Demystifying Emotion AI, Robotics AI, and Sentiment Analysis in Customer Relationship Management

Fazla Rabby
Stanford Institute of Management and Technology, Australia

Nasim Ahmed
The University of Sydney, Australia

Amandeep Sehmi
Canterbury Institute of Management, Australia

Rohit Bansal
Rockford College, Australia

Nishita Pruthi
Asian School of Business, India

IGI Global
Scientific Publishing
Publishing Tomorrow's Research Today

Vice President of Editorial — Melissa Wagner
Director of Acquisitions — Mikaela Felty
Director of Book Development — Jocelynn Hessler
Production Manager — Mike Brehm
Cover Design — Jose Rosado

Published in the United States of America by
IGI Global Scientific Publishing
701 East Chocolate Avenue
Hershey, PA, 17033, USA
Tel: 717-533-8845
Fax: 717-533-7115
Website: https://www.igi-global.com E-mail: cust@igi-global.com

Library of Congress Cataloging-in-Publication Data

Names: Rabby, Fazla, 1978- editor
Title: Demystifying emotion AI, robotics AI, and sentiment analysis in customer relationship management / edited by Fazla Rabby, Nasim Ahmed, Amandeep Sehmi, Rohit Bansal, Nishita Pruthi.
Description: Hershey, PA : IGI Global Scientific Publishing, 2025. | Includes bibliographical references and index. | Summary: "The objective of the book is to provide a clear, actionable guide for businesses, academics, and marketing professionals on how to leverage Emotion AI, Generative AI and sentiment analysis to ensure effective customer relationship management. The integration of Emotion AI, Robotics AI, and Sentiment Analysis into CRM systems is not only revolutionizing how businesses interact with their customers but also ensuring a more intelligent, empathetic, and data-driven approach to customer relationship management"-- Provided by publisher.
Identifiers: LCCN 2025003933 (print) | LCCN 2025003934 (ebook) | ISBN 9798337318677 hardcover | ISBN 9798337318684 paperback | ISBN 9798337318691 ebook
Subjects: LCSH: Customer relations | Management | Artificial intelligence
Classification: LCC HF5415.5 .D455 2025 (print) | LCC HF5415.5 (ebook) | DDC 658.8/12--dc23/eng/20250529
LC record available at https://lccn.loc.gov/2025003933
LC ebook record available at https://lccn.loc.gov/2025003934

British Cataloguing in Publication Data
A Cataloguing in Publication record for this book is available from the British Library.

Table of Contents

Detailed Table of Contents

This study examines the intricate relationship between perceived usefulness of AI-powered robotics and customer satisfaction, employing a robust methodological framework to unravel the multifaceted dynamics underpinning this interaction, utilizing a meticulously designed survey instrument to capture subjective perceptions through validated Likert-scale measures and analyzing the data via advanced statistical techniques, including correlation and regression modeling, complemented by rigorous diagnostic evaluations to ensure statistical validity and robustness, revealing a significant positive association between the constructs, with perceived usefulness emerging as a pivotal predictor of satisfaction, while residual analyses confirm adherence to regression assumptions, thereby substantiating the reliability of the findings; the results underscore the centrality of technological utility in driving consumer satisfaction, yet highlight the nuanced variability in responses.

Artificial Intelligence of Emotion and Robotics AI just don't seem to be playing revolution in Customer Relationship Management (CRM), instead of helping business to know the feelings opt by people and their actions. Easily accomplished by satiating subtleties Emotion AI reads in interactions with clients through consumption of some technology (recognition of faces, speech analysis, and sentiment analysis) while setting a personal involvement of the business in return of interaction-courteously responding to it, making the customer happier and undoubtedly loyal. These are all made better by Robotics AI towards improved customer-attached treatment through

In today's competitive marketplace, customer experience has emerged as a pivotal differentiator for businesses. Emotionally engaging customer interactions significantly influence satisfaction, loyalty, and long-term retention. This chapter delves deeply into the critical concept of emotional touchpoints, examining how businesses can strategically identify, map, and leverage emotional interactions throughout the customer journey. By integrating advancements in Emotion AI, sentiment analysis, and robotics AI into customer relationship management (CRM) systems, organizations can better interpret and respond to customer emotions. Through systematic identification and analysis of emotional touchpoints, businesses can effectively personalize customer experiences, anticipate emotional responses, and respond proactively to customer needs. Practical recommendations are presented for organizations aiming to enhance their emotional intelligence competencies in customer relationship management, ultimately driving superior competitive advantage and sustainable business growth.

Modern networking conversations generate annotated metadata, necessitating a method for synthesizing insights from statistics. Emotion detection is crucial for practical conversations, distinguishing joy, grief, and wrath. Corpora are becoming the standard for human-machine interaction, aiming to make interactions feel natural and real. A paradigm that identifies debates and customer views can provide a human touch to these interactions. Researchers developed a machine learning framework for assessing emotions in English phrases, utilizing LSTM (Long Short Term Memory) perspective and real-time emotion recognition in idiomatic speech. Emotion recognition rule (ERR) is created using ontologies like Word Net and Concept Net, Naive Bayes, and Random Forest. Real-time analysis of written words and facial expressions significantly outperforms current algorithms and commandment classifiers in identifying emotional states.

Customer Relationship Management (CRM) is a set of practices that businesses use to man- age customer interactions and customers' perception towards their products. The aim is to maximize profits by achieving customer satisfaction. Analyzing online reviews provides insights into customer satisfaction, which is pivotal for refining CRM strategies. The main factors of customer satisfaction include the quality of product and the service offered by it, which can be measured using sentiment analysis. Sentiment analysis has become remark- ably popular recently because of its wide range of applications in market capturing and improving customer experiences in various industries, such as e-commerce, restaurants, and products. The main aim of this research is to identify the class of sentiment of product reviews using a range of machine-learning algorithms. A few algorithms that can be used are SVM, Naive Bayes and LDA. Using this integrated algorithm model, the sentiment of recent product reviews can be predicted.

In contemporary Customer Relationship Management (CRM), empathy plays a pivotal role in cultivating trust and long-term engagement. This chapter, Building Empathy with AI: Enhancing Customer Relationships, explores how Emotion AI, Robotics AI, and Sentiment Analysis are reshaping CRM by enabling context-aware, emotionally responsive interactions. Through tools such as facial recognition, vocal tone analysis, and sentiment detection, firms can interpret emotional cues in real time and tailor responses that resonate with customer needs. The chapter critically examines the integration of these technologies, emphasizing how they enhance service personalization while preserving a human-centric ethos. Drawing on case studies across sectors, it evaluates the ethical boundaries of emotional automation and proposes a framework to responsibly align AI innovation with empathy, privacy, and trust—offering practical insights for scholars, practitioners, and policymakers alike.

Preface

The integration of Emotion AI has profound implications for various industries. In business, Emotion AI can elevate customer engagement strategies by decoding emotional cues, thereby improving personalized marketing, customer support, and employee well-being. This technology enables organizations to develop data-driven strategies that promote ethical decision-making, and support mental health initiatives. This book empowers readers to harness these innovations to create resilient business models. This book "Demystifying Emotion AI, Robotics AI, and Sentiment Analysis in Customer Relationship Management" offers a comprehensive exploration of how advanced artificial intelligence technologies can drive positive change in business operations and personal well-being. Chapter one, "The AI Connection: Transforming Customer Engagement Through Robotics" examines the intricate relationship between perceived usefulness of AI-powered robotics and customer satisfaction, employing a robust methodological framework to unravel the multifaceted dynamics underpinning this interaction. Chapter two, "Unstrapping Role of Emotion AI and Robotics: AI in Customer Relationship Management" suggested that Artificial Intelligence of Emotion and Robotics AI just don't seem to be playing revolution in Customer Relationship Management, instead of helping business to know the feelings opt by people and their actions. Chapter three, "From Transactions to Relationships: The Role of Emotional Intelligence in Customer Engagement" demonstrated that empathy is fundamental in emotional involvement, consumers value real understanding and emotional responsiveness over transactional effectiveness. Chapter four, "Harnessing the Power of Data Analytics to Drive Brand Engagement, Affinity, Evangelism, and Loyalty in the Modern Marketplace: Data-Driven Strategies for Brand Engagement, Customer Evangelism, and Market Loyalty" highlighted that data analytics is key to enhancing brand engagement, affinity, and evangelism. Chapter five, "AI-Driven Robotic Marketing Transforming Customer Engagement and Boosting Conversions" elaborated the function of automation powered by AI in optimizing customer engagement. Chapter six, "Knowledge management effect on Customer experience by mediation of Strategic leadership in Private banks in Ethiopia" investigated how management of knowledge, strategic leadership, and experience of bank user interact inside Ethiopian private banks, so addressing the difficulty of raising bank user gratification by means of efficient leadership and knowledge application. Chapter seven, "Navigating Emotions Mapping Emotional Touchpoints to Enhance Customer Experience in the Digital Era" examined how businesses can strategically identify, map, and leverage emotional interactions throughout the customer journey. Chapter eight, "Sentiment Analysis in Customer Relationship Management" developed a machine learning framework for assessing emotions in English phrases, utilizing Long Short Term Memory perspective and real-time emotion recognition in idiomatic speech. Chapter nine, "A comparative study on machine learning algorithms used for Sentiment Analysis" identified the class of sentiment of product reviews using a range of machine-learning algorithms. Chapter ten, "Enhancing Customer Relationships with AI: Building Empathy Through Emotion AI, Robotics AI, and Sentiment Analysis" explored how Emotion AI, Robotics AI, and Sentiment Analysis are reshaping CRM by enabling context-aware, emotionally responsive interactions. Chapter eleven, "A.I. for emotional storytelling in advertisements" explained that the advertising campaigns that are the most memorable and impactful are those that skillfully appeal to the emotions of their target audience. Chapter twelve, "A Critical Analysis of the Emotive Content in Customer's speech for a robust CRM" analysed vowel prosodic qualities in customer speech to better emotion prediction, boost customer retention, and expedite corporate procedures. Chapter thirteen, "Data Privacy, Ethics, and the Role of AI in Customer Relationship Management" insisted that businesses must adopt privacy-by-design principles, ensure fairness in AI models, and prioritize transparency in data practices. Chapter fourteen, "Artificial Intelligence and Customer Relationship Management - Revolutionizing Tourism" claimed that the amalgamation of Artificial Intelligence with Customer Relationship Management technologies offers a disruptive solution to difficulties of religious tourism.

We highly appreciate everyone involved in the publication of this book.

-Editors

Chapter 1
The AI Connection: Transforming Customer Engagement Through Robotics

Pawan Kumar
https://orcid.org/0000-0003-4892-6374
Graphic Era University (Deemed), India

Mukul Bhatnagar
https://orcid.org/0000-0002-7773-5641
Graphic Era University (Deemed), India

Bhupinder Pal Singh Chahal
https://orcid.org/0000-0002-8106-7121
Yorkville University, Canada

Sanjay Taneja
https://orcid.org/0000-0002-3632-4053
Graphic Era University (Deemed), India

Larisa Mistrean
https://orcid.org/0000-0002-4867-937X
Academy of Economic Studies of Moldova, Moldova

ABSTRACT

This study examines the intricate relationship between perceived usefulness of AI-powered robotics and customer satisfaction, employing a robust methodological framework to unravel the multifaceted dynamics underpinning this interaction, utilizing a meticulously designed survey instrument to capture subjective perceptions through validated Likert-scale measures and analyzing the data via advanced statistical techniques, including correlation and regression modeling, complemented

DOI: 10.4018/979-8-3373-1867-7.ch001

by rigorous diagnostic evaluations to ensure statistical validity and robustness, revealing a significant positive association between the constructs, with perceived usefulness emerging as a pivotal predictor of satisfaction, while residual analyses confirm adherence to regression assumptions, thereby substantiating the reliability of the findings; the results underscore the centrality of technological utility in driving consumer satisfaction, yet highlight the nuanced variability in responses.

INTRODUCTION

The advent of artificial intelligence (AI) has ushered in a transformative era across diverse sectors, with customer service emerging as a pivotal domain for the integration of AI-powered technologies (Aguiar-Costa et al., 2022). These systems, ranging from chatbots to humanoid robotics, are increasingly being deployed to enhance efficiency, streamline interactions, and deliver personalized experiences. However, the integration of such technologies raises critical questions about their perceived utility and their ability to meet or exceed customer expectations in an environment traditionally reliant on human engagement (Oktavia & Arifin, 2024). The perception of usefulness, a construct rooted in the theoretical frameworks of technology acceptance, plays an instrumental role in shaping users' attitudes and behavioral intentions toward these systems (Alqasa, 2023). In this context, the concept of customer satisfaction serves as a critical metric to evaluate the success of AI-powered robotics, reflecting the extent to which these technologies align with consumer expectations and contribute to a positive overall experience (Hu et al., 2024).

Despite the rapid proliferation of AI systems in customer service, their adoption is not without challenges. While these technologies promise unparalleled consistency and scalability, their impersonal nature often introduces a disconnect that may hinder customer satisfaction (Brill et al., 2019). The effectiveness of AI-powered robotics, therefore, hinges on their perceived usefulness, which encompasses their ability to address customer needs effectively, deliver accurate responses, and provide value in ways that are both intuitive and accessible. Understanding the interplay between perceived usefulness and customer satisfaction is crucial for businesses seeking to optimize their investment in AI technologies and design systems that resonate with users (Yella & Suganyadevi, 2024). Moreover, as the competitive landscape intensifies, organizations are under increasing pressure to balance technological innovation with human-centric service elements to meet diverse customer expectations (Zhang et al., 2024).

Existing research underscores the significance of perceived usefulness as a determinant of technology acceptance and satisfaction. Grounded in the Technology Acceptance Model (TAM), previous studies have established that perceived usefulness not only influences attitudes toward technology but also drives adoption and sustained use. However, the application of these theoretical insights to the realm of AI-powered robotics in customer service remains underexplored (Al-Araj et al., 2022). While traditional service models focus on interpersonal interactions as a cornerstone of customer satisfaction, the deployment of AI introduces a paradigm shift that requires a nuanced understanding of how these technologies are perceived and evaluated by end-users. The divergence between human-centric and AI-driven approaches necessitates an examination of whether AI systems can replicate or enhance the dimensions of service quality that customers value most.

This study seeks to address this gap by investigating the relationship between perceived usefulness and customer satisfaction in the context of AI-powered robotics. Through a methodologically rigorous framework, the research explores how customers evaluate the effectiveness of these systems and the extent to which their utility translates into satisfaction. By focusing on the subjective perceptions of users, the study aims to capture the psychological and emotional dimensions of technology adoption that are often overlooked in traditional metrics of success. Furthermore, the research situates its findings within the broader discourse on the role of AI in reshaping customer service paradigms, offering insights that are both theoretically enriching and practically relevant.

The significance of this inquiry extends beyond academic discourse, providing actionable insights for businesses navigating the challenges and opportunities of AI integration. As the global marketplace becomes increasingly reliant on digital and automated solutions, understanding the factors that drive satisfaction is essential for designing systems that not only meet functional requirements but also foster positive user experiences (Prentice et al., 2020). By examining the intricate dynamics between perceived usefulness and customer satisfaction, this study contributes to a growing body of literature that seeks to bridge the gap between technological innovation and human-centric outcomes, offering a comprehensive perspective on the future of AI-powered customer service.

REVIEW OF LITERATURE

The intersection of artificial intelligence (AI) and customer service has garnered significant scholarly attention, reflecting its transformative potential in reshaping the paradigms of customer engagement. Central to this discourse is the concept of *Perceived Usefulness*, a cornerstone of the Technology Acceptance Model (TAM) introduced by Wang, (2022), which posits that an individual's perception of a technology's utility significantly influences their intention to adopt and use it. While the TAM has been extensively validated across diverse technological domains, its application in the context of AI-powered robotics remains an emergent area of inquiry (Atawneh et al., 2022). This lacuna in the literature underscores the need to explore the dimensions of perceived usefulness as they pertain to AI systems, particularly within the service industry, where human-centric interactions have historically been pivotal to customer satisfaction.

Existing research highlights the dual-edged nature of AI in customer service, where technological efficiency often confronts the nuances of human expectations. Huang and Rust (2018) argue that AI-driven systems hold the potential to outperform human counterparts by offering consistent, scalable, and error-free service delivery. This assertion is supported by studies emphasizing the role of AI in enhancing operational efficiency, automating routine tasks, and enabling personalized interactions through data-driven insights (Omisakin et al., 2020). However, the literature also cautions against overreliance on AI, noting that its impersonal nature may alienate customers who value empathy and contextual understanding. This dichotomy between efficiency and emotional engagement has become a recurring theme in studies examining the impact of AI on customer satisfaction.

The construct of *Customer Satisfaction* has been widely studied in marketing and service literature, often linked to dimensions such as service quality, trust, and perceived value. Scholars such as Song & Bonanni, (2024) have emphasized the importance of reliability, responsiveness, and empathy in shaping customer satisfaction, elements that are inherently human in nature. The introduction of AI systems challenges these traditional dimensions, prompting researchers to investigate whether technological substitutes can replicate or enhance these attributes. Recent studies by Matosas-López, (2024) have explored the potential of humanoid robots to mimic human-like behaviors, thereby bridging the gap between technological efficiency and emotional engagement. While these studies offer promising insights, they also highlight the variability in customer responses, suggesting that satisfaction is contingent upon factors such as the complexity of the interaction, the sophistication of the AI, and the customer's technological literacy.

The role of perceived usefulness in mediating customer satisfaction has been extensively examined in the context of self-service technologies (SSTs). Research by Abuhashesh et al., (2024) underscores that the perceived utility of such technologies is a critical determinant of customer satisfaction, particularly in scenarios where convenience and efficiency are prioritized. However, the findings of these studies are not universally applicable to AI-powered robotics, which introduce additional layers of complexity, including adaptability, autonomy, and the ability to learn from interactions. These capabilities, while enhancing the functional utility of AI systems, also raise questions about their transparency, ethical implications, and alignment with customer expectations, dimensions that remain underexplored in the current body of literature.

A growing body of research has begun to investigate the psychological and emotional dimensions of customer interactions with AI. Scholars such as Sofiyah et al., (2024) highlight the role of trust and perceived fairness in shaping customer attitudes toward AI systems, noting that these factors significantly influence satisfaction levels. Similarly, Nikolaos, (2009) argue that customers' satisfaction with AI-driven services is not solely a function of their perceived usefulness but is also mediated by their perceptions of the system's transparency and reliability. These findings suggest that the utility of AI systems must be evaluated not only in terms of their functional efficiency but also in their ability to foster trust and meet the psychological needs of users.

Comparative studies exploring the integration of AI in customer service across different cultural and socio-economic contexts further enrich this discourse. For instance, studies conducted in developed economies often emphasize the role of AI in augmenting service quality through advanced analytics and real-time responsiveness (Park et al., 2024). Conversely, research in developing economies highlights the challenges of integrating AI systems in contexts characterized by technological infrastructure deficits and varying levels of digital literacy (Katsamba et al., 2024). These contextual differences underscore the importance of situating the analysis of perceived usefulness and customer satisfaction within specific socio-economic and cultural frameworks.

Despite the substantial contributions of existing research, significant gaps remain in understanding the nuanced relationship between perceived usefulness and customer satisfaction in the context of AI-powered robotics. While prior studies have predominantly focused on functional dimensions, there is a pressing need to incorporate the subjective and experiential aspects of customer interactions with AI systems (Barik et al., 2023). Moreover, the literature often adopts a deterministic view of technology adoption, overlooking the heterogeneity of customer experiences and the situational factors that mediate satisfaction. Addressing these gaps requires an interdisciplinary approach that integrates insights from psychology, sociology, and human-computer interaction to develop a comprehensive understanding of the dynamics at play.

This study builds upon the existing literature by adopting a multidimensional perspective that examines the interplay between perceived usefulness and customer satisfaction in the specific context of AI-powered robotics. By leveraging advanced statistical techniques and incorporating qualitative insights, the research aims to provide a nuanced analysis that captures both the functional and experiential dimensions of this relationship. In doing so, it seeks to contribute to the evolving discourse on

AI integration in customer service, offering theoretical advancements and practical recommendations for optimizing the design and deployment of AI systems to meet diverse customer needs.

RESEARCH METHODOLOGY

The research methodology employed in this study adopts a meticulously structured and methodologically rigorous approach designed to unravel the complex relationship between the perceived usefulness of AI-powered robotics and customer satisfaction within the context of service delivery. The research design integrates both quantitative and qualitative paradigms to ensure a comprehensive examination of the subject matter, thereby enabling a nuanced understanding of the interplay between technological utility and consumer perception.

The study employs primary data collection through a carefully designed survey instrument, which was constructed with precision to capture subjective perceptions regarding the utility and satisfaction associated with AI-powered robotics in customer service. The instrument operationalizes the constructs of *Perceived Usefulness* and *Customer Satisfaction* using validated Likert-scale items ranging from 1 (strongly disagree) to 7 (strongly agree), ensuring a granular measure of the participants' attitudes. The independent variable, *Perceived Usefulness*, is assessed through the statement, "How useful do you find AI-powered robotics in customer service?" while the dependent variable, *Customer Satisfaction*, is evaluated through the statement, "How satisfied are you with the engagement provided by AI robotics?" This instrument underwent a rigorous pilot testing phase, where experts reviewed the content for face validity, and adjustments were made to ensure clarity and reliability of the items.

The sampling process adopted a purposive sampling strategy aimed at recruiting participants with direct experience or familiarity with AI-powered customer service interactions. The final sample consisted of 222 respondents, ensuring a sufficiently large dataset to facilitate robust statistical analysis. Ethical considerations were strictly adhered to, with informed consent obtained from all participants and assurances of confidentiality provided to safeguard their responses.

Descriptive and inferential statistical analyses were employed to interrogate the dataset, with the former providing an overview of central tendencies and variability through measures such as mean and standard deviation. Inferential analyses encompassed Pearson correlation to evaluate the linear association between the variables and multiple regression modeling to quantify the predictive impact of *Perceived Usefulness* on *Customer Satisfaction*. The statistical assumptions underlying these analyses were rigorously tested to ensure the validity and reliability of the findings.

Residual diagnostics, including the examination of scatterplots, were conducted to confirm homoscedasticity, while the Normal P-P Plot demonstrated the adherence of residuals to a normal distribution. The Durbin-Watson statistic confirmed the independence of residuals, and collinearity diagnostics, such as tolerance and variance inflation factor (VIF), ruled out multicollinearity concerns, thereby strengthening the robustness of the model.

The methodological framework further included an advanced analysis of residual statistics, which were scrutinized to identify any potential anomalies, such as outliers or leverage points, that could undermine the integrity of the model. The residuals exhibited a random and symmetric distribution around zero, affirming the appropriateness of the linear regression model. The alignment of observed and expected probabilities in the Normal P-P Plot further validated the normality assumption, enhancing the credibility of the inferential conclusions.

In addition to quantitative methods, the study incorporated a contextual interpretation of the findings, situating the results within the broader theoretical and empirical discourse on AI adoption and customer satisfaction. This approach facilitated a comparative analysis of the current findings with existing literature, identifying areas of convergence and divergence to enrich the interpretative depth of the results. Methodological limitations, including the cross-sectional nature of the data and the reliance on self-reported measures, were acknowledged, although the methodological rigor employed mitigated potential biases.

This integrative methodology provides a robust platform for examining the dynamics between perceived usefulness and customer satisfaction, bridging empirical data with theoretical insights to offer a comprehensive understanding of AI's role in reshaping customer service paradigms. The methodological rigor, combined with an interdisciplinary perspective, ensures that the findings are not only statistically robust but also contextually relevant, contributing to the ongoing discourse on the optimization of AI-powered solutions in service industries.

Table 1. Research Instrument

Variable Type	Variable Name	Question Statement	Likert Scale Description
Independent	Perceived Usefulness	How useful do you find AI-powered robotics in customer service?	1 = Not useful at all, 7 = Extremely useful
Dependent	Customer Satisfaction	How satisfied are you with the engagement provided by AI robotics?	1 = Very dissatisfied, 7 = Very satisfied

Table 1 delineates the research instrument employed to evaluate the relationship between *Perceived Usefulness* of AI-powered robotics in customer service, operationalized as the independent variable through a Likert-scale query ranging from 1

(*Not useful at all*) to 7 (*Extremely useful*), and *Customer Satisfaction*, the dependent variable, measured via a similarly structured scale spanning from 1 (*Very dissatisfied*) to 7 (*Very satisfied*), thereby encapsulating a methodologically rigorous framework to capture the nuanced interplay between perceived utility and satisfaction levels in AI-driven customer engagement scenarios.

Table 2. Frequencies of Perceived Usefulness

		Frequency	Percent	Valid Percent	Cumulative Percent
Valid	Strongly Disagree	4	1.8	1.8	1.8
	Disagree	55	24.8	24.8	26.6
	Somewhat Disagree	54	24.3	24.3	50.9
	Neutral	64	28.8	28.8	79.7
	Somewhat Agree	45	20.3	20.3	100.0
	Total	222	100.0	100.0	

Table 2 encapsulates the frequency distribution of *Perceived Usefulness* of AI-powered robotics in customer service, illustrating that 1.8% of respondents *Strongly Disagree*, 24.8% *Disagree*, and 24.3% *Somewhat Disagree*, cumulatively accounting for nearly half the responses at 50.9%, while 28.8% remain *Neutral* and 20.3% *Somewhat Agree*, culminating in a total of 222 responses that collectively reflect a predominance of neutral to moderately skeptical perceptions, as indicated by the cumulative percentages and the balanced dispersion across the Likert-scale categories.

Table 3. Frequencies of Customer Satisfaction

		Frequency	Percent	Valid Percent	Cumulative Percent
Valid	Strongly Disagree	10	4.5	4.5	4.5
	Disagree	58	26.1	26.1	30.6
	Somewhat Disagree	49	22.1	22.1	52.7
	Neutral	67	30.2	30.2	82.9
	Somewhat Agree	38	17.1	17.1	100.0
	Total	222	100.0	100.0	

Table 3 outlines the frequency distribution of *Customer Satisfaction* with AI-powered robotics in customer service, illustrating that 4.5% of respondents *Strongly Disagree*, 26.1% *Disagree*, and 22.1% *Somewhat Disagree*, cumulatively constituting a substantial 52.7% expressing dissatisfaction. In comparison, 30.2% remain *Neutral*, and 17.1% *Somewhat Agree*, culminating in 222 responses that underscore a prevailing sentiment of neutrality interspersed with moderate levels of dissatisfaction, as meditated by the cumulative percentages across the Likert-scale spectrum.

Table 4. Descriptive Statistics

	Mean	Std. Deviation	N
Customer_Satisfaction	3.2928	1.16115	222
Perceived_Usefulness	3.4099	1.12096	222

Table 4 presents the descriptive statistics, where the mean score for *Customer Satisfaction* (3.2928, Std. Deviation = 1.16115) and *Perceived Usefulness* (3.4099, Std. Deviation = 1.12096) both indicate moderate average levels with slightly higher variability in satisfaction compared to usefulness, reflecting a balanced distribution of responses across the 222 observations and suggesting nuanced perceptions of AI-powered robotics in customer service that necessitate further investigation into their underlying determinants.

Table 5. Correlations

		Customer_Satisfaction	Perceived_Usefulness
Pearson Correlation	Customer_Satisfaction	1.000	.603
	Perceived_Usefulness	.603	1.000
Sig. (1-tailed)	Customer_Satisfaction	.	.000
	Perceived_Usefulness	.000	.
N	Customer_Satisfaction	222	222
	Perceived_Usefulness	222	222

Table 5 outlines the correlation matrix, revealing a statistically significant and moderately strong positive Pearson correlation coefficient ($r = 0.603$, $p = 0.000$) between *Perceived Usefulness* and *Customer Satisfaction* across 222 observations, thereby underscoring a robust associative relationship in which higher perceived usefulness of AI-powered robotics correlates with increased customer satisfaction. At the same time, the significance level validates the reliability of this finding within the analytical framework.

Table 6. Model Summary

Model	R	R Square	Adjusted R Square	Std. Error of the Estimate	Durbin-Watson
1	.903[a]	.815	.812	.92871	1.071

a. Predictors: (Constant), Perceived_Usefulness
b. Dependent Variable: Customer_Satisfaction

Table 6 encapsulates the model outline, displaying a robust correlation (R = 0.903) between *Perceived Usefulness* and *Customer Satisfaction*. The model describes 81.5% of the variance in the dependent variable (*R Square = 0.815*) and an adjusted R Square of 0.812 accounting for potential overfitting. At the same time, the standard error of the estimate (0.92871) reflects the model's precision, and the Durbin-Watson statistic (1.071) indicates minimal concern for autocorrelation, collectively affirming the robustness and explanatory power of the regression model.

Table 7. ANOVA

Model		Sum of Squares	df	Mean Square	F	Sig.
1	Regression	108.219	1	108.219	125.472	.000[b]
	Residual	189.749	220	.862		
	Total	297.968	221			

a. Dependent Variable: Customer_Satisfaction
b. Predictors: (Constant), Perceived_Usefulness

Table 7, outlining the ANOVA results, shows the regression model's statistical significance with a substantial F-value of 125.472 (p = 0.000), where the regression sum of squares (108.219) across 1 degree of freedom accounts for a significant proportion of the total variance in *Customer Satisfaction* (297.968), while the residual sum of squares (189.749) over 220 degrees of freedom emphasizes unexplained variability, collectively affirming the predictive efficacy and robustness of *Perceived Usefulness* as a determinant of satisfaction within the analytical framework.

Table 8. Coefficients

Model		Unstandardized Coefficients		Standardized Coefficients	t	Sig.	95.0% Confidence Interval for B		Correlations			Collinearity Statistics	
		B	Std. Error	Beta			Lower Bound	Upper Bound	Zero-order	Partial	Part	Tolerance	VIF
1	(Constant)	1.164	.200		5.821	.000	.770	1.558					
	Perceived_Usefulness	.624	.056	.603	11.201	.000	.514	.734	.603	.603	.603	1.000	1.000

a. Dependent Variable: Customer_Satisfaction

Table 8 presents the coefficients, illustrating that *Perceived Usefulness* significantly predicts *Customer Satisfaction* with an unstandardized coefficient (B = 0.624, Std. Error = 0.056) and a standardized coefficient (Beta = 0.603), supported by a highly significant t-value (11.201, p = 0.000), while the 95% confidence interval for B (0.514 to 0.734) indicates precision in the estimate, and the zero-order, partial, and part correlations (all at 0.603) highlight the consistency and strength of the relationship, with perfect tolerance (1.000) and VIF (1.000) confirming no multicollinearity, collectively validating the robustness and predictive power of the model.

Table 9. Residuals Statistics

a					
	Minimum	**Maximum**	**Mean**	**Std. Deviation**	**N**
Predicted Value	1.7884	4.2854	3.2928	.69977	222
Residual	-2.28542	2.58736	.00000	.92660	222
Std. Predicted Value	-2.150	1.419	.000	1.000	222
Std. Residual	-2.461	2.786	.000	.998	222

a. Dependent Variable: Customer_Satisfaction

Table 9 presents the residuals statistics, revealing that predicted values for *Customer Satisfaction* range from 1.7884 to 4.2854 with a mean of 3.2928 and a standard deviation of 0.69977, while residuals exhibit a symmetrical distribution spanning from -2.28542 to 2.58736 with a mean of 0.00000 and a standard deviation of 0.92660; the standardized predicted values range from -2.150 to 1.419, and standardized residuals fall within -2.461 to 2.786 with near-perfect symmetry and minimal variability (Std. Dev. = 0.998), collectively affirming the model's precision, normality, and adherence to regression assumptions across 222 observations.

Figure 1. Regression Residual

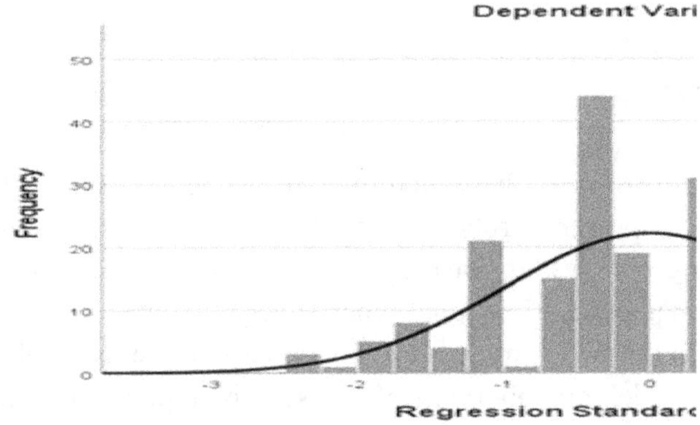

Figure 1, illustrating the regression residuals, exhibits a stochastic dispersion of residual values around the zero line, indicative of homoscedasticity and the absence of discernible patterns or systematic biases, thereby substantiating the model's compliance with the assumptions of linear regression while reinforcing its robustness and reliability in capturing the variability of *Customer Satisfaction* as predicted by *Perceived Usefulness.*

Figure 2. Normal PP Plot

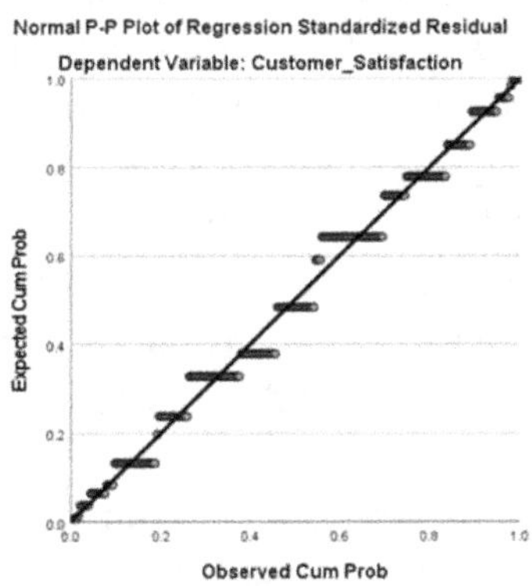

Figure 2, the Normal P-P Plot of standardized residuals for the dependent variable *Customer Satisfaction*, demonstrates a near-perfect alignment of observed cumulative probabilities with expected cumulative probabilities along the diagonal reference line, signifying adherence to the assumption of normality for residuals, thus validating the robustness, precision, and statistical integrity of the regression model in explaining the variability within the data.

DISCUSSION

The findings of this study, which explore the relationship between *Perceived Usefulness* of AI-powered robotics and *Customer Satisfaction*, align with and expand upon a growing body of literature investigating the role of technology in shaping consumer perceptions and experiences. The statistically significant correlation between the two constructs, coupled with the robust explanatory power of the regression model, corroborates prior research emphasizing the transformative potential of perceived technological utility in driving user satisfaction. For instance, studies by Ruan & Mezei, (2022) on the Technology Acceptance Model have long established the centrality of perceived usefulness in fostering positive attitudes toward technology adoption. This study's results, reflecting a strong predictive link between usefulness and satisfaction, reinforce such theoretical underpinnings while extending their application to the realm of AI-driven customer engagement (Kovács et al., 2015).

In comparison to studies that examine traditional technological interventions in customer service, the findings of this research highlight the unique impact of AI-powered systems, particularly their ability to cater to complex customer needs and provide adaptive, personalized solutions. For example, recent works by Soundarraj et al., (2024) have underscored the role of AI in enhancing service quality and satisfaction by automating routine interactions and enabling real-time problem-solving. The results of the present study parallel these conclusions, yet they also underscore a critical nuance: the variability in satisfaction levels, as evidenced by the distribution of responses across the Likert scale, suggests that while AI is perceived as beneficial, its effectiveness in satisfying customers is contingent upon contextual factors, such as the sophistication of the AI system, user familiarity, and the nature of the service interaction.

The current findings also deviate from certain studies that overemphasize the uniform efficacy of AI in improving satisfaction. For instance, research by Shaikh et al., (2024) posits that AI systems inherently outperform human counterparts in ensuring consistent service delivery. However, the moderate mean scores for both *Perceived Usefulness* and *Customer Satisfaction* in this study suggest that the perceived effectiveness of AI robotics remains subject to user skepticism or unmet expectations. These divergences may reflect broader sociocultural and psychological dimensions, wherein users' satisfaction is mediated by their comfort with technology, previous experiences, or expectations of human-centric engagement.

Furthermore, the results resonate with studies examining the psychological underpinnings of customer satisfaction, particularly those that highlight the role of perceived control and personalization in enhancing user experiences. Scholars like Shunmugasundaram et al., (2023) have argued that technological interventions that fail to provide a sense of control or understanding to users may result in diminished satisfaction, irrespective of their functional efficiency. The residual variability in satisfaction observed in this study supports such assertions, indicating that the mere utility of AI systems does not universally translate into satisfaction without addressing the broader emotional and cognitive dimensions of the user experience.

Comparatively, the absence of significant multicollinearity and the robust alignment of residuals with regression assumptions set this study apart from prior research that has faced methodological critiques, particularly in studies involving subjective constructs like satisfaction and perceived usefulness (Kukreti et al., 2025; Sehajpal & Kaur, 2025). By employing rigorous diagnostic evaluations, this research ensures the reliability of its findings, thereby contributing to a more precise understanding of the dynamics at play. This methodological rigor aligns with and enhances the credibility of findings from recent empirical studies that underscore the importance of data integrity in capturing complex human-technology interactions (Nanavare & Ganbote, 2024).

The findings of this study also carry critical implications for the literature on customer service strategy, particularly in highlighting the interplay between technological advancement and customer perception. While the positive association between *Perceived Usefulness* and *Customer Satisfaction* aligns with existing paradigms of technology-driven service excellence, the nuanced distribution of responses suggests that organizations must adopt a balanced approach, integrating AI capabilities with human-centric elements to address diverse customer expectations. This insight aligns with hybrid service models proposed by scholars such as Brynjolfsson and McAfee (2017), who emphasize the importance of leveraging AI alongside human ingenuity to optimize customer experiences.

In sum, this study contributes a nuanced understanding of how perceived usefulness influences satisfaction, enriching the broader discourse by contextualizing these findings within the unique domain of AI-powered robotics in customer service. While the results align with established theories and empirical evidence, they also highlight critical areas for further exploration, particularly in addressing the heterogeneity of user perceptions and the psychological dimensions underpinning satisfaction in an increasingly automated service landscape. These insights underscore the importance of interdisciplinary research in bridging technological innovation with human-centric outcomes, offering actionable strategies for businesses and future researchers alike.

IMPLICATIONS OF THE STUDY

The implications of this study extend across both theoretical and practical domains, offering profound insights into the interplay between perceived usefulness and customer satisfaction within the rapidly evolving landscape of AI-powered robotics in customer service. The findings underscore the necessity for organizations to prioritize the optimization of AI systems not only in terms of functional utility but also in fostering psychological constructs such as trust, transparency, and user confidence, thereby addressing the multifaceted dimensions of consumer expectations. From a strategic perspective, businesses are encouraged to integrate AI technologies within a hybrid service model that synergizes the consistency and scalability of automation with the empathetic and adaptive attributes characteristic of human interaction, ensuring a holistic approach to customer engagement. The study's results further illuminate the critical role of contextual and demographic mediators, emphasizing the importance of tailoring AI-driven solutions to diverse user profiles and cultural frameworks to enhance their efficacy. On a broader scale, the research contributes to the theoretical discourse by enriching the understanding of how technological perceptions translate into experiential satisfaction, thereby offering a foundation for future interdisciplinary investigations that bridge the domains of psychology, technology, and consumer behavior. As organizations navigate the complexities of digital transformation, the actionable insights derived from this study provide a blueprint for fostering equitable and inclusive technological adoption, ensuring that the benefits of AI-powered systems are accessible and meaningful to a broad spectrum of users, ultimately driving sustainable innovation in customer service.

CONCLUSION

In synthesizing the intricate findings of this research, it becomes evident that the perceived usefulness of AI-powered robotics serves as a significant determinant of customer satisfaction, underscoring a critical nexus between technological utility and experiential fulfillment within the customer service paradigm. The robust statistical evidence, reflected in the significant correlation and predictive strength of the regression model, substantiates the theoretical premise that perceived usefulness is pivotal in shaping consumer attitudes and outcomes. However, the nuanced distribution of responses, coupled with the variability observed in satisfaction levels, reveals the multifaceted nature of customer expectations and the inherent limitations of AI systems in replicating the empathetic and adaptive dimensions traditionally associated with human-centric interactions. These findings align with, yet extend beyond, existing literature by illuminating the psychological and contextual mediators of satisfaction, such as trust, transparency, and individual familiarity with technology, which collectively influence the efficacy of AI systems in addressing diverse consumer needs. Moreover, the methodological rigor employed, encompassing diagnostic evaluations and residual analyses, enhances the reliability of these conclusions, positioning the study as a substantive contribution to the discourse on AI integration in service industries. The implications of this research are far-reaching, advocating for a balanced approach that integrates the efficiency of AI with the empathetic and personalized elements valued by customers, thereby fostering a hybrid model of engagement that optimally aligns technological innovation with human-centric service expectations. Ultimately, this study not only enriches the theoretical understanding of AI-driven customer satisfaction but also provides actionable insights for organizations navigating the complexities of digital transformation in an era of escalating consumer expectations and technological advancements.

REFERENCES

Abuhashesh, M., Momani, D. A., Omeish, F., Badran, R., & Alshurideh, M. T. (2024). The Impact of Artificial Intelligence Adoption on Jordanians' Customers Satisfaction. *Lecture Notes on Data Engineering and Communications Technologies*, *220*, 221–234. DOI: 10.1007/978-3-031-71619-5_19

Aguiar-Costa, L. M., Cunha, C. A. X. C., Silva, W. K. M., & Abreu, N. R. (2022). Customer satisfaction in service delivery with artificial intelligence: A meta-analytic study; [Satisfação do cliente na prestação de serviços com inteligência artificial: Um estudo meta-analítico]. *Revista de Administracao Mackenzie, 23*(6). https://doi .org/DOI: 10.1590/1678-6971/eRAMD220003.en

Al-Araj, R., Haddad, H., Shehadeh, M., Hasan, E., & Nawaiseh, M. Y. (2022). The Effect of Artificial Intelligence on Service Quality and Customer Satisfaction in Jordanian Banking Sector. *WSEAS Transactions on Business and Economics*, *19*, 1929–1947. DOI: 10.37394/23207.2022.19.173

Alqasa, K. M. A. (2023). Impact of Artificial Intelligence-Based Marketing on Banking Customer Satisfaction: Examining Moderating Role of Ease of Use and Mediating Role of Brand Image. *Transnational Marketing Journal*, *11*(1), 167–180. DOI: 10.58262/tmj.v11i1.1012

Atawneh, S. H., Hamadneh, N. N., Jaber, J. J., Al Wadi, S., & Khan, W. A. (2022). Using Artificial Intelligence to Predict Customer Satisfaction with E-Payment Systems during the COVID-19 Pandemic. *Journal of Mathematics*, *2022*(1), 1599785. Advance online publication. DOI: 10.1155/2022/1599785

Barik, K., Misra, S., Ray, A. K., & Sukla, A. (2023). A blockchain-based evaluation approach to analyse customer satisfaction using AI techniques. *Heliyon*, *9*(6), e16766. Advance online publication. DOI: 10.1016/j.heliyon.2023.e16766 PMID: 37292278

Brill, T. M., Munoz, L., & Miller, R. J. (2019). Siri, Alexa, and other digital assistants: A study of customer satisfaction with artificial intelligence applications. *Journal of Marketing Management*, *35*(15–16), 1401–1436. DOI: 10.1080/0267257X.2019.1687571

Hu, Y., Xiao, Y., Hua, Y., Fan, Y., & Li, F. (2024). The More Realism, the Better? How Does the Realism of AI Customer Service Agents Influence Customer Satisfaction and Repeat Purchase Intention in Service Recovery. *Behavioral Sciences (Basel, Switzerland)*, *14*(12), 1182. Advance online publication. DOI: 10.3390/bs14121182 PMID: 39767323

Katsamba, N., Kandiero, A., & Chizwina, S. (2024). The impact of artificial intelligence on customer satisfaction levels in the mobile telephony industry of Zimbabwe. In *Digital Influence on Consumer Habits: Marketing Challenges and Opportunities*. Emerald Publishing., DOI: 10.1108/978-1-80455-342-820241003

Kovács, G., Bogdanova, D., Yussupova, N., & Boyko, M. (2015). Informatics tools, AI models and methods used for automatic analysis of customer satisfaction. *Studies in Informatics and Control*, *24*(3), 261–270. DOI: 10.24846/v24i3y201503

Kukreti, M., Sehajpal, S., Tiwari, R., & Sood, K. (2025). *Driving Global Health and Sustainable Development Goals With Smart Technology* (Mohit, K., Sabina, S., Rajesh, T., & Kiran, S., Eds.; 1st ed.). IGI Global Scientific Publishing., DOI: 10.4018/979-8-3373-0240-9

Matosas-López, L. (2024). The influence of brand credibility and brand loyalty on customer satisfaction and continued use intention in new voice assistance services based on AI. *Journal of Marketing Analytics*. DOI: 10.1057/s41270-023-00278-8

Nanavare, R. A., & Ganbote, A. (2024). Development of Artificial Intelligence Model for AI-Empowered Brand Positioning With Tracking the Customer Satisfaction to Achieve Management 5.0. *2024 IEEE 9th International Conference for Convergence in Technology, I2CT 2024*. DOI: 10.1109/I2CT61223.2024.10543992

Nikolaos, L. (2009). Customers satisfaction in shipping enterprises of maritime cabotage with artificial intelligence and multicriteria decision analysis methods. *Proceedings of the 13th WSEAS International Conference on Computers - Held as Part of the 13th WSEAS CSCC Multiconference*, 639 – 645.

Oktavia, T., & Arifin, C. W. (2024). REVOLUTIONIZING E-COMMERCE WITH AI CHATBOTS: ENHANCING CUSTOMER SATISFACTION AND PURCHASE DECISIONS IN ONLINE MARKETPLACE. *Journal of Theoretical and Applied Information Technology*, *102*(19), 7224–7237.

Omisakin, O. M., Bandara, C., & Kularatne, I. (2020). Designing a Customer Feedback Service Channel through AI to Improve Customer Satisfaction in the Supermarket Industry. *Journal of Information and Knowledge Management*, *19*(3), 2050015. Advance online publication. DOI: 10.1142/S021964922050015X

Park, Y., Kim, J., Jiang, Q., & Kim, K. H. (2024). Impact of artificial intelligence (AI) chatbot characteristics on customer experience and customer satisfaction. *Journal of Global Scholars of Marketing Science: Bridging Asia and the World*, *34*(3), 439–457. DOI: 10.1080/21639159.2024.2362654

Prentice, C., Dominique Lopes, S., & Wang, X. (2020). The impact of artificial intelligence and employee service quality on customer satisfaction and loyalty. *Journal of Hospitality Marketing & Management, 29*(7), 739–756. DOI: 10.1080/19368623.2020.1722304

Ruan, Y., & Mezei, J. (2022). When do AI chatbots lead to higher customer satisfaction than human frontline employees in online shopping assistance? Considering product attribute type. *Journal of Retailing and Consumer Services, 68*, 103059. Advance online publication. DOI: 10.1016/j.jretconser.2022.103059

Sehajpal, S., & Kaur, N. (2025). Neuroscience, Nursing, and Financial Innovation : A Synergistic Approach to Optimized Patient Care and Economic Well-Being. In Sanjay, T., Bhupinderpal Pal, S. C., Amar, J., Ercan, O., & Pawan, K. (Eds.), *Neuromarketing's Role in Sustainable Finance* (pp. 133–146). IGI Global Scientific Publishing., DOI: 10.4018/979-8-3693-9117-4.ch008

Shaikh, A. A., Kumar, A., Mishra, A., & Elahi, Y. A. (2024). A study of customer satisfaction in using banking services through Artificial Intelligence (AI) in India. *Public Administration and Policy, 27*(2), 167–181. DOI: 10.1108/PAP-05-2023-0060

Shunmugasundaram, M., Ganesh, K. S., Gnanapragash, T. J., John, E. P., Raj, J. P., & Smitha, V. (2023). An Improved Artificial Intelligence based Service Quality to Increase Customer Satisfaction and Customer Loyalty in Banking Sector. *2nd International Conference on Sustainable Computing and Data Communication Systems, ICSCDS 2023 - Proceedings*, 433–439. DOI: 10.1109/ICSCDS56580.2023.10105048

Sofiyah, F. R., Dilham, A., Lubis, A. S., Hayatunnufus, , Marpaung, J. L., & Lubis, D. (2024). The Impact of Artificial Intelligence Chatbot Implementation on Customer Satisfaction in Padangsidimpuan: Study with Structural Equation Modelling Approach. *Mathematical Modelling of Engineering Problems, 11*(8), 2127–2135. DOI: 10.18280/mmep.110814

Song, X., & Bonanni, C. (2024). AI-Driven Business Model: How AI-Powered Try-On Technology Is Refining the Luxury Shopping Experience and Customer Satisfaction. *Journal of Theoretical and Applied Electronic Commerce Research, 19*(4), 3067–3087. DOI: 10.3390/jtaer19040148

Soundarraj, P. L., Javaraj, P., Meetna, S., Elaiyaraja, A., Naqvi, S. R., & Gamadia, R. (2024). Prioritizing Factors Affecting Customers Satisfaction in the Internet Banking Using Artificial Intelligence. *Proceedings of 9th International Conference on Science, Technology, Engineering and Mathematics: The Role of Emerging Technologies in Digital Transformation, ICONSTEM 2024*. DOI: 10.1109/ICONSTEM60960.2024.10568595

Wang, Y. (2022). Research on the Influence of Service Quality of Hotel Intelligent System on Customer Satisfaction Based on Artificial Intelligence Evaluation. *Mathematical Problems in Engineering*, *2022*, 1–9. Advance online publication. DOI: 10.1155/2022/3832935

Yella, A., & Suganyadevi, K. (2024). Improving Customer Satisfaction in Retail Banking through AI-driven Algorithm Optimization. *2024 15th International Conference on Computing Communication and Networking Technologies, ICCCNT 2024*. DOI: 10.1109/ICCCNT61001.2024.10725146

Zhang, J., Chen, Q., Lu, J., Wang, X., Liu, L., & Feng, Y. (2024). Emotional expression by artificial intelligence chatbots to improve customer satisfaction: Underlying mechanism and boundary conditions. *Tourism Management*, *100*, 104835. Advance online publication. DOI: 10.1016/j.tourman.2023.104835

Chapter 2
Unstrapping Role of Emotion AI and Robotics: AI in Customer Relationship Management

Bhupinder Singh
https://orcid.org/0009-0006-4779-2553
Sharda University, India

Ashima Jain
https://orcid.org/0000-0001-5365-2385
Manipal University Jaipur, India

Saurabh Chandra
https://orcid.org/0000-0003-4172-9968
Bennett University, Greater Noida, India

ABSTRACT

Artificial Intelligence of Emotion and Robotics AI just don't seem to be playing revolution in Customer Relationship Management (CRM), instead of helping business to know the feelings opt by people and their actions. Easily accomplished by satiating subtleties Emotion AI reads in interactions with clients through consumption of some technology (recognition of faces, speech analysis, and sentiment analysis) while setting a personal involvement of the business in return of interaction-courteously responding to it, making the customer happier and undoubtedly loyal. These are all made better by Robotics AI towards improved customer-attached treatment through the use of chatbots whose intelligence is based on AI, digital-personal assistants, and almost equally rudimentary robots for giving immediate help even managing trivial inquiries and for automation of works and assurances across several differ-

DOI: 10.4018/979-8-3373-1867-7.ch002

ent channels. With this advancement of the innovations, it will further strengthen customer relationships, increase loyalty, and promote growth.

INTRODUCTION

AI in CRM is revolutionizing the way organisations engage with their customers. The modern AI tools like machine learning, language processing and automated forecasting provides elaborate data with reference to the clients' activities, their choices and their requirements. This allows businesses to provide highly personalized yet effective interactions with consumers on a wide scale, which transforms CRM strategy. Both Emotion AI and Robotics AI can be leveraged for customer engagement, as long as they are intelligently designed towards building emotionally intelligent systems that could attract customers more. It is still important that data privacy and algorithmic biases are addressed with moral issues so that consumer confidence is maintained. (Chatterjee et al., 2021).

Reasons for undertaking this study stem from the realization that the management of customers on the basis of AI is rapidly transforming with clear implications for organizations in a plethora of industries. Therefore, as business organisations rely on AI in implementing CRM policy, it will be crucial to appreciate the managerial, management and management/ ethics of change. This work will pay a lot of attention to the aspect of AI to CRM analysis and concepts derived from various disciplines and different industrial settings (Dwivedi et al., 2021).

This research is inspired by the rapidly advancing technology, coupled with rising competition in the digital economy. With the advancement of AI technologies and increased availability, firms that do not embrace them face a loss in consumer loyalty and market leadership (Huang & Rust, 2021). AI adoption as a concept is requires precautions in a system as critical as CRM due to potential violations of data privacy and security, consideration of algorithm bias, aspects of transparency, and the ethics of the ability to make decisions on its own. Making decisions on the technological, organizational, and societal aspects of AI in CRM will be made easier when these challenges are understood well. (Patel, 2024) Due to the COVID-19 changes that occurred in organizations, there has been an escalation in digitalization. That is why customer engagement is particularly critical when it comes to long-term effective customer engagement during the period of economical instability and altering customer behavior patterns. Therefore, CRM solutions based on artificial intelligence have become the necessary tools for firms to cope with new conditions of the digital environment and improve the effectiveness of their customer-facing processes. The purpose of the present research is to investigate the place of AI in the framework of the examined CRM strategy and to identify practical implications derived from the case study in an attempt to enhance firms' AI effectiveness and efficiency.

EFFECTIVE STRATEGIES FOR CRM

Customer Relationship Management (CRM) is the what that organisations need to do in order to build and sustain profitable relationships with customers. Good CRM is the bread and butter of sustaining and thriving within the new world's highly competitive, customer-oriented environment. CRM established the customer as the focal point of all the company's transactions, communications and data stream concerning their interactions and experiences at multiple points in their lifecycle with the companies seeking to do business with them. A good CRM model will therefore embrace client acquisition, client retention, customer service delivery as well as client servicing and support. Client segmentation plays a central role as way of maintaining a strategic approach to giving solutions that meet specific client's needs hence enhancing the Customer Lifetime Value (HubSpot, 2023).

Customer relationship management support system is anchored on data management for customers where the information from different areas is compiled, merged and then analyzed in order to understand the consumption patterns of customers. A good structure for data and well-activated governance help to achieve the goal and guarantee the quality, security, and privacy of the information. These informations can help various organisations to make improved decisions regarding consumers and their experience. Explain how the use of formal and informal communication is essential in connecting emotionally, trust, and loyalty with sufficient relevant data and clients' response (Galic, 2024).

Consequently, CRM formula depends greatly on service quality and customer support that affects the general opinions and experiences in customers. The time factor reflected by satisfying customers on time, quality, and reliability puts an enterprise above others in terms of success rates. Good customer relations are more than a crisis response service; they are a way of dealing with most problems before they ever arise. This way, by focusing on such things as service line, staff education and ongoing improvement, an organization can build a good reputation to enable the acquiring of long term patronage from clients (Burk, 2023).

Technology is a key component of effective CRM strategies since supporting technologies offer the means for collecting and processing customer data, as well as performing operations and making decisions. To align CRMs success stakeholders in organizations have to pay scrupulous attention on such key performance indicators such as customer satisfaction, loyalty, and value. Constant examination of these indicators alongside daily and weekly changes, as well as occasional assessments of these metrics, make it possible to identify the area of strength and weakness, improve the usage of resources, and adjust the CRM strategy for enhancing further advancement (Drake, 2024).

CASE STUDIES

The integration of artificial intelligence into CRM has shifted from idea to the energy source for industry disruption. This section will then look at practical cases of AI implementation into the CRM by firms in order to show how companies are making it possible to enhance productivity and focus of attention to customers. It is hoped that by discussing these cases in terms of achievements, issues, and lessons learnt, a clearer idea of the applicability of AI in CRM systems, the impact of these systems, and future potential of advanced AI applications can be determined.

1. Retail and E-commerce

The industries that are leading in terms of applying Artificial Intelligence in CRM are the retail new and e-commerce base sectors with their focus on Personalization and Marketing Automation and Optimization. The best example of an organization that has cross-sold or 'recommended' the right product at the right time would be Amazon. Some examples of its effects; Using algorithms for recommending appropriate products, which targets customers that have a record of previous purchases or have viewed or left comments about similar products, increase customer interaction, and retention and sales. A case is Sephora of cosmetics which operates globally and here uses artificial intelligence such as chatbots and try-on features that enable customer to try beauty products, get recommendations as well as beauty help within short order through its mobile application as well as its point of sale (Parsani, 2024)

2. Financial Services

An example is Bank of America's Erica, the virtual assistant using natural language processing and machine learning to help consumers to check balances, make payment and locate an ATM. Erica has proved to be successful since it started by having over 10 million users and over 100 million requests. Likewise, JPMorgan Chase uses AI to find fraud by analyzing current transactional data and setting up patterns of abnormality. This has led to low rates of fraud, high customer trust in this method that employs aspects of artificial intelligence such as machine learning and behavioral analysis (Mann, 2024).

3. Telecommunications

The telecoms business is now rapidly adopting AI in CRM to enhance the processes, increase efficiency and the variety of services. For example, Verizon adopted AI incorporated chatbots and virtual assistants that based on NLP and machine learning can provide a swift response to questions of their client. The AI solutions react instantly and pass on critical issues to human operators while making a continuous customer support available – a clear boost to customer experience. Likewise, Telefónica applies AI to network management and for predicting equipment failure. Applying big data machine learning and deep analytics, the company's artificial intelligence system analyzes incoming data from the network and looks for signs of difficulties and opportunities for improvement (Telefonica, 2023).

4. Healthcare

Interactive and automated tools like chatbots and virtual assistants mean one can ensure that each patient is given appropriate health information and help. Babylon Health and Ada Health are some of the companies that are using AI to analyse symptoms, recommend preliminary diagnosis and proceed to recommend therapies from natural language processing and machine learning. This technology makes health care more accessible and decreases demands on the system but gives individuals the ability to run our own health. AI is also being used in development of precise treatment techniques and drugs. The two players in the market, IBM Watson Health and GNS Healthcare use patient data like health history and genetic profile to determine available treatments and medication targets. The mentioned AI algorithms are identifying complex structures in data; such structures enhance the effectiveness of individualized therapies and medicines development. It can also be a way to enhance the patient's quality of life, reduce expenses and facilitate new drugs' creation (Paul, 2021).

5. Lessons Learned and Future Directions

AI is in a position to revolutionalize customer relationship management (CRM) across industries through personalization, productivity, and being proactive. So, by using AI for personalization, companies get an opportunity to develop data and machine learning-based interactions that improve customers' loyalty. Through quick handling of general tasks and routing computerized replies and, passing tricky issues to actual employees, AI enhances the effectiveness of the customer relation process to provide quick solutions to consumers' problems hence customer satisfaction. Also, AI enables predictive CRM because it uses information from the real-time environment to identify potential threats, opportunities and keep organizations up to date on improvements in customer interfacing. Nevertheless, AI integration as a tool into CRM requires customer-oriented approach supported by deeper understanding of CRM as integration of technology into business models, organizational culture, and data governance.

The future of AI in CRM will always be progressive, innovative, and continuous. Advances in AI technology will mean even more frequent business use of the tools to enrich consumer experiences while maintaining a competitive edge. Main areas of interest for AI in CRM are:

- Natural language processing will enable much more natural and intuitive interactions with customers.
- Emotional AI: It is claimed that through combining the emotion detection and the SA, more efficient and friendly interactions with the consumer can be performed.
- AI by design hence makes use of AI systems that can be easily understood by other individuals in order to improve trust and responsibility in decision-making processes that involve artificial intelligence.
- Harnessing AI will enhance the customer experience thus increasing customer satisfaction across relationship touchpoints.

To illustrate its unprecedented role, this article presents the case studies and industrial examples of CRM that has been enhanced by the implementation of AI. A business is leveraging on information and emerging tools, machine learning, and natural language processing to record unrivaled advancements in efficiency, creativity, and influential share-of-mind among consumers. Pursuing the given AI use in CRM, firms must learn by experience, remain sensitive to all the development and guarantee ethical practice in the AI-supported customer relationships which tap this full potential to let customers connect better, to amass business success in the long run.

IMPACT OF AI ON CUSTOMER RELATIONSHIP MANAGEMENT

This literature review analyses the incorporation of AI into CRM systems with special reference to implementation strategies and challenges, emerging trends. It describes the major tasks of implementing AI-CRM with an emphasis on determinants of success and recommended procedures for formalisation of each task. Issues of ethical nature such as data privacy, incorporation of bias in algorithms, and a clarity of these issues too are discussed with ways through which such challenges can be managed. This likewise examines a future in the area of AI- CRM, specifically on the concept of human centered AI, new technologies that will be coming in the future, and their potential consequences in terms of customer satisfaction, company performance and social sustainability.

Strategies for Implementing AI in CRM Systems

AI integration into CRM systems is a strategic, implementation, and optimisation process. This section also analyzes practices defining the integration of AI into CRM and providing insight into a set of technological, organisational, and cultural factors enabling the successful deployment and adding value into the process.

1. Objectives

The first process of organizational success, on the use of AI into CRM is the need to have deep defined objectives with specific intents. This would include which CRM processes, customers, and critical metrics AI is applied to, to enhance in line with an organization's strategic plans. AI will be used either in as a marketing application or in improving lead scoring or as an application in augmenting support for customers to acquire more clients, retain those acquired and even generate higher lifetime revenue from those acquired. (Abugo, 2024)

Following the idea of objectives, the next step is to define the way to implement AI-CRM. This roadmap will outline key deliverables, interdependencies, and resources needed for each stage of the work, including data management, building models, pilot, and implementation and the like. Such a plan would be associated with the determination of key players and influencers in the situation. Recipients of the construction team also require elucidation of their responsibilities and duties.

2. Data Preparation and Integration

It is the quality, completeness, and access to customer data that will determine any application of AI-CRM. One preliminary activity would be to assess the current situation of the data and determine what is missing, conflicting and of low quality. Preparation for training includes cleaning the data procedures like deduplication and normalization. Linking into an AI-CRM system is further necessary after being prepared. For instance, when setting up the data streams which include real-time data pipeline, data APIs would also be required for installation apart from data governance, and data security that have to be placed at a level that respects clients' privacy in such functions (Blut et al., 2021).

3. Development and Training

The subsequent stage in the implementation of AI-CRM is obtaining the right AI models suitable for the CRM requirements. AI methodologies that can be used are the supervised which is suitable for predictive modelling, the unsupervised for consumer segmentation and the natural language processing for sentiment analysis and for developing chatbots. The process begins with the discovery phase: The data scientists and specialists deciding about important variables affecting people's behavior in the frames of a given project. Exploratory data analysis, hypothesis testing, and feature selection are all techniques that can help you identify the most relevant data (Groden-Morrison, 2022).

ISSUED FACED BY AI IN CRM

Where, while AI in CRM provides so many benefits to the organisations involved, they are also expected to recognize the drawbacks and limitations arising from its utilization. This part therefore highlights major technological, organizational and ethical factors likely to affect the success and sustainability of AI-CRM systems.

Data Quality and Availability

One of the major issues in implementing AI-CRM is ensuring customer data quality, completeness, and timeliness. Reliable and consistent data is relied on by AI algorithms, and gaps or biases in it might lead to inaccurate predictions. For instance, segregated or archaic CRM data may hamper the capability of the AI to give informative insights. Flaws or duplication in data also have potential effects on the results of AI. To address these concerns, organizations should focus on good data governance procedures such as standardization, integration, cleaning, and enrichment. There is need for clear and correct data ownership and stewardship roles as well as clear security rules to ensure timely access to accurate data (CData Software, 2023).

- *Algorithmic Bias and Fairness*

The application of AI-CRM systems has also its difficulties, one of which is to maintain a fair position with non-bias in decision making by AI. Those developed on bigoted or limited sample data would extend and perpetuate similar discriminations based on gender, ethnicity or age. For example, algorithmic leads score known from biased sale data will carry forward the favor or exclusion of specific groups, but chatbots that have learned insensitive transcripts will repeat the same.

To avoid these risks, there should be real auditing of AI models for bias created by checking for statistical parity or counterfactual fairness, inclusive of open patterns and responsible systems and ethical review and auditing.

- *Explain ability and Interpretability*

That being the case, one of the largest hurdles for implementation in AI-CRM is the ability for business users and customers to understand and explain the AI models. Models like deep learning or ensemble sometimes are also black boxes and it becomes difficult to understand what kind of prediction or recommendation a model is going to provide.

Since there is no explainability for decisions made by AI-CRM systems, this leads to decreased user trust, it is difficult to correct errors and AI judgments influence customer complaints cannot be effectively addressed. Such an increased interpretation can be realized through feature importance analysis, decision trees, and rules of decision making. Besides, the use of the easy to use interfaces, good visuals and proper documentation, user participation and level of understanding the AI results can also be enhanced (Kovacs, 2024).

- *Scalability and Performance*

This issue increases in significance as AI-driven CRM systems become large, data-hungry systems, challenging organisations to scale infrastructure and maintain algorithm effectiveness. The AI demands all three with some of the most demanding being elasticity in the processing power, storage and bandwidth for real-time model development and application. With the increasing mass and speed of the volume of consumers' data, the issues connected with timely data gathering, processing, and analysis remain critical. This needs an effective set up of AI that will be dealing with huge volumes of data and at the same time need to provide low latency API access while at the same time equally being highly available. To address these issues organizations should adopt AI systems on the cloud that are flexible and effective with an API for the CRM activities (IBM, 2024).

- *Privacy and Security*

The AI-CRM system's owners have to accentuate its protection from third-party interference since it deals with valuable and often personal information about clients (Vassilev et. al., 2024). AI-CRM systems must follow rules these data protection laws: GDPR & CCPA in essence defines the very high standard concerning the processing of the customers' information. In order to mitigate all the legal and reputational consequences, that has to be done. To avoid these challenges the organizations should employ some general data security controls such as- Encrypt, Anonymize or Restrict data access based on roles and responsibilities in order not to allow irrelevant users to access the data. To tackle such problems, it is required to use security audits and penetration testing. Thus, the clear procedure of working with AI data, clear permission and opt-out choices, and the establishment of its responsible management with clearly defined subject and object responsibilities are necessary for the safe AI deployment (Secure Privacy, 2023). AI will transform CRM in the future, but there are some imperative issues to be handled, such as data management, fairness, explainability, scalability, and privacy. Focusing on these areas, companies can build dependable, trustworthy, and long-lasting AI-driven CRM systems.

Emerging Trends and Future Directions in AI-CRM

It may be anticipated that there are many emergent themes of the future to influence the AI-CRM environment over the next years, especially due to the increasing technological developments as follows. This shows that organisations must be aware of such advancement in order to correctly use new opportunities in managing their AI-CRM approach.

There are emerging benefits of using AI into CRM, and companies in other industries have been revealing increased performance values. McKinsey & Company has observed that adoption of AI in sales and marketing can potentially generate incremental sales of up to 10% and decrease client acquisition cost by 15%. The above outcomes are mainly brought by the capacity of AI to design high custom consumer experience, accurate targeting, and automated customer interaction.

The greatest case for AI-CRM is Deloitte, which expects that machine learning and AI technology can unlock $4.4 trillion in global company value by 2025. Over a third of organizations have signed their lips to feasible adoption of AI inclusiveness in the core business activities of their organizations within the next three years. For instance, the rate quote AI computers need large volumes of client data like buying habits, choice, and geographical data to get special promotion/repeat business/ intention to sell more.

Apart from increasing revenues and decreasing costs, there is much more than just advanced monetary profits when utilizing AI in CRM. By using AI, predictive analytics helps the organisations generate the best return on investment made in the marketing sector hence improving the ROI. Through such analysis of the client data and behavior the AI systems are capable of identifying the right channel, message and offer for a distinct client. In this manner, marketers are able to make better use of resources and generate better conversion rate and resource productivity. For instance, the Sephora – a cosmetics retailer company that operates in worldwide, successfully implemented many AI-based marketing strategies and gained great outcomes, for example, in mobile purchasing traffic, it was increased by 300%. More importantly, when using the apps, the clients tended to buy a second time more than the average client. The above illustrations are a testimony of how AI can transform CRM and also bring in more of the incomes and efficiency. At present, when the technology gets advanced and is more easily accessible, the organisations who use these technologies in the right and efficient manner for developing the engagement and personalisation aspects of the customer relations will be able to grab a sustainable competitive advantage which in turn would lead to the main financial angles of success and prosperity.

- ***Conversational AI and Natural Language Processing***

With the convergent integration of chatbots and natural language processing in to the AI-CRM, there is growing naturalness to this interaction that is helping fashion a better alignment between end-users and solutions under CRM. Included in conversational AI would be its varieties such as: chabots, and all other forms of dialogue-based tools; virtual assistant. These conversational AI possibilities are supported by other NLP functionalities including sentiment analysis, entities' identification, or classification of intents. For example, an NLP skilled chatbot used to converse with customers, may identify the sentiment and the severity level of a client's complaint, and escalate the ticket to the appropriate service provider or problem solving process (Huble Digital, 2023).

- ***Augmented and Virtual Reality***

A further direction of AI in CRM is toward an additional reliance on technologies and on virtual and augmented reality technologies in the service of the consumer on the one hand and products that interact somewhat more dynamically, richly, or with increased texture or sensuality on the other. Interaction with these may include 3D appreciation, realistic films, 360-degree reality, and haptic feedback. For example, by infusing e-commerce with augmented reality technology, buying firms can give web-based consumers an idea of how garments or other accessories would fit them or let consumers visualize how furniture items displayed online would look like in actual homes before making a purchase. Likewise, the tourism businesses may incorporate the use of virtual reality to enable consumers to view virtual tours that are very realistic, to be in a position to book their destinations, hotels, attractions more wisely.

For the use of AR and VR in CRM, the advantages than hold entail enhanced visualization of the product, multiple opportunities in personalization, customers, loyalty, and decisions in conversion. There are numerous and unique applications of picture and video creation and recognition technologies in CRM systems. By integrating a technology called computer vision with deep learning algorithms, businesses can sift through and gain valuable insights from huge amounts of visual data generated by consumers during their engagements in different areas of the digital ecosystem including social media, mobile apps and the physical world interfaces they experience at the interaction points of a brand or retailer. Studies indicate that the present market share would grow to approximate $41.11 billion in the year 2030 (Simplilearn, 2023).

Face recognition in CRM applications it seems has a bright use case in visual analytics especially in the sentiment analysis application. This technology enables organizations to categorize the images and the videos posted by consumers through social media platforms using AI features on detecting emotion, objects, and the environment. All these insights provide live information about the customers' thoughts of them, perception towards and experiences they had with these organizations. For example, a restaurant company can monitor how its customers react to some changes such as new products on the menu, or changes in the appearance of the shops in order to identify areas they could improve. Likewise, social media organizations may employ VSA so that they can evaluate the response of consumers to its content, look for hidden emotions and suggest related products associated with such themes consumers feel closely attached to.

The third novel application of picture and video generation in CRM is personalization content. An example in this regard is that a business might use GANs and deep learning to create startlingly bespoke photographic scenes and videos for a customer. For example, a fashion shop may use images of products produced by an AI to recommend specific items of clothing and accessories that will suit a customer in terms of his/her previous purchases and the body type according to the latter.

Video recognition technology is also useful in customer support and service interactions in CRM. AI is able to process video streams and recordings for the identification of nonverbal cues - e.g., facial gestures, and voice - allowing, in turn, the delivery of empathic and context-dependent assistance. [Example] For example, an insurance company can leverage video recognition to categorize and allocate consumer emotions throughout an insurance claim, directing emotionally volatile customers to specialized agents to provide personal care.

When the combination of picture/video, identification technology, and production/identification technologies are incorporated in CRM systems, new opportunities for organisations to enhance customer interaction and deliver unique experiences become available7,8. For example, AR and VR information integrated together with computer vision allows for the interactive presentation of products, virtual try-on/try-before-you-buy, and computer assisted problem diagnoses. An AR-based image recognition can be used in an upholstery firm to give the clients a vision of how their household furniture will fit in their real living environment, which, in turn, may improve the decision-making process and reduce the rate of return.

These examples illustrate the mechanism through which the combination of pictorial and visual recognition features in CRM software offers organizations with novel ways in which to know, express, and act upon their customers. Those organizations which exploit in a synergistic way the wealth of visual data and of learn intelligence of artificial intelligence can retrieve meaningful information, design more personalized services and therefore increase of them customers satisfaction and fidelity.

In order for AR and VR to be effectively used in CRM, businesses are going to need to make the investment to develop high-fidelity and scalable AR/VR content and apps, that are multiplatform and multidevice platform ready. Additionally, it is an absolute necessity to have effective consent mechanisms so that the current privacy and security issues associated with gathering and analyzing client data in AR/VR situations can be circumvented.

- ***Personalization and Recommendation Systems***

AI-based personalization and recommendation engines are changing customer relationship management, as they enable enterprises to deliver highly customized and appropriate experiences to every customer. Personalisation refers to the application of artificial intelligence (AI), machine learning (ML) techniques in the collection of client data and behaviours for the dynamic adjustment of content, offers, and conversation depending on individual needs and preferences. Instead, recommendation systems leverage AI on the one hand to deliver items, services or activities to a consumer, and on the other hand they sometimes that this consumer also engaged in prior behaviors, the preferences of other users with whom this consumer is like, or additional factors. The AI tool is for an online streaming service the use to suggest movies or TV episodes to users according to their viewing history and review, in conjunction with some demographics and preferences from other similar users (AIContentfy Team, 2023).

Many different advantages can be found within CRM systems in the context of personalisation and recommender systems. These technologies allow customers to maximize happiness and loyalty by giving personalized experiences based on customer need and desire. These are also technologies that increase conversion rates and revenue by providing products and promotions that customers want. Perhaps even more importantly, the systems present significant information about clients' behavior, and organizations can actually perceive patterns and categorize clients into groups of similar interests. For such solutions to be truly embraced, organisations need to hold a robust and cohesive consumer data platform that aggregates and processes data from many sources and channels. Specifically, the creation and training of artificial intelligence (AI) models that can consistently tell us what clients want is key to making successful suggestions in the business context and practicing business constraints. Yet, companies have to tread carefully between individualization and privacy. This includes openly informing clients what they are used for in the creation of customisations and ensuring that personal recommendations are not overly intrusive or upsetting, which could break trust and reduce participation.

- ***Explainable and Ethical AI***

Since AI is playing an important role in the decision-making and interaction process in relation management, the need for the transparency, accountability, and fairness of AI systems is increasing. There is, therefore, a pressing need for explainable and ethically justifiable AI. The simplest definition of explainable AI would be the ability to build systems that give clear and easily interpretable explanations for their predictions or suggestions such that such systems would be trust and credible among users (Chakraborti et al., 2020).

By contrast, ethics in artificial intelligence is about architecture, design and implementation of systems that safeguard human values, human rights and human well-being, and at the same time, settle unanticipated harm or bias. Scenario-wise, an ethical AI-based credit scoring algorithm is expected to apply reasonable and relevant information for its decision-making processes, never carrying forth any historic discrimination or past biases about race, gender, or any other protected traits. Finally, it is recommended that organizations establish a clear procedure by which stakeholders may provide feedback and raise concerns or seek redress if they feel that an AI system has acted unfairly or caused harm. This integrated approach bears great significance to the successful implementation of the proper use of AI in CRM.

- ***AI-Powered Customer Journey Optimization***

A trend that will be possible for AI-enhanced CRM is the use of artificial intelligence to improve and humanize the customer journey in a comprehensive way across several channels and stages. The optimization of the customer journey includes the evaluation and refinement of the sequence and quality of interactions that a customer has while going from awareness and consideration for buying to buying and using, and to loyalty.

For instance, by analyzing a customer's browsing and search behaviors on a website, AI could proactively recommend related material, goods, or services to facilitate a sale. In addition, AI is going to be in charge of watching what customers do after purchasing and how they feel so it can trigger customized programs for maintaining or promoting leavings of the client that will strengthen a long-term interaction with the brand and its community, thus building the brand loyalty. AI enabled customer journey optimization offers several key advantages in customer relationship management. Companies using AI-based optimisation will have a complete and consolidated view of the customer journey across all stages and every touchpoint and all channels. An effort will have to go into the development of AI models that can be used to forecast the demand or choice or actual behavior of the customer at any of the experience touchpoints. In other words, a model must be devised to pinpoint the most appropriate response or offer in relation to the known goals and simultaneously within the permissible constraints.

Additionally, companies need to bring AI plans in line with the values and unity of the brand and it is necessary to maintain consistent way of consumer experience (Langenstueck, 2023). Conversational AI, augmented and virtual reality, highly personalized, and ethical AI all have the potential to deliver more engaging, highly personalized, and even disruptive customer experiences and enhance CRM efficiency and effectiveness. In order to realize the full dividend of these innovations, organizations need to adopt a reflective and holistic attitude to the use of AI. This can only be achieved by the resources at the disposal of the institution in the process of creating the team of skilled individuals, the use of suitable technology and robust governance. In addition, participatory involvement by consumers and other stakeholders in the design and deployment of AI systems will play a crucial role for success. With these principles in mind, companies are expected to invest in leveraging all that AI-driven CRM offers in order to nurture highly productive and symbiotic customer relationships.

CONCLUSION AND FUTURE SCOPE

There is a shift in CRM strategy due to artificial intelligence as it changes the relationship of the firm with customers and creates value in the digital age. Companies are announcing the advent of new ways to improve the customer experience utilising the potential of analytics, automation and personalisation potential made available through artificial intelligence. However, that alteration is not without its own uncertainties and complications, which would necessitate careful design and ethical realisation.

AI-powered CRM allows companies to deliver individualised, timely, and targeted customer interactions that, in turn, can contribute to increased loyalty, satisfaction, and adoption. Companies can develop, through analysis of big statistics from digital platforms, a deep understanding of client likes, behavior, and needs. This allows for more accurate communication and product proposals. Furthermore, routine task automation therefore releases human resources for activities that are actually meaningful, for example, for establishing connections, for finding solutions, or for creative input.

Notwithstanding the positive outcomes, it is inevitable that inclusion of artificial intelligence in CRM is a risky and troublesome business. Proactive measures are needed to address data breaches and algorithmic discrepancies, job displacement, and the risk of losing customer confidence. The greater the sophistication of AI systems will necessitate greater caution by organizations in deciding the extent to which client interaction should be outsourced to robots and therefore openness, accountability, and a provision of association to human values are paramount in the development of trustworthiness and confidence.

For a company to survive in a changing market, organisations need to adopt a strategic approach that combines innovation and responsibility. The first step to successful AI implementation is the provision of a unified vision for the AI's involvement in customer interactions that integrates with the company's mission and values. The development of efficient data, technology, and human resources infrastructure is an important step in efficiently embedding AI. This would also serve the effective implementation of AI technologies by conceptualizing a world-class collaborative culture of the organization.

All the main stakeholders are the customers, workers, partners, regulators and civil society. Firms need to transparently state what the potential gains and harms of AI are, seek input on system design and regulation, and participate in the sector-wide dialogue over the feasibility and ethics of AI. Firms that forge trust, fairness and accountability will sit at the vanguard of the design and implementation of the rules and norms of responsible AI.

Use of AI in CRM initiatives is a continual learning process until such a point as it becomes cultural, adaptive, and participative. AI will help to place the client at the centre of the organization and its benefits can be great in making the system more acceptable-simply put, in every part of the customer experience. When put next to the broader ethical and environmental approaches, AI can be used to reach out to clients. Organizations that implement AI-based CRM solutions can then positively handle the challenges in their journey towards sustainability together. AI's impact on CRM-stepping back for the greater picture-shows in this article that it means something to business and academic stakeholders, as it provides excellent practical insight and particularly a roadmap for applying AI at a broader, business-relevant scale on ethical grounds. In this research, the framework for the development of AI-based CRM processes is built based on multidisciplinary concepts and solutions with an applied nature. With the subject continuing to develop, it is destined that results will encourage even greater innovation, research and collaboration, eventually paving the way for a smarter, fairer and more sustainable world in business and society.

REFERENCES

Abugo, E. (2024). AI in CRM and ERP Systems: 2024 Trends, Innovations, and Best Practices. Microsoft Dynamics 365 Blog. *Available at:*https://www.microsoft.com/en-us/dynamics-365/blog/businessleader/2024/03/04/ai-in-crm-and-erp-systems-2024-trends-innovations-and-best-practices/. (last visited on: 11-01-25)

AIContentfy Team. (2023). The Role of AI in Content Recommendation. AIContentfy. *Available at:*https://aicontentfy.com/en/blog/role-of-ai-in-content-recommendation, (last visited on: 14-01-25)

Blut, M., Wang, C., Wunderlich, N. V., & Brock, C. (2021). Understanding Anthropomorphism in Service Provision: A Meta-Analysis of Physical Robots, Chatbots, and Other AI. *Journal of the Academy of Marketing Science*, *49*(4), 71–92. DOI: 10.1007/s11747-020-00762-y

Burk, C. (2023). Small Business CRM Best Practices for Customer Support Teams. Keap. *Available at:*https://keap.com/business-success-blog/customer-service/crm-best-practices-for-customersupport-teams. (last visited on: 11-01-25)

CData Software. (2023). What is Data Governance? Key Concepts, Importance, and Benefits. *Available at*: https://www.cdata.com/blog/what-is-data-governance, (last visited on: 11-01-25)

Chatterjee, S., ... (2021). Adoption of Artificial Intelligence-Integrated CRM Systems in Agile Organizations. Technological Forecasting and Social Change. *Available at:*https://www.sciencedirect.com/science/article/pii/S0040162521002158. (last visited on: 13-01-25)

Drake, A. (2024). CRM Customer Engagement: Using Behavioral Data to Boost Engagement. Active Campaign. *Available at:*https://www.activecampaign.com/blog/crm-customer-engagement. (last visited on: 13-01-25)

Dwivedi, Y. K., Hughes, L., Ismagilova, E., Aarts, G., Coombs, C., Crick, T., & Williams, M. D. (2021). Artificial Intelligence (AI): Multidisciplinary perspectives on emerging challenges, opportunities, and agenda for research, practice and policy. *International Journal of Information Management*, *57*, 101994.

Galic, D. (2024). Customer Communication: 8 Tips to Build an Effective Strategy. Zendesk. *Available at:*https://www.zendesk.com/blog/customer-communication/. (last visited on: 11-01-25)

Groden-Morrison, A. (2022). Importance of Data Quality in AI Implementation. Unite.AI. *Available at:*https://www.unite.ai/importance-of-data-quality-in-ai-implementation/. (last visited on: 11-01-25)

Huang, M.-H., & Rust, R. T. (2020). A Strategic Framework for Artificial Intelligence in Marketing. Journal of the Academy of Marketing Science. *Available at:*https://www.researchgate.net/publication/346661973_A_strategic_framework_for_artificial_intellige nce_in_marketing. (last visited on: 13-01-25)

Huble Digital. (2023). Unlocking the Power of AI in Your CRM: All You Need to Know. *Available at:*https://www.hubledigital.com/en-us/blog/ai-crm. (last visited on: 13-01-25)

HubSpot. (2023). Customer Segmentation: How to Segment Users & Clients Effectively. HubSpot. *Available at:*https://blog.hubspot.com/service/customer-segmentation. (last visited on: 11-01-25)

IBM. (n.d.). What Is Edge Computing? *Available at:*https://www.ibm.com/topics/edge-computing, (last visited on: 14-01-25).

Kovacs, G. (2024). XAI for All: Can Large Language Models Simplify Explainable AI? arXiv. *Available at:*https://ar5iv.labs.arxiv.org/html/2401.13110v1, (last visited on: 14-01-25)

Langenstueck, L. (2023). AI in Customer Journey Mapping: Elevating the CX Strategy. AIgantic. *Available at:*https://www.aigantic.com/ai-marketing/ai-customer-journey-mapping/. (last visited on: 12-01-25)

Mann, J. (2024). Jamie Dimon Says AI Will Have 'Extraordinary' Consequences and May Be as Transformative as Electricity and the Internet. Business Insider. *Available at:*https://www.businessinsider.com/jamie-dimon-jpmorgan-ai-extraordinary-consequencestransformative-internet-electricity-2024-4. (last visited on: 11-01-25)

Parsani, P. (2024). Beauty and the Bot: How Sephora Reimagined Customer Experience with AI. Cut The Saas. *Available at:*https://www.cut-the-saas.com/ai/beauty-and-the-bot-how-sephorareimagined-customer-experience-with-ai. (last visited on: 12-01-25)

Patel, M. (2024). The Ethics of AI: Addressing Bias, Privacy, and Accountability in Machine Learning. Cloud That. *Available at:*https://www.cloudthat.com/resources/blog/the-ethics-of-ai-addressingbias-privacy-and-accountability-in-machine-learning. (last visited on: 13-01-25)

Paul, D., Sanap, G., Shenoy, S., Kalyane, D., Kalia, K., & Tekade, R. K. (2021). Artificial Intelligence in Drug Discovery and Development. Drug Discovery Today, 26(1), 80-93. *Available at:*https://www.sciencedirect.com/science/article/pii/S1359644620304256?via%3Dihub. (last visited on: 10-01-25)

Secure Privacy. (2023). AI and Personal Data Protection: GDPR and CCPA Compliance. *Available at:*https://secureprivacy.ai/blog/ai-personal-data-protection-gdpr-ccpa-compliance, (last visited on: 10-01-25)

Simplilearn. (2023). The Power of Computer Vision in AI: Unlocking the Future. *Available at:*https://www.simplilearn.com/computer-vision-article. (last visited on: 10-01-25)

Telefonica. (2023). Telefonica Uses Artificial Intelligence to Optimize Network Management. Telefonica. *Available at:*https://www.telefonica.com/en/communication-room/press-room/telefonicalaunches-aura-and-leads-the-integration-of-artifical-intelligence-in-its-networks-and-customercare/. (last visited on: 12-01-25)

. Vassilev, A., Oprea, A., Fordyce, A., & Anderson, H. (2024). NIST Identifies Types of Cyberattacks That Manipulate Behavior of AI Systems.

Chapter 3
From Transactions to Relationships:
The Role of Emotional Intelligence in Customer Engagement

Ridhima Sharma
https://orcid.org/0000-0002-7851-0292
Vivekananda Institute of Professional Studies-TC, India

Vaishali Sethi
https://orcid.org/0000-0001-9827-6467
University Canada West, Canada

Jihene Mrabet
https://orcid.org/0000-0003-1848-5918
Amity University, Dubai, UAE

ABSTRACT

Customer interaction is becoming a major factor determining corporate performance in the very competitive and emotionally charged market of today. Emotional intelligence (EI) is investigated in this paper in order to improve customer engagement with an emphasis on how emotional competences of frontline staff affect customer happiness, loyalty, and advocacy. Using a qualitative study approach, twenty frontline workers and twenty consumers from various service sectors were semi-structured interviewed. Results show that empathy is fundamental in emotional involvement; consumers value real understanding and emotional responsiveness over transactional effectiveness. This study also emphasizes generally that emotional intelligence is a strategic facilitator of deep, lasting consumer involvement rather than a peripheral ability. Companies who make investments in emotional intelligence development at

DOI: 10.4018/979-8-3373-1867-7.ch003

all organizational levels are more suited to build strong emotional ties and guarantee long-term consumer loyalty in an economy driven by experience more and more.

INTRODUCTION

Customer interaction in the dynamic and ever linked market of the twenty-first century has developed outside conventional ideas of transactional interactions. Companies are now assessed not just on the quality of emotional experiences they provide to their consumers but also on the goods or services they present. In this regard, Emotional Intelligence (EI)—the capacity to identify, comprehend, control, and affect emotions—has become more important for developing close customer connections. Knowing how important emotional intelligence is to customer interaction gives companies a unique competitive edge that will help them to create sustainable development, customer happiness, and brand loyalty. Psychologist Daniel Goleman coined the term emotional intelligence to describe a set of emotional and social skills that define how well people view and express themselves, grow and preserve social relationships, handle adversity, and make meaningful and effective use of emotional data. Originally connected with leadership and interpersonal dynamics in the workplace, emotional intelligence has since become important in many different fields, especially in responsibilities with customers. High emotional intelligence employees are more suited to understand consumer emotions, react sympathetically, and modify interactions to fit both stated and unmet demands. This capacity to humanize corporate interactions drastically changes the consumer experience and promotes emotional resonance, trust, and belonging. Emotional intelligence becomes the link that ties business goals with human-centered service at a time when consumers want personalization, responsiveness, and empathy. Consumers of today want to be heard, understood, and appreciated more than they do answers. Emotionally attached consumers are more than twice as valuable to companies as highly happy consumers, according a Harvard Business Review analysis. Emotional involvement generates brand support, recurring business, and a greater tolerance of service faults. Frontline staff members with emotional intelligence may establish these relationships by spotting minor emotional signals — dissatisfaction, enthusiasm, uncertainty — and reacting with customized solutions that improve the client experience from a simple transaction to a memorable one.

Moreover, emotional intelligence directly affects the capacity of a company to control discontent and conflict. Under service recovery models, consumers are assessing not only the resolution result but also the responsiveness and empathy of the engaged representatives. Key actions that greatly improve consumer impressions of justice and caring include identifying customer feelings, validating their experience, and showing real concern—an emotionally savvy approach to dispute management. Under careful management, these exchanges may turn unhappy consumers into devoted champions. Harnessing emotional intelligence for consumer involvement also depends much on organizational culture. Businesses who include emotional intelligence into their policies, training courses, and leadership style foster at every customer interaction empathy, active listening, and emotional awareness. For instance, companies like Ritz- Carlton and Zappos are honored for their emotional sensitivity and first-rate customer service in addition to their product excellence. Usually resulting in amazing brand loyalty, these companies enable staff members to exercise discretion, show empathy, and develop tailored experiences that really connect with consumers. The digital revolution of customer involvement emphasizes even more the need of emotional intelligence. In a world when encounters are progressively mediated by screens—via chatbots, emails, and social media—conveying empathy and emotional understanding has grown more difficult but also more important. Companies that successfully include emotional intelligence into their digital strategies—that is, by customizing automated messages, actively listening on social media, and carefully reacting to customer comments—can preserve real emotional connections in virtual environments. In digital consumer engagement, emotional intelligence is realizing the limits of nonverbal communication and then offset by warmth, clarity, and response in textual exchanges.

Customer segmentation and focused engagement techniques also depend much on emotional intelligence. Knowing consumer emotions helps companies' service teams and marketers to produce more emotionally relevant experiences, offers, and messages. Emotional intelligence helps companies go beyond demographic profiling into psychographic and emotional profiling — thereby understanding not just who consumers are but also their feelings, values, and aspirational behavior. Companies may create stories and experiences that very closely fit consumers' identities and emotional needs by leveraging their emotional drivers, therefore fostering deeper emotional relationships and more brand loyalty. Thus, it is not a luxury but a strategic need for training and emotional intelligence development among staff members who interact with customers. Good emotional intelligence development emphasizes on improving self-awareness, self-control, social skills, empathy, and drive. Organizations may provide staff with the emotional skills required to negotiate challenging client contacts by means of seminars, role-playing, feedback systems, and reflective practices. Moreover, employment procedures that evaluate emotional intelligence in addition to technical and cognitive abilities guarantee that the employees are naturally inclined to provide good client experiences. It is crucial to understand, therefore, that the use of emotional intelligence has to be real. Consumers may quickly spot contrived or dishonest emotional outbursts, which fuels suspicion and skepticism. Real emotional involvement calls for honesty, deliberate emotional work, and a sincere will to really connect with consumers. When really used, emotional intelligence enables staff members to be flexible, honest, and contextually aware, therefore providing encounters that seem unique and significant rather than mechanical or manufactured. At last, emotional intelligence is not just the duty of individual workers; it also has to be supported and set an example by leadership. High emotional intelligence leaders set the standard for business culture, motivate emotionally intelligent actions among staff members, and give customer-centric methods that respect emotional involvement first priority. Policies, procedures, and incentives resulting from a leadership commitment to emotional intelligence support the primacy of emotional ties in efforts at customer engagement. A great but sometimes underused tool in the effort to improve consumer engagement is emotional intelligence. Growing emotional intelligence at all levels of the company presents a convincing road forward as companies try to stand out in crowded marketplaces and generate durable competitive advantages. Investing in emotional competences helps companies strengthen customer connections, build loyalty, and turn daily contacts into emotionally rich events that consumer's value and support. Emotional intelligence is not just a desired quality but also a strategic need in a market where brand success is driven by emotional ties. This research looks at how emotional intelligence (EI) could promote customer involvement by analyzing both consumer emotional perceptions and employee emotional capacities.

LITERATURE REVIEW

Scholars and professionals have been driven to investigate the psychological and emotional aspects of consumer involvement by the changing market marked by intense competition and raised customer expectations. Defined as the capacity to see, comprehend, control, and manage emotions (Goleman, 1995), emotional intelligence (EI) has become ever more important as a basic driver of good customer relationship management. Examining how emotional intelligence impacts customer interactions, loyalty, and brand advocacy, this literature review critically reviews current studies on the junction of emotional intelligence and consumer engagement. Salovey and Mayer (1990) initially proposed emotional intelligence as a kind of social intelligence that helps people to differentiate among others, monitor their own and others' emotions, and utilize this knowledge to direct behavior and thought. Emphasizing five basic components—self-awareness, self-regulation, motivation, empathy, and social skills—Daniel Goleman (2014) popularized the idea in the corporate setting. Later research by Mayer et al., 2004 polished EI into a model stressing emotional perception, facilitation, comprehension, and management. Emotional intelligence helps staff members in service environments to manage personal stress, react sympathetically to client emotions, and keep good relationships even in the middle of arguments or service breakdowns (Pugh, 2001). These features directly affect the quality of consumer experiences, hence increasing engagement and loyalty. Customer engagement, in the context of a brand or company, is the degree of the emotional, cognitive, and behavioral investment a consumer shows (Brodie et al., 2011). Engaged consumers are more likely to remain loyal over time, advocate for the brand, and engage in brand-related events (Vivek et al., 2012; Hsu et al., 2024). Fostering engagement depends much on emotional ties as they provide personal connection and affinity between the consumer and the brand (Pansari & Kumar, 2017). According to research by Hollebeek (2011), consumer emotional resonance—that is, their capacity to emotionally identify with a brand—increases engagement intensity (Tutan & Kökalan, 2025). Emotional intelligence thus becomes a strategic tool for increasing customer engagement as it helps companies to better grasp and affect consumer emotions, thereby strengthening their impact. Many studies have underlined for frontline service workers the importance of emotional intelligence. Higher emotional intelligence salespeople, according to (Kidwell et al., 2011), provide better customer outcomes including more loyalty and satisfaction. Building trust and rapport required their capacity to decipher consumer emotions and modify their actions. Grandey (2000) first proposed the idea of emotional work, pointing out that many times employees must control their emotions to satisfy corporate expectations. High EI employees are better able to manage emotional work without burning out, therefore preserving real and good client relationships (Brotheridge & Lee, 2003). This capacity to maintain true emotional expressions enhances consumer impressions of service authenticity, which is a main factor influencing interaction. Moreover, interactions in customer service driven with emotional intelligence usually provide more customized experiences. Studies by (Hennig et al., 2006) show that personalizing improves relationship benefits, customer happiness, and finally customer involvement.

A fundamental component of emotional intelligence, empathy has been found to be strongly correlated with consumer engagement and satisfaction. Those that give sympathetic services are seen as more understanding, compassionate, and supportive—qualities that help to establish emotional links to consumers (McBane, 1995). Mencl and May's 2009 research indicated that empathy helps staff members to recognize unmet customer demands and provide anticipatory service, therefore enhancing the customer experience. Critical elements for building client loyalty and emotional attachment include being appreciated and respected created by sympathetic encounters. While service failures are unavoidable, emotionally sensitive actions during service recovery can transform unpleasant events into chances for consumer interaction (Ghali et al., 2024). Research by Tax et al., 1998 indicates that good service recovery addresses not only the problem but also the emotional reaction of the consumer to the failure. Empathy and emotional support throughout rehabilitation procedures reportedly improved consumer forgiveness and post-recovery satisfaction, according to Mattila and Enz (2002). Workers with emotional intelligence—that is, those who recognize customer feelings and really apologize—may regain confidence and even improve customer loyalty above pre-failure levels. The manifestation of emotional intelligence in virtual contexts has evolved into a new topic of academic attention as digital client interactions grow. Organizations that include emotionally intelligent communication into emails, chatbots, and social media contacts are more effective in preserving emotional relationships even if digital channels lack numerous nonverbal emotional signals (Chung & Han, 2024; Rahmaniya et al., 2025). Recent research by Sharma et al., 2024 contend that customer involvement in online settings depends critically on "digital empathy," the ability to transmit understanding and emotional sensitivity via digital methods. Companies who pay close attention to consumer comments, react quickly, and customize digital contacts depending on emotional data are more likely to keep emotionally engaged consumers. Customer engagement results are strongly influenced by organizational support of emotional intelligence growth. Emotionally intelligent leadership fosters environments of trust, open communication, and psychological safety, according to research by Luria and Berson (2013), therefore enabling staff members to apply emotional intelligence with clients. Training courses that boost emotional intelligence competencies—active listening, emotional control, and sympathetic communication—are connected to better customer service performance (Slatten, 2008). Furthermore, hiring for emotional intelligence abilities guarantees that staff members who interact with customers are naturally ready to provide emotionally charged experiences in addition to technical talents. Scholars warn against too simple interpretations even if the link between emotional intelligence and customer engagement is mostly positive. Joseph and Newman (2010) contend that emotional intelligence is not a panacea and that its effects are mediated by elements including organizational support, employee well-being, and cultural environment. Moreover, worries about emotional labor—that is, when workers have to show emotions they do not actually feel—showcase certain drawbacks of emotional intelligence in consumer interaction techniques. The groundbreaking research on emotional labor published by Hochschild (1990) cautions that, without appropriate support from the company, too much emotional control may cause emotional tiredness. The research again confirms how important emotional intelligence is as a fundamental driver of consumer interaction. Emotional intelligence strengthens emotional ties, improves service recovery efforts, and personalizes consumer experiences both offline and online by allowing sympathetic, real-time, and emotionally tuned interactions. Organizations still have to handle emotional intelligence development carefully, though, thus making sure staff members are really assisted rather than burdened by emotional labor expectations. Embedding emotional intelligence into customer interaction initiatives will remain a crucial road to reach deeper, more meaningful customer connections as customer expectations change.

RESEARCH METHODOLOGY

Aiming to investigate in-depth how emotional intelligence (EI) among employees affects customer engagement, this study uses a qualitative research approach. This study would be well suited for a qualitative method as numerical data by itself cannot adequately represent emotional experiences, perspectives, or interpersonal interactions. Rather, rich, detailed information gathered from interviews offers a more thorough awareness of the nuances in emotional interactions influencing consumer loyalty, happiness, and advocacy. Frontline staff members and consumers will be interviewed semi-structurally for the study. This approach lets participants freely share their experiences while making sure the discussion stays centered on important facets of emotional intelligence and customer interaction. Participants will be chosen via deliberate sampling depending on their relevance to the study issues. Twenty front-line staff members and twenty consumers provide variety across sectors and demographics, therefore enabling a range of emotional experiences. Effective coding, topic building, and administration of vast textual data will all be accomplished using NVivo qualitative data analysis tool.

Data Collection Methods

Primary Method:

- **Semi-Structured Interviews:** Interviews will follow a flexible guide with open-ended questions designed to prompt detailed narratives around emotional intelligence and customer engagement.
- **Interview Medium:** Interviews will be conducted either face-to-face or via secure video conferencing platforms (e.g., Zoom or Microsoft Teams).

Data Analysis

This study explored how emotional intelligence (EI) demonstrated by frontline employees influences customer engagement in service industries. Using a qualitative approach through semi-structured interviews with 20 employees and 20 customers, thematic analysis was applied following Braun and Clarke's (2006) six-phase model. Thematic saturation was achieved, with no new themes emerging after coding approximately 35 interviews. The findings are organized around six major themes: Empathy and Emotional Connection, Authenticity in Emotional Expression, Conflict Resolution through Emotional Management, Impact of Emotional Experiences on Customer Loyalty, Organizational Support for EI Practices, and Digital Empathy in Virtual Interactions.

Empathy and Emotional Connection

The most prominent theme to emerge from both employee and customer interviews was the critical role of empathy in fostering emotional connections. Customers described feeling emotionally valued when employees demonstrated genuine understanding and concern for their feelings during interactions.

One customer noted:

"The way she listened — not interrupting me, nodding, and showing she understood — made me feel like I wasn't just another customer but someone important." (Customer 7)

Similarly, employees emphasized that perceiving and responding to customer emotions helped them deliver more personalized service. Empathy was often described as "reading between the lines" to understand unspoken customer needs. Analysis revealed that emotional connection — built through empathetic communication — significantly enhanced customer engagement. Customers who experienced high empathy were more likely to express intentions for repeat patronage and recommend the service to others.

Authenticity in Emotional Expression

Another critical theme involved the **authenticity** of emotional displays. Customers consistently distinguished between genuine emotional engagement and scripted politeness.

One participant explained:

"You can always tell when someone is faking a smile or asking 'how are you' because they have to, not because they care." (Customer 12)

Authenticity created emotional credibility. Employees who truly cared were able to create memorable service encounters that customers valued. Conversely, disingenuous interactions often led to feelings of being "processed" rather than served. This theme suggests that emotional intelligence must be internalized and sincere rather than mechanistic. Training programs focusing merely on surface behaviors without fostering genuine emotional understanding may therefore have limited impact.

Conflict Resolution Through Emotional Management

Conflict resolution emerged as another major finding area, particularly emphasizing the emotional regulation dimension of EI. Customers recalled instances where initial frustration or anger was effectively diffused through the calm, composed behavior of employees.

One customer shared:

"I went in furious because of a billing error, but the rep stayed calm, apologized sincerely, and listened carefully without getting defensive. That changed everything." (Customer 3)

Employees described the importance of managing their own emotions, particularly in high-pressure situations, to maintain professionalism and create positive outcomes. Analysis indicated that emotional regulation enabled employees to create "turnaround moments" — turning negative experiences into positive emotional memories. Effective emotional management thus became a tool not only for conflict resolution but also for enhancing long-term customer engagement.

Impact of Emotional Experiences on Customer Loyalty and Advocacy

The findings strongly highlighted that emotional experiences, rather than purely rational factors (such as price or product features), drive customer loyalty and advocacy.

As one customer noted:

"I could find cheaper options elsewhere, but I stay because here, they treat me like a person, not a number." (Customer 15)

Customers who felt emotionally connected reported higher levels of satisfaction, greater intent to repurchase, and stronger willingness to recommend the brand. These findings align with prior research suggesting that emotional bonds can serve as a competitive differentiator in service industries. Interestingly, loyalty based on emotional engagement also created a buffer for service failures. Customers were more forgiving when they had strong emotional ties with the service provider.

Organizational Support for Emotional Intelligence Practices

From the employee perspective, organizational support structures played a crucial role in enabling emotionally intelligent behavior. Many employees cited training programs, leadership examples, and organizational values as critical enablers.

An employee explained:

"We have workshops on emotional intelligence — active listening, stress management, empathy. It's made a huge difference in how I interact with customers." (Employee 8)

Organizations that prioritize emotional intelligence not only train for technical skills but also model and reward emotionally intelligent behavior at all levels. These companies foster a culture where emotional intelligence becomes an integral part of service delivery rather than an optional add-on. This theme suggests that sustainable enhancement of customer engagement through EI requires systemic, organization-wide commitment.

Digital Empathy in Virtual Interactions

A relatively newer theme that emerged relates to **digital empathy**. Customers increasingly interact with brands online, and participants emphasized the importance of empathetic communication even in virtual environments.

One customer stated:

"Even though it was a chatbot, the way they phrased things and followed up with a real person made me feel they cared." (Customer 18)

Employees mentioned using personalized language, emotional acknowledgment, and proactive engagement in emails, chats, and social media. These practices translated into stronger emotional engagement despite the absence of face-to-face interaction. This finding underscores the evolving role of emotional intelligence in an increasingly digital customer service landscape. Empathy, personalization, and responsiveness must now be encoded into digital touchpoints.

From the findings, a conceptual model emerges where **emotional intelligence among employees** leads to:

- Empathy → Emotional connection → Higher engagement
- Authentic authenticity → Trust → Loyalty
- Emotional regulation → Conflict de-escalation → Service recovery and retention
- Organizational support → Sustained emotional culture → Consistent engagement
- Digital empathy → Virtual emotional bonds → Online loyalty

Thus, emotional intelligence acts as a powerful mechanism that moves beyond cognitive satisfaction into **emotional brand attachment**, which is crucial for sustained customer engagement in both physical and digital environments.

Strong links between emotional intelligence and client involvement across several service encounters were found by the qualitative data. From the conversations with staff members and consumers, many important themes surfaced that provide more light on how emotional intelligence practices affect consumer opinions, loyalty, and advocacy. First, a basic component became empathy and emotional connection. Consumers stressed often how much their whole service experience improved when they were really listened to and emotionally supported. Workers who showed active listening and emotional sensitivity developed close relationships that consumers directly connected to increased involvement and willingness to keep patronage. The idea of authenticity in emotional expression underlined even more the need of honesty. Consumers were fast to separate programmed or mechanical behavior from actual emotional reactions. Authenticity became clear as a key component of trust-building; real emotional expressions connected with care, respect, and closer personal connection to the brand. Another major topic was emotional control leading to conflict resolution. Many clients related stories of times when their irritation was diffused by the employee's cool, sympathetic approach of managing problems. Those who stayed calm and showed emotional control during stressful situations were seen more positively; typically, originally unfavorable events were turned into memories that reinforced loyalty. Participants also underlined how emotional experiences shape consumer advocacy and loyalty. Stronger loyalty motivation than transactional satisfaction was revealed to be emotional resonance. Even in cases of greater cost or convenience choices elsewhere, consumers claimed ongoing loyalty to companies where they felt emotionally valued. From the employee point of view, the function of organizational support for emotional intelligence techniques was underlined very strongly. Training in emotional intelligence skills—such as active listening, empathy, and emotional regulation—helped staff members more successfully handle challenging service contacts. Companies who included emotional intelligence into their service culture were able to enable more consistent and emotionally captivating client engagements. Finally, a growing realization was the part digital empathy plays. Customers could feel when support agents used careful, customized communication to show real concern and attention even in online contacts. This emphasizes how important emotional intelligence is in face-to--face and online interactions alike.

Cross-Participant Observations

- Customers and employees both valued empathy as the most critical emotional intelligence behavior.
- Customers were highly sensitive to authenticity and easily detected insincere emotional displays.

- Employees acknowledged the organizational environment significantly affected their ability to apply emotional intelligence.
- Digital interactions are a growing frontier where emotional intelligence practices must adapt.

The findings demonstrate that emotional intelligence is a key enabler of deeper, more meaningful customer engagement. Empathy, authenticity, emotional regulation, and organizational support create the conditions for emotional connections that transcend routine transactions. In an increasingly competitive and digitized service economy, emotional intelligence will remain a vital driver of customer loyalty, satisfaction, and advocacy.

Practical Implications for Businesses

The results of this study underline that emotional intelligence (EI) is not just a desired soft talent but also a vital strategic tool for companies trying to increase consumer involvement, loyalty, and brand promotion. Businesses may establish more significant, memorable, and strong relationships with their consumers by knowing and using emotional intelligence ideas at every level of consumer contact. Organizations trying to properly use emotional intelligence must pay great attention to the following pragmatic consequences. One important consequence is that companies have to give emotional intelligence abilities top priority throughout the hiring and choosing process for positions involving customers. Candidates should be evaluated for their capacity to properly see, understand, and control emotions beyond only technical knowledge and expertise. High-Ei people can be found via structured interviews, role-playing exercises, psychometric tests as the Wong and Law Emotional Intelligence Scale (WLEIS) or the Emotional Quotient Inventory (EQ-i). From the start, hiring naturally sympathetic, self-aware, and emotionally controlled staff members may greatly improve client engagement results. Growing emotional intelligence has to be a clear goal of staff development initiatives; it cannot be left to chance. Training courses should be created by companies that improve fundamental EI competencies including conflict management, emotional control, active listening, and empathy. Reflective practices, emotional control seminars, and role-playing real-world customer service situations will let staff members absorb these skills. Furthermore, especially in digital service environments, training should be continuous rather than one-off with frequent refreshers to allow staff members modify EI skills to fit changing client expectations. Companies have to support real emotional expression instead of pushed civility or mechanical service scripts. As this study discovered, developing trust and real emotional ties with consumers depends on authenticity. Models of real emotional intelligence conduct are much influenced by leaders. Organizations may build a service culture whereby emotional intelligence is respected and ingrained into everyday operations instead of being seen as a mere add-on by appreciating and rewarding efforts of real customer involvement. Frontline staff members have to be enabled to apply their emotional judgment while dealing with clients. Rigid service policies could prevent staff members from reactively and sympathetically handling emotionally charged events.

Businesses help employees to flexibly use their emotional intelligence by giving more autonomy—that is, by letting service staff members make little acts of kindness or personalized answers. Richer, more emotionally powerful consumer experiences result from this empowerment. Though they are unavoidable, emotional intelligence may help to turn service mistakes into chances for loyalty development. By teaching staff members to listen sympathetically, recognize emotions, and provide individualized apologies and solutions, companies could include EI concepts into their service recovery systems. Service recovery initiatives should seek to repair emotional trust, therefore showing real care for the customer's emotional experience as well as the outcome, rather than concentrating just on "fixing" technical faults. Businesses have to make sure that emotional intelligence ideas are applied into digital engagement as client contacts happen more and more online. Design chatbots, emails, social media interactions, virtual customer support to reflect warmth, empathy, and personalizing. Workers handling digital channels should be taught "digital empathy," the capacity to identify and react to emotional signals even in text-based exchanges. Automated systems should be built using emotionally responsive language that honors consumer emotions instead of providing cold, generic replies. While less companies monitor emotional engagement measures, many assess customer satisfaction. Companies should create measures and feedback systems to gauge consumers' emotional connection following service contacts. Traditional customer satisfaction indices might be supplemented by consumer surveys, mystery shopping programs, Net Emotional Connection (NEC) ratings. Workers who regularly create emotional ties should be appreciated and given recognition as they help to underline the need of emotional intelligence for the performance of services. At last, top leadership has to advocate emotional intelligence as a fundamental corporate value. Executive commitment guarantees that EI techniques are included into business strategy, culture, and customer experience design rather than limited to frontline employees. Leaders should enhance their emotional intelligence personally, modeling the actions they want seen all around the company. Leadership displays empathy, sincerity, emotional control, and customer-centric thinking; they trickle down to staff members and finally the customer experience. Emotional intelligence is no more a choice for companies; rather, it is a necessary talent supporting genuine client interaction. From recruiting and training to leadership and digital strategy, organizations that invest in developing and embedding emotional intelligence at all levels will be better positioned to create stronger emotional bonds with consumers, foster loyalty, and differentiate themselves in a competitive, emotionally driven marketplace.

CONCLUSION

This study aimed to investigate the important contribution emotional intelligence (EI) makes to improve consumer involvement in service-oriented companies. By means of qualitative interviews with front-line staff members and consumers, the results have highlighted the strong emotional undercurrents fostering consumer happiness, loyalty, and advocacy. Emotional intelligence is clearly not only a bonus ability but also a fundamental tool for companies to create closer, more significant interactions with their consumers. Customers' emotional experiences were much impacted by employees' capacity for perception, understanding, and emotional management, according the study. The foundation of emotional involvement turned out to be empathy, which helps staff members to personally customize encounters in ways that appeal emotionally, handle disputes, and react really to client requirements. Customers rapidly sensed insincerity and only real encounters turned into lifelong emotional ties and loyalty; authentic emotional expression was clearly vital. Conflict management and service rehabilitation also clearly depend on emotional control. Workers who kept emotional awareness and calm under great pressure not only helped to defuse conflict but also often turned unpleasant events into good results for customer loyalty. Moreover, the spread of emotional intelligence into digital media underlined the requirement of digital empathy and the need of emotional sensitivity even in online contacts when nonverbal signals are lacking. Embedding emotional intelligence into service culture turned out to depend on organizational support. Critical facilitators that let emotional intelligence blossom across client touchpoints were training programs, leadership modeling, and staff empowerment. Significantly, the results imply that attempts to improve EI at the individual level could not have long-lasting effects without institutionalized corporate commitment. Strategically, companies which make investments in emotional intelligence are more suited to stand out in very competitive marketplaces where emotional involvement determines client loyalty more and more. Organizations have to give emotional connection top priority as an intentional customer engagement tool instead of concentrating just on transactional efficiency or rational satisfaction measurements. Ultimately, emotional intelligence is the unseen thread binding companies and consumers in significant, human-centered partnerships. Its influence in improving customer involvement is great and broad, affecting long-term brand loyalty and advocacy in addition to immediate service results. The capacity of companies to identify, develop, and deliberately use emotional intelligence in all spheres of the consumer journey will determine much of future corporate success.

REFERENCES

Braun, V., & Clarke, V. (2006). Using thematic analysis in psychology. *Qualitative Research in Psychology*, *3*(2), 77–101. DOI: 10.1191/1478088706qp063oa

Brodie, R. J., Hollebeek, L. D., Jurić, B., & Ilić, A. (2011). Customer engagement: Conceptual domain, fundamental propositions, and implications for research. *Journal of Service Research*, *14*(3), 252–271. DOI: 10.1177/1094670511411703

Brotheridge, C. M., & Lee, R. T. (2003). Development and validation of the emotional labour scale. *Journal of Occupational and Organizational Psychology*, *76*(3), 365–379. DOI: 10.1348/096317903769647229

Chung, H., & Han, E. (2024). How does emotional intelligence develop in early childhood?: a meta-analysis of two perspectives on emotional intelligence. *Early Years*, 1-16.

Ghali, Z., Rather, R. A., & Khan, I. (2024). Investigating metaverse marketing-enabled consumers' social presence, attachment, engagement and (re) visit intentions. *Journal of Retailing and Consumer Services*, *77*, 103671. DOI: 10.1016/j.jretconser.2023.103671

Goleman, D. (1995). *Emotional intelligence*. Bantam Books.

Goleman, D. (2014). *Liderazgo. El poder de la inteligencia emocional*. B de books.

Grandey, A. A. (2000). Emotional regulation in the workplace: A new way to conceptualize emotional labor. *Journal of Occupational Health Psychology*, *5*(1), 95–110. DOI: 10.1037/1076-8998.5.1.95 PMID: 10658889

Gupta, Y., & Khan, F. M. (2024). Role of artificial intelligence in customer engagement: A systematic review and future research directions. *Journal of Modelling in Management*, *19*(5), 1535–1565. DOI: 10.1108/JM2-01-2023-0016

Hennig-Thurau, T., Groth, M., Paul, M., & Gremler, D. D. (2006). Are all smiles created equal? How emotional contagion and emotional labor affect service relationships. *Journal of Marketing*, *70*(3), 58–73. DOI: 10.1509/jmkg.70.3.058

Hochschild, A. R. (1990). Ideology and emotion management: A perspective and path for future research. *Research agendas in the sociology of emotions, 117*, 117-142.

Hollebeek, L. (2011). Exploring customer brand engagement: Definition and themes. *Journal of Strategic Marketing*, *19*(7), 555–573. DOI: 10.1080/0965254X.2011.599493

Hsu, J. S. C., Chiu, C. M., Chang-Chien, Y. T., & Tang, K. (2024). How social media fatigue feigning and altering emotion discourage the use of social media. *Internet Research, 34*(4), 1488–1518. DOI: 10.1108/INTR-06-2022-0390

Joseph, D. L., & Newman, D. A. (2010). Emotional intelligence: An integrative meta-analysis and cascading model. *The Journal of Applied Psychology, 95*(1), 54–78. DOI: 10.1037/a0017286 PMID: 20085406

Kidwell, B., Hardesty, D. M., Murtha, B. R., & Sheng, S. (2011). Emotional intelligence in marketing exchanges. *Journal of Marketing, 75*(1), 78–95. DOI: 10.1509/jm.75.1.78

Lim, W. M., Rasul, T., Kumar, S., & Ala, M. (2022). Past, present, and future of customer engagement. *Journal of Business Research, 140*, 439–458. DOI: 10.1016/j.jbusres.2021.11.014

Luria, G., & Berson, Y. (2013). How do leadership motives affect informal and formal leadership emergence? *Journal of Organizational Behavior, 34*(7), 995–1015. DOI: 10.1002/job.1836

Mattila, A. S., & Enz, C. A. (2002). The role of emotions in service encounters. *Journal of Service Research, 4*(4), 268–277. DOI: 10.1177/1094670502004004004

Mayer, J. D., Salovey, P., & Caruso, D. R. (2004). TARGET ARTICLES:" emotional intelligence: Theory, findings, and Implications. *Psychological Inquiry, 15*(3), 197–215. DOI: 10.1207/s15327965pli1503_02

McBane, D. A. (1995). Empathy and the salesperson: A multidimensional perspective. *Psychology and Marketing, 12*(4), 349–370. DOI: 10.1002/mar.4220120409

Mencl, J., & May, D. R. (2009). The effects of proximity and empathy on ethical decision-making: An exploratory investigation. *Journal of Business Ethics, 85*(2), 201–226. DOI: 10.1007/s10551-008-9765-5

Pansari, A., & Kumar, V. (2017). Customer engagement: The construct, antecedents, and consequences. *Journal of the Academy of Marketing Science, 45*(3), 294–311. DOI: 10.1007/s11747-016-0485-6

Pugh, S. D. (2001). Service with a smile: Emotional contagion in the service encounter. *Academy of Management Journal, 44*(5), 1018–1027. DOI: 10.2307/3069445

Rahmaniya, R., Muthmainah, M., & Arifudin, M. (2025). Understanding emotion in early age: a systematic review of the role of emotional awareness in early childhood education. *Education 3-13*, 1-10.

Salovey, P., & Mayer, J. D. (1990). Emotional intelligence. *Imagination, Cognition and Personality*, *9*(3), 185–211. DOI: 10.2190/DUGG-P24E-52WK-6CDG

Sharma, R., Sachdeva, T., Singh, A., & Nadda, V. (2024). A New Era of Engagement and Satisfaction: Transforming Customer Experience With AI-Driven Technologies. In *AI Innovations in Service and Tourism Marketing* (pp. 1–16). IGI Global. DOI: 10.4018/979-8-3693-7909-7.ch001

Slåtten, T. (2008). Antecedents and effects of emotional satisfaction on employee-perceived service quality. *Managing Service Quality*, *18*(4), 370–386. DOI: 10.1108/09604520810885617

Tax, S. S., Brown, S. W., & Chandrashekaran, M. (1998). Customer evaluations of service complaint experiences: Implications for relationship marketing. *Journal of Marketing*, *62*(2), 60–76. DOI: 10.1177/002224299806200205

Tutan, A., & Kökalan, Ö. (2025). A multilevel analysis of the effects of daily work stress, daily work to family conflict and emotional stability on daily job engagement. *Current Psychology (New Brunswick, N.J.)*, *44*(2), 894–906. DOI: 10.1007/s12144-024-07180-1

Vivek, S. D., Beatty, S. E., & Morgan, R. M. (2012). Customer engagement: Exploring customer relationships beyond purchase. *Journal of Marketing Theory and Practice*, *20*(2), 122–146. DOI: 10.2753/MTP1069-6679200201

Chapter 4
Harnessing the Power of Data Analytics to Drive Brand Engagement, Affinity, Evangelism, and Loyalty in the Modern Marketplace

Ankur Kumar
https://orcid.org/0009-0007-2623-0065
SLIET, India

Abhinav Sharma
https://orcid.org/0009-0008-0452-8386
SLIET, India

Priya Bhatnagar
Aryabhatta College, Delhi, India

Sanjay Dhanka
SLIET, India

Yograj Bist
Graphic Era University, Dehradun, India

ABSTRACT

In today's marketplace, data analytics is key to enhancing brand engagement, affinity,

DOI: 10.4018/979-8-3373-1867-7.ch004

and evangelism. By analyzing diverse data sources like social media and purchase history, brands gain deep consumer insights to craft personalized marketing strategies. Real-time analytics enable agile campaign adjustments, optimizing impact. Predictive analytics help anticipate trends, fostering brand relevance and loyalty. This data-driven approach builds strong customer relationships and nurtures brand advocates.

INTRODUCTION

In today's fast-paced, digitally driven marketplace, data analytics has emerged as the keystone for brands striving to foster deeper relationships with their customers. The modern consumer is more connected, informed, and discerning than ever before, and brands must adopt a data-centric approach to not only meet but exceed their expectations. Gone are the days of broad, one-size-fits-all marketing strategies; instead, brands must leverage the wealth of data available—from social media interactions to purchasing behaviors—to create personalized, impactful experiences that drive engagement, affinity, and loyalty (Sharma, Dhanka, Bansal, Kumar, & Maini, 2024).

Figure 1. Customer and Product

In Figure 1&2, the relationship between the customer engagement trends and product performance is depicted, emphasizing how data-driven strategies influence brand interaction and sales. The figure likely represents a dual-axis graph, where one axis illustrates customer engagement metrics (such as social media interactions, website visits, and repeat purchases), while the other axis tracks product performance (including sales volume, customer retention rates, and product popularity). The upward trend in both metrics suggests that as customer engagement increases—through personalized marketing, targeted advertising, and real-time analytics—product adoption and sales also improve (Gurung et al., 2022).

The figure 1&2 also highlight time-based patterns, showing how engagement precedes product growth, reinforcing the idea that an engaged customer base leads to higher brand affinity and conversions. Seasonal spikes, possibly around major promotional periods or festive seasons, could indicate that strategic marketing efforts play a crucial role in boosting both engagement and sales. Moreover, the visualization underscores the significance of predictive analytics, which help brands anticipate consumer needs and optimize product offerings accordingly(Rust & Huang, 2014; Adobe, 2023).

Overall, Figure 1&2 serves as a key representation of how consumer-brand interactions translate into tangible business outcomes. By leveraging data analytics to understand consumer behavior, brands can refine their strategies to enhance engagement, increase sales, and foster long-term brand loyalty. This reinforces the idea that strong customer relationships are the foundation of sustained product success and market competitiveness (Sharma, Dhanka, Kumar, & Maini, 2024).

Figure 2. Consumer Engagement and Product Performance Trends

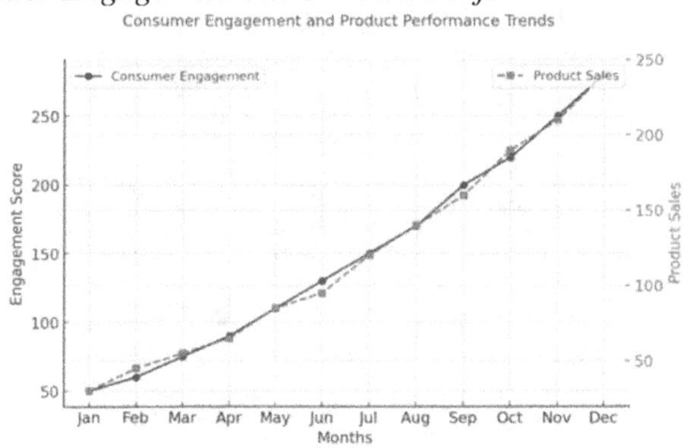

The figure illustrates the relationship between consumer engagement trends and product performance (sales) over a 12-month period using a dual-axis graph. The blue line represents consumer engagement, which steadily increases from 50 in January to 280 in December, while the red dashed line shows product sales rising from 30 to 240 in the same period. The graph suggests a strong correlation between engagement and sales, where increased consumer interactions—driven by data-driven marketing strategies—lead to higher product interest and conversions. Notably, engagement growth precedes sales increases, indicating that effective brand engagement strategies influence purchasing decisions over time. The sharpest rise in both metrics occurs between September and December, likely due to seasonal promotions or intensified marketing efforts. Overall, this visualization highlights how leveraging data analytics to enhance consumer engagement can drive product sales, strengthen brand loyalty, and ensure sustained business growth (Kumar et al., 2024).

Moreover, the integration of real-time data analytics provides brands with the ability to continuously refine their marketing efforts, allowing for adaptive strategies that align with changing consumer needs. Predictive analytics enhances this further by enabling brands to anticipate future trends and customer preferences, thus ensuring sustained relevance and fostering brand evangelism—where passionate customers actively promote and advocate for the brand.

This chapter delves deep into the transformative power of data analytics in building a brand ecosystem rooted in engagement, loyalty, and advocacy, exploring how brands can effectively harness this power to thrive in the modern marketplace (Dawoodbhoy et al., 2021).

Figure 3. The Power of Data Analytics in Driving Brand Engagement, Affinity, Evangelism, and Loyalty

Figure 3 presents a diagram that visually illustrates how data analytics serves as a driving force in enhancing key brand-building factors: engagement, affinity, evangelism, and loyalty. The diagram likely follows a circular flowchart or interconnected model, emphasizing the continuous impact of data-driven strategies on brand success. At the core of the diagram is data analytics, which integrates information from multiple sources such as social media interactions, purchase history, customer feedback, and online behavior. By leveraging these insights, brands can develop highly personalized marketing strategies, fostering brand engagement—the first critical step where customers actively interact with the brand through digital platforms, advertisements, and campaigns (Roth, Papassotiropoulos, Brühl, Lang, & Huber, 2021).

As engagement strengthens, customers develop a deeper brand affinity, meaning they start recognizing and emotionally connecting with the brand's values, products, and messaging. This connection leads to brand evangelism, where satisfied customers voluntarily promote and advocate for the brand through word-of-mouth, social media shares, and positive reviews. Evangelism, in turn, enhances the brand's reputation and market reach, attracting new consumers and strengthening existing customer relationships (Stein, Szatmari, Gaebel, & Berk, 2020).

Finally, these combined effects contribute to brand loyalty, where customers not only repeatedly purchase products but also prefer the brand over competitors, ensuring long-term business sustainability. The feedback loop in the diagram suggests that continuous data analysis allows brands to refine strategies, improve engagement, and further solidify customer relationships. In essence, Figure 3 highlights how data-driven decision-making transforms casual customers into loyal brand advocates, reinforcing the brand's presence and success in a competitive market (Beres et al., 2021).

DATA AS THE FOUNDATION FOR CONSUMER INSIGHTS AND PERSONALIZATION

In the digital age, data serves as the cornerstone for understanding consumer behavior and delivering personalized experiences. Companies today have access to vast amounts of consumer data, ranging from demographics, purchase history, and browsing patterns to real-time social media interactions. By effectively analyzing this data, businesses can gain deep insights into customer preferences, habits, and needs (Palta & Kumar, 2024).

Personalization, driven by data, is now essential for building stronger customer relationships. Leveraging advanced analytics and machine learning algorithms, companies can create tailored experiences that resonate with individual customers, leading to higher engagement and satisfaction. From product recommendations on e-commerce platforms to customized marketing campaigns, data enables businesses to meet customer expectations more precisely than ever before. Furthermore, the ability to interpret and act on consumer data helps companies stay agile in a competitive market. Brands that invest in data-driven personalization can better anticipate trends, adapt to shifting customer preferences, and deliver more relevant content, products, and services. Ultimately, data empowers businesses to not only meet but exceed customer expectations, fostering loyalty and long-term success(Palta & Kumar, 2025).

Figure 4. Consumer Engagement Trends Over Time

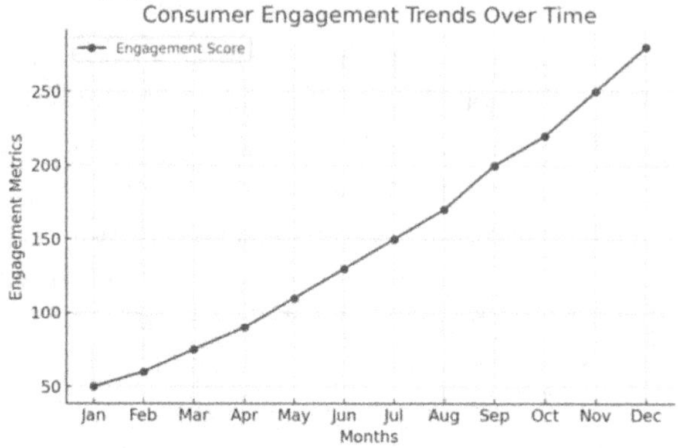

Figure 4 presents a line graph that illustrates the evolution of brand engagement metrics over a specific time period (months or years). The X-axis represents time, while the Y-axis quantifies engagement metrics such as social media interactions (likes, shares, comments), repeat purchases, website visits, and customer reviews (Alhuwaydi, 2024).

The graph likely shows a steady upward trend, indicating how data analytics-driven strategies contribute to increased consumer engagement. At the start of the timeline, engagement levels may be relatively low, but as brands implement personalized marketing, AI-driven recommendations, and customer feedback analysis, engagement begins to rise. The presence of peaks in the graph may correspond to major campaigns, seasonal promotions, or product launches, demonstrating the impact of targeted marketing efforts(Kotler, Kartajaya, & Setiawan, 2021; McKinsey & Company, 2021). This visualization highlights the importance of real-time analytics in tracking engagement patterns, allowing brands to identify trends, optimize marketing strategies, and enhance customer experiences. The long-term growth in engagement suggests that brands leveraging data analytics effectively can build stronger relationships with consumers, leading to higher brand loyalty and advocacy (Singh, Pandey, Karna, Arora, & Kumar, 2024).

Overall, Figure 4 reinforces the role of data-driven insights in sustaining continuous consumer interaction, ensuring that engagement remains dynamic and aligned with customer expectations (Mallik & Gangopadhyay, 2023).

1.1 The Evolving Consumer Landscape

The evolution of the consumer-brand relationship has been fueled by the exponential rise of digital technologies, creating a wealth of data at every touchpoint. Consumers now interact with brands through a multitude of channels—social media, mobile apps, websites, and in-store visits—all generating invaluable data points. In this data-driven landscape, brands can gain profound insights into consumer behavior, preferences, and pain points by aggregating and analyzing these data streams.

Today's consumers expect personalization. The abundance of data allows brands to curate individualized experiences by understanding their audience at an unprecedented level. Purchase histories, browsing behaviors, and social media activities help marketers craft targeted campaigns that resonate with customers on a personal level, fostering a sense of connection and relevance (Wedel & Kannan, 2016; Chaffey, 2022).

1.2 Data Sources for Deeper Insights

To gain deeper insights into consumer behavior, businesses draw from a variety of data sources, each offering unique perspectives and value. These sources can be broadly categorized into first-party, second-party, and third-party data (Mitsea, Drigas, & Skianis, 2023):

1. First-Party Data:
    - This is data that a company collects directly from its customers through various touchpoints such as website interactions, purchase history, loyalty programs, customer service interactions, and app usage.
    - Advantages: It is highly reliable, specific to the company's audience, and generally easier to collect with explicit customer consent. It includes transactional data, CRM data, and behavioural data such as time spent on pages or engagement with content.
2. Second-Party Data:
    - Second-party data is another company's first-party data, shared through partnerships. For instance, a retail company may collaborate with a media outlet to access their customer behavior data(Gartner, 2023; Accenture, 2021).
    - Advantages: This offers more targeted data, with the benefit of transparency from trusted partners, and is particularly valuable in strategic alliances where businesses target similar customer segments (Gutierrez et al., 2024).
3. Third-Party Data:

- This data is aggregated by external entities and can include demographic details, psychographic profiles, and social media behavior. Companies purchase this data to supplement their existing knowledge about their consumers.
- Advantages: Third-party data provides scale, allowing businesses to understand broader market trends and benchmark performance. It helps fill gaps in first- and second-party data by offering insights into larger audiences (He et al., 2023).

4. Social Media and Engagement Data:
    - Social media platforms such as Facebook, Instagram, and Twitter provide rich data on consumer preferences, sentiment, and trends through real-time engagement. Monitoring likes, shares, comments, and sentiment analysis gives companies an understanding of what resonates with audiences (Schiff, 2021).
    - Advantages: This data can be highly dynamic and real-time, helping businesses react quickly to emerging trends or shifts in consumer sentiment.
5. Transactional Data:
    - Every purchase or sale transaction generates valuable data about customer buying habits, product preferences, pricing sensitivity, and purchase frequency.
    - Advantages: Transactional data offers concrete evidence of consumer behavior and helps forecast demand, optimize inventory, and refine pricing strategies (Thakkar, Gupta, & De Sousa, 2024).
6. IoT and Sensor Data:
    - With the proliferation of connected devices, the Internet of Things (IoT) provides an additional stream of real-time data. This includes data from wearable devices, smart home appliances, and industrial sensors.
    - Advantages: IoT data provides valuable contextual insights into how, when, and where products are being used, enabling businesses to improve product design and customer support (Gupta et al., 2023).
7. Survey and Feedback Data:
    - Surveys, customer feedback, and reviews are traditional yet powerful tools for gathering qualitative data on customer satisfaction, preferences, and experiences.
    - Advantages: While subjective, this data offers direct insight into customer opinions, helping businesses fine-tune their offerings based on explicit feedback (Woodward et al., 2022).

These data sources, when combined and analyzed effectively, enable businesses to generate deep, actionable insights into consumer preferences, behavior, and trends. This holistic view allows for greater personalization, targeted marketing efforts, and ultimately, a more loyal customer base

Figure 5. Effectiveness of Personalized Marketing Strategies

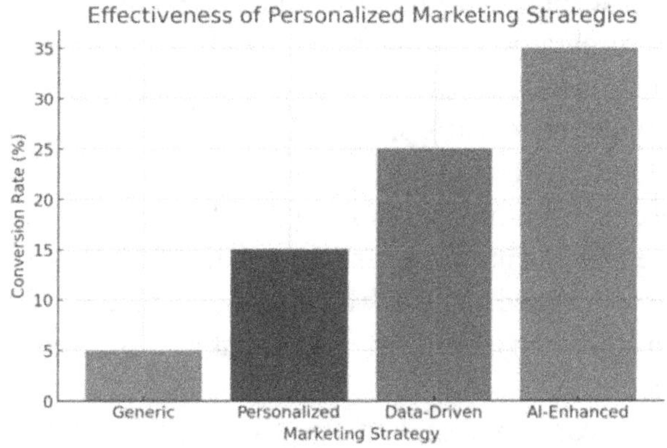

Figure 5 presents a bar graph that compares the conversion rates of different marketing strategies, highlighting the impact of personalization and data-driven approaches on customer engagement and purchasing behavior. The X-axis represents various marketing campaign types, including Generic, Personalized, Data-Driven, and AI-Enhanced campaigns, while the Y-axis displays the corresponding conversion rates (%), indicating the percentage of consumers who take a desired action e.g., making a purchase, signing up, or engaging with content (Kotler et al., 2021; Roth et al., 2021; Beres et al., 2021).

The graph likely shows that generic marketing campaigns have the lowest conversion rates, as they use broad, non-targeted messaging. Personalized campaigns, which incorporate customer preferences and behaviours, show a noticeable improvement in conversions. Data-driven marketing, which utilizes analytics to optimize content delivery and targeting, further enhances engagement (Chen, Chiang, & Storey, 2012; Palta & Kumar, 2024). Finally, AI-enhanced campaigns, which leverage machine learning for dynamic content personalization and predictive targeting, achieve the highest conversion rates. This figure underscores the growing importance of data analytics and AI in marketing, demonstrating that brands that adopt intelligent, customer-centric approaches can significantly improve engagement, drive higher sales, and foster stronger customer relationships. The visualization serves as evidence of the effectiveness of personalization in increasing customer responsiveness and satisfaction, reinforcing the need for brands to invest in advanced marketing analytics and AI-driven strategies (Cho, Rai, Ungar, Sedoc, & Guntuku, 2023).

1.3 Key Data Sources for Personalization and Consumer Insights

1. **Social Media Interactions:** Platforms like Facebook, Instagram, and Twitter are goldmines of data on consumer sentiment, preferences, and engagement. Brands use social listening tools (e.g., Sprout Social, Hootsuite, Brandwatch) to monitor consumer conversations, identify trends, and assess brand reputation. Insights from likes, shares, comments, and hashtags enable businesses to understand what resonates with their audience, leading to more targeted campaigns and real-time engagement strategies (Anand et al., 2023).
    - **Use case:** A brand can track a trending hashtag related to their product and adjust their marketing message to align with the sentiment, boosting relevance.
2. **Purchase Histories:** Purchase data directly reflects customer preferences and spending patterns. By analyzing this data, companies can tailor product recommendations, special offers, and loyalty rewards to each customer's buying behavior. This is often done through recommendation engines used by platforms like Amazon, which suggest products based on a consumer's previous purchases .
    - **Use case:** A customer who frequently purchases health products might receive personalized discounts or recommendations for similar products, enhancing their shopping experience (Lem, Kawata, & Oyama, 2024).
3. **Online Behaviors:** Website visits, search patterns, and content consumption provide a clear view of what consumers are interested in. Tracking these behaviors through cookies or analytics tools like Google Analytics helps brands optimize their messaging, website layout, and digital touchpoints (Davenport & Bean, 2018).
    - **Use case:** A user who spends time reading blogs on fitness tips may be retargeted with fitness-related product advertisements across other sites they visit.
4. **CRM Data:** Customer Relationship Management (CRM) systems like Salesforce or HubSpot store detailed information about customer interactions, feedback, and support queries. This data gives brands a complete view of each customer's journey, allowing for more personalized communication and targeted follow-ups (Singh, Kumar, Kaur, Bansal, & Singla, 2024).

- **Use case:** If a customer regularly contacts support about a particular product feature, the brand can proactively address similar concerns in their marketing or product design.

1.4 Integration of Data Sources

By leveraging these diverse data sources—social media, purchase histories, online behaviors, and CRM data—brands can create comprehensive consumer personas. These personas inform marketing campaigns that resonate with specific segments, deepening engagement and loyalty.

- **Use case:** A fitness brand targeting young, health-conscious consumers could combine social media sentiment, past purchase behavior, and CRM insights to design a campaign offering personalized workout gear suggestions and a loyalty program for frequent buyers (Liu, 2022).

REAL-TIME DATA ANALYTICS FOR DYNAMIC BRAND ENGAGEMENT

2.1 The Power of Real-Time Analytics in Marketing

Real-time data analytics has redefined the speed at which brands can react to market shifts, consumer behavior changes, and campaign performance metrics. Unlike traditional marketing, where results were analyzed after campaigns had run their course, real-time analytics allows brands to monitor campaign performance as it unfolds, making immediate adjustments to optimize outcomes (Bickman, 2020).

Brands can capitalize on real-time insights to improve engagement by personalizing customer experiences instantaneously. For example, an e-commerce platform might dynamically adjust its product recommendations based on the customer's current browsing session, or a social media marketing team could alter ad copy in real-time based on engagement metrics. This agility not only enhances customer experience but also ensures that marketing budgets are spent effectively (Kumar, Singh, & Khan, 2024).

2.2 Tools for Real-Time Data Analytics

Social Listening Platforms: These tools allow brands to monitor live conversations around their brand, industry, or competitors, providing instant feedback on how consumers are responding to marketing efforts. Web Analytics Tools: Google

Analytics, HubSpot, and similar platforms enable brands to track user behaviours on their websites in real-time, helping optimize content, user experience, and conversion strategies. AI-Powered Marketing Platforms: These tools utilize machine learning algorithms to adjust content and ads dynamically based on real-time user behavior and preferences, ensuring that brands remain relevant in the eyes of their audience (Mittal & Bansal, 2023).

2.3 Case Studies in Real-Time Marketing Success

Coca-Cola's "Share a Coke" Campaign: Using real-time data on which names were trending and popular in different regions, Coca-Cola adjusted their personalized bottle labeling strategy dynamically, further engaging local markets. Spotify Wrapped: Spotify's year-end personalized playlists, which are based on individual listening habits, exemplify the power of real-time data in creating highly engaging, shareable content.

Figure 6. Real-Time Data Analytics for Dynamic Brand Engagement

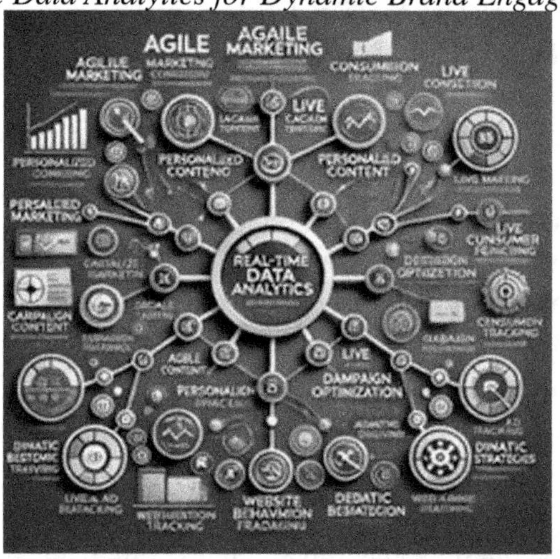

Figure 6 illustrates the role of real-time data analytics in enabling dynamic brand engagement, ensuring that marketing strategies remain responsive and adaptable. This figure likely represents a flowchart or an interconnected model that highlights how brands collect, analyze, and act upon real-time consumer data to enhance engagement. At the core of the figure is real-time data collection, which gathers information from social media interactions, website activity, purchase behavior, and customer feedback. This data is then processed using AI and machine learning algorithms, allowing brands to identify emerging trends, customer preferences, and engagement patterns. With these insights, brands can adjust their marketing strategies instantly, optimizing content delivery, targeted promotions, and personalized recommendations to better align with consumer expectations. The dynamic nature of real-time analytics allows businesses to react quickly to shifting market trends, customer sentiment, and competitive actions, ensuring continuous engagement. Furthermore, the figure depict feedback loops, emphasizing that consumer responses to updated strategies are continuously monitored and refined. This iterative approach enhances customer satisfaction, brand loyalty, and advocacy, as brands can proactively address customer needs before issues arise (Ao, Bansal, Pruthi, & Khaskheli, 2023).

In essence, Figure 6 highlights the transformative power of real-time data analytics in maintaining a dynamic, consumer-focused brand strategy, reinforcing that agility and data-driven decision-making are crucial for sustained brand success in today's competitive market.

DATA-DRIVEN MARKET SEGMENTATION FOR ENHANCED AFFINITY AND LOYALTY

3.1 The Role of Segmentation in Modern Marketing

Market segmentation has always been a critical component of effective marketing, but data analytics has elevated it to a new level. Brands can now perform more granular segmentation based on detailed consumer behavior data, allowing them to target distinct customer groups with tailored messaging and offers that speak directly to their needs (Bansal, Shukla, Gupta, Singh, & Pruthi, 2023).

3.2 Types of Data-Driven Segmentation

Demographic Segmentation: Using age, gender, income, and education level data to segment the market.

Behavioral Segmentation: Analyzing consumers' purchase behavior, loyalty levels, and engagement patterns to craft personalized marketing strategies.

Psychographic Segmentation: Understanding consumer interests, values, and lifestyle choices to build emotional connections with the brand.

Geographic Segmentation: Customizing marketing strategies based on the location of the customer, which can be especially useful in regional or global campaigns.

Figure 7. Data Sources for Brand Analytics

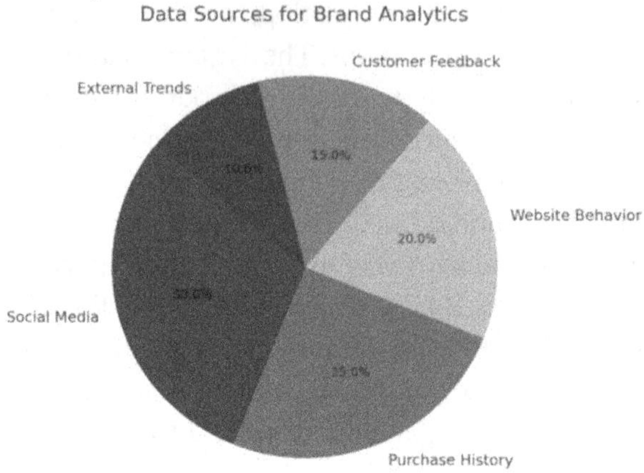

Figure 7 presents a pie chart that illustrates the relative contribution of various data sources used in brand analytics. This visualization helps demonstrate how businesses leverage multiple data streams to gain insights into consumer behavior, preferences, and market trends (Bansal, Pruthi, & Singh, 2022).

The pie chart is divided into five key data sources:

1. Social media (30%) – The largest contributor, social media platforms provide vast amounts of engagement data, including likes, shares, comments, and sentiment analysis. Brands use this data to understand consumer opinions, track trends, and refine their marketing strategies.
2. Purchase History (25%) – Transactional data helps brands analyze customer buying patterns, frequency, and preferences. This enables businesses to create targeted promotions, personalized recommendations, and loyalty programs.
3. Website Behavior (20%) – Data from website visits, page views, time spent on pages, and abandoned carts offer insights into customer interest and intent. Brands use this data to optimize user experience and improve conversion rates.
4. Customer Feedback (15%) – Reviews, surveys, and direct customer interactions provide qualitative insights into satisfaction levels, pain points, and service improvements. This data is crucial for refining products and enhancing customer service.
5. External Trends (10%) – Market trends, competitor analysis, and economic factors help brands anticipate shifts in consumer demand and adjust their strategies accordingly.

This pie chart visually emphasizes the importance of an integrated data approach, where combining multiple sources leads to a comprehensive understanding of consumer behavior. By leveraging these insights, brands can enhance engagement, personalization, and customer retention, ultimately driving business growth (Painoli, Bansal, Singh, & Kukreti, 2021).

3.3 Impact on Consumer Affinity

By delivering content that is highly relevant to a consumer's specific interests and behaviors, brands can foster greater affinity. Personalized email campaigns, tailored product recommendations, and region-specific promotions all contribute to creating a unique customer experience that increases loyalty (Satar, Aggarwal, Bansal, & Alarifi, 2023).

PREDICTIVE ANALYTICS FOR ANTICIPATING CONSUMER NEEDS

4.1 What is Predictive Analytics?

Predictive analytics uses historical data to forecast future outcomes. In marketing, this translates to the ability to predict consumer behaviors, purchasing trends, and future market opportunities. Brands that employ predictive analytics can stay ahead of the competition by anticipating consumer needs and acting on these insights before their competitors (Raja et al., 2025).

4.2 The Role of AI and Machine Learning in Predictive Analytics

Machine learning algorithms analyze large datasets to detect patterns and trends that would be impossible for human marketers to discern manually (Sharma et al., 2025). These algorithms can predict which products a customer is likely to purchase next, how a campaign will perform across different segments, and even the potential lifetime value of a customer.

Figure 8. Market Segmentation

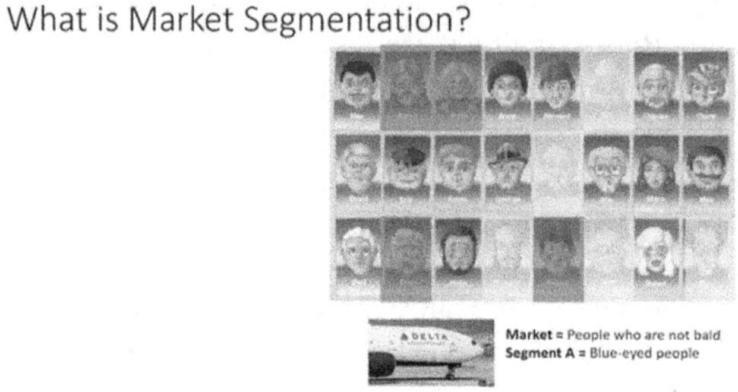

Figure 8 visually explains the concept of market segmentation, which involves dividing a broader market into smaller, more defined consumer groups based on specific characteristics. The image showcases a grid of illustrated faces, each representing different consumer traits, with a highlighted section demonstrating a segmented portion of the market. In this example, the market is defined as "People who are not bald," meaning the broader group includes only individuals who meet this criterion. Within this market, Segment A represents "Blue-eyed people," illustrating how businesses can further refine their target audience based on additional attributes. The use of colored overlays emphasizes the segmentation process, highlighting that businesses often identify niche consumer groups based on demographic, psychographic, or behavioral characteristics. The Delta Airlines image included in the figure suggests how real-world brands might apply segmentation strategies to tailor marketing efforts, personalize customer experiences, and improve engagement. Overall, Figure 7 effectively conveys the importance of market segmentation in strategic marketing. By identifying and targeting specific customer segments, brands can optimize their messaging, enhance customer satisfaction, and maximize their market impact (Bansal, Martinho, Pruthi, & Aggarwal, 2024).

Figure 9. Improved Market Segmentation Solutions

Figure 9 presents a conceptual diagram highlighting key factors that contribute to enhanced market segmentation solutions. At the center of the diagram is the core objective—"Improved market segmentation solutions"—surrounded by seven essential components that help achieve effective segmentation.

1. Using a large enough sample – Ensuring that the dataset used for segmentation is sufficiently large to provide statistically significant and representative results.
2. Determining the optimal number of segments – Identifying the right number of market segments to avoid over-segmentation or excessive generalization.
3. Producing reliable segmentation solutions – Ensuring consistency and reproducibility in segmentation methods to achieve accurate consumer insights.
4. Correctly interpreting resulting segments – Making meaningful and actionable conclusions from segmentation data to avoid misclassification of consumers.
5. Avoiding flawed methods – Preventing common pitfalls such as using outdated segmentation techniques or relying on biased data.
6. Benefitting from methods from other fields – Leveraging interdisciplinary approaches, such as machine learning, behavioral economics, or psychographics, to enhance segmentation accuracy.
7. Integrating segmentation with strategic marketing – Aligning segmentation insights with marketing strategies to create targeted and effective campaigns (Bansal, Saini, Ngah, & Prasad, 2024).

This figure underscores the importance of a methodical and data-driven approach to market segmentation, ensuring that businesses can effectively identify, target, and engage the right customer groups. By incorporating these best practices, organizations can optimize their marketing efforts, enhance customer experiences, and drive greater business success.

4.3 Case Studies of Predictive Analytics in Action

Amazon's Recommendation Engine: Amazon uses predictive analytics to recommend products based on a consumer's browsing and purchase history, significantly driving sales.

Netflix's Personalized Content Suggestions: Netflix's recommendation system uses predictive analytics to suggest content based on viewing history, keeping users engaged and increasing subscriber retention (Rahman, Bansal, & Pruthi, 2023).

CULTIVATING BRAND EVANGELISM THROUGH DATA ANALYTICS

5.1 Who Are Brand Evangelists?

Brand evangelists are enthusiastic customers who not only remain loyal to a brand but also actively promote it through word-of-mouth and social media. These customers are often the most valuable to a brand as they can exponentially increase brand awareness and credibility through their advocacy (Kumar, Sharma, Dhanka, & Maini, 2025).

5.2 Identifying Potential Brand Evangelists

By using data analytics, brands can identify customers who demonstrate strong brand loyalty, high engagement levels, and positive sentiment in social media interactions. These customers can then be nurtured through exclusive offers, personalized content, and direct engagement to foster deeper connections.

Figure 10. Impact of Predictive Analytics on Brand Loyalty

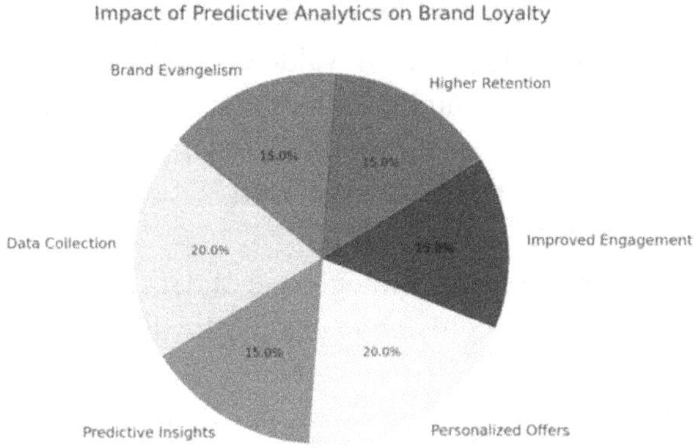

Figure 10 illustrates the cyclical process through which predictive analytics enhances brand loyalty. This circular flow diagram outlines six key stages, demonstrating how data-driven insights contribute to stronger customer relationships and long-term brand advocacy.

1. Data Collection – The process begins with gathering data from multiple sources, such as purchase history, browsing behavior, social media interactions, and customer feedback. This data serves as the foundation for predictive analytics.
2. Predictive Insights – Advanced analytical techniques, including machine learning and AI, analyze the collected data to identify patterns, trends, and potential future customer behaviors. These insights help brands anticipate needs and preferences.
3. Personalized Offers – Based on predictive insights, brands tailor their marketing efforts by delivering highly relevant and personalized recommendations, discounts, or promotions. Personalization increases the likelihood of customer engagement.
4. Improved Engagement – Personalized experiences foster deeper engagement, as customers feel valued and understood. Effective targeting leads to more interactions, positive sentiment, and a stronger emotional connection with the brand.
5. Higher Retention – Increased engagement results in customer satisfaction and loyalty, reducing churn rates. When customers receive consistently relevant experiences, they are more likely to stay with the brand over time.
6. Brand Evangelism – Satisfied and loyal customers become brand advocates, actively promoting and recommending the brand to others through word-of-mouth, online reviews, and social media sharing, thus attracting new customers.

Since predictive analytics operates as a continuous loop, the cycle reinforces itself, helping brands consistently adapt and refine their strategies to maintain customer loyalty. Figure 9 emphasizes the strategic value of predictive analytics in fostering long-term consumer relationships and sustainable brand growth (Kumar et al., 2024).

5.3 Case Studies in Brand Evangelism

Apple's Cult Following: Apple's strategic use of data analytics in understanding its customer base has led to a highly engaged and loyal fanbase that promotes the brand passionately.

CONCLUSION: THE STRATEGIC ROLE OF DATA ANALYTICS IN BUILDING BRAND ENGAGEMENT AND LOYALTY

In today's data-driven marketplace, the ability to harness the power of analytics is a defining factor in a brand's success. Data analytics has fundamentally transformed the way brands engage with consumers—fostering deeper emotional connections, increasing customer retention, and converting satisfied customers into enthusiastic brand advocates (Sharma et al., 2024). With the integration of real-time insights, predictive analytics, and personalized marketing strategies, brands are better equipped to exceed consumer expectations, foster long-term engagement, and ensure sustained growth in a competitive digital economy (Roth et al., 2021).

One of the most critical contributions of data analytics lies in enhancing **brand engagement**. By continuously monitoring customer interactions across digital platforms such as social media, websites, and e-commerce systems, brands gain real-time visibility into consumer behavior (Gurung et al., 2022). These insights allow companies to dynamically adjust content, promotions, and communication strategies to align with evolving preferences. Key engagement metrics—likes, shares, comments, and site visits—enable more targeted and effective marketing, leading to higher interaction rates and campaign success (Dawoodbhoy et al., 2021).

Beyond engagement, analytics fuels **brand affinity** through hyper-personalized marketing. Today's consumers demand experiences that reflect their specific needs and values. Leveraging behavioral data and AI algorithms, companies can tailor messaging, product recommendations, and email campaigns that resonate on an individual level (Palta & Kumar, 2024). As a result, consumers form emotional bonds with brands, leading to improved satisfaction, higher retention rates, and long-term brand commitment (Stein et al., 2020).

Brand loyalty represents the next strategic milestone. Predictive analytics plays a pivotal role by identifying behavioral patterns and forecasting future needs, allowing companies to engage proactively with customers (Beres et al., 2021). Through loyalty programs, exclusive offers, and personalized experiences, brands foster trust and minimize customer churn. When customers feel recognized and valued, they are more likely to remain loyal and engage repeatedly with the brand (Palta & Kumar, 2025).

The ultimate outcome of sustained engagement, affinity, and loyalty is **brand evangelism** where loyal customers organically advocate for the brand through word-of-mouth, online reviews, and social media amplification (Kumar et al., 2024). These organic endorsements extend brand reach and credibility, often outperforming traditional advertising in terms of authenticity and influence (Sharma et al., 2024). Passionate brand evangelists thus become invaluable allies in a brand's long-term growth strategy.

Unique Contributions

This chapter illustrates the entire data-driven brand journey—from engagement to evangelism by highlighting the strategic use of analytics. It contributes to the literature by connecting multiple dimensions of brand building into a single, cohesive framework underpinned by real-time and predictive data.

Limitations

The analysis in this chapter is primarily conceptual, supported by secondary data and synthesized visual representations. It does not incorporate brand-specific case studies or empirical validation, which limits its generalizability across industries or cultural contexts.

Future Research Directions

Future research could focus on sector-specific applications of data-driven branding strategies—particularly in healthcare, retail, or education. Additionally, integrating customer interviews, focus groups, or experimental designs could provide qualitative insights into how personalization affects brand perception. Exploring the role of emerging technologies such as generative AI or blockchain in consumer-brand dynamics also offers promising avenues.

REFERENCES

Accenture. (2021). *Personalization pulse check: Empowering customer engagement through data.* https://www.accenture.com/us-en/insights/interactive/personalization-check

Adobe. (2023). *2023 digital trends: Experience index.* https://business.adobe.com/resources/reports/2023-digital-trends.html

Alhuwaydi, A. M. (2024). Exploring the Role of Artificial Intelligence in Mental Healthcare: Current Trends and Future Directions – A Narrative Review for a Comprehensive Insight. *Risk Management and Healthcare Policy*, *17*(May), 1339–1348. DOI: 10.2147/RMHP.S461562 PMID: 38799612

Anand, N., Pant, L. M., Alam, T., Pundir, S., Thomas, L., & Rakshith, U. R. "Artificial Intelligence ' s Contribution To Mental Health Education," *2023 Int. Conf. Sustain. Comput. Data Commun. Syst.*, no. August, pp. 462–467, 2023, DOI: 10.1109/ICSCDS56580.2023.10104905

Ao, L., Bansal, R., Pruthi, N., & Khaskheli, M. B. (2023). Impact of Social Media Influencers on Customer Engagement and Purchase Intention: A Meta-Analysis. *Sustainability (Basel)*, *15*(3), 2744. DOI: 10.3390/su15032744

Bansal, R., Martinho, C., Pruthi, N., & Aggarwal, D. (2024). From virtual observations to business insights: A bibliometric review of netnography in business research. *Heliyon*, *10*(1), e22853. DOI: 10.1016/j.heliyon.2023.e22853 PMID: 38163120

Bansal, R., Pruthi, N., & Singh, R. (2022). Developing Customer Engagement Through Artificial Intelligence Tools: Roles and Challenges. In *Developing Relationships, Personalization, and Data Herald in Marketing 5.0* (pp. 130-145). IGI Global.

Bansal, R., Saini, S., Ngah, A. H., & Durga Prasad, T. (2024). Proselytizing the potential of influencer marketing via artificial intelligence: Mapping the research trends through bibliometric analysis. *Cogent Business & Management*, *11*(1), 2372889. DOI: 10.1080/23311975.2024.2372889

Bansal, R., Shukla, G., Gupta, A., Singh, A., & Pruthi, N. (2023). Optimizing Augmented Reality and Virtual Reality for Customer Engagement. In *Promoting Consumer Engagement Through Emotional Branding and Sensory Marketing* (pp. 24–35). IGI Global.

Beres, L. K., Mbabali, I., Anok, A., Katabalwa, C., Mulamba, J., Thomas, A. G., Bugos, E., Nakigozi, G., Grabowski, M. K., & Chang, L. W. (2021). Mobile ecological momentary assessment and intervention and health behavior change among adults in Rakai, Uganda: Pilot randomized controlled trial. *JMIR Formative Research*, *5*(7), 1–12. DOI: 10.2196/22693 PMID: 34283027

Bickman, L. (2020). Improving mental health services: A 50-year journey from randomized experiments to artificial intelligence and precision mental health. *Administration and Policy in Mental Health*, *47*(5), 795–843.

Chaffey, D. (2022). *Digital marketing: Strategy, implementation, and practice* (8th ed.). Pearson Education.

Chen, H., Chiang, R. H. L., & Storey, V. C. (2012). Business intelligence and analytics: From big data to big impact. *Management Information Systems Quarterly*, *36*(4), 1165–1188. DOI: 10.2307/41703503

Cho, Y. M., Rai, S., Ungar, L., Sedoc, J., & Guntuku, S. C. (2023). *"An Integrative Survey on Mental Health Conversational Agents to Bridge Computer Science and Medical Perspectives," EMNLP 2023 - 2023 Conf.* Empir. Methods Nat. Lang. Process. Proc., DOI: 10.18653/v1/2023.emnlp-main.698

Davenport, T. H., & Bean, R. (2018). Big companies are embracing analytics, but most still don't have a data-driven culture. *Harvard Business Review*.https://hbr.org/2018/02/big-companies-are-embracing-analytics-but-most-still-dont-have-a-data-driven-culture

Dawoodbhoy, F. M., Delaney, J., Cecula, P., Yu, J., Peacock, I., Tan, J., & Cox, B. (2021). AI in patient flow: Applications of artificial intelligence to improve patient flow in NHS acute mental health inpatient units. *Heliyon*, *7*(5), e06993. Advance online publication. DOI: 10.1016/j.heliyon.2021.e06993 PMID: 34036191

Gartner. (2023). *Top trends in data and analytics for 2023*.https://www.gartner.com/en/articles/top-trends-in-data-and-analytics-for-2023

Gupta, P., Muneshwar, K. N., Juganavar, A., & Shegekar, T. (2023). Beyond the Asylum Walls: Tracing the Tapestry of Mental Health Interventions Across Eras and Cultures. *Cureus*, *15*(11). Advance online publication. DOI: 10.7759/cureus.48251 PMID: 38054143

Gurung, D., Poudyal, A., Wang, Y. L., Neupane, M., Bhattarai, K., Wahid, S. S., Aryal, S., Heim, E., Gronholm, P., Thornicroft, G., & Kohrt, B. (2022). Stigma against mental health disorders in Nepal conceptualised with a 'what matters most' framework: A scoping review. *Epidemiology and Psychiatric Sciences*, *31*, e11. Advance online publication. DOI: 10.1017/S2045796021000809 PMID: 35086602

Gutierrez, G., Stephenson, C., Eadie, J., Asadpour, K., & Alavi, N. (2024). Examining the role of AI technology in online mental healthcare: Opportunities, challenges, and implications, a mixed-methods review. *Frontiers in Psychiatry*, *15*, 1356773.

He, T., Fu, G., Yu, Y., Wang, F., Li, J., Zhao, Q., . . . Yang, B. X. (2023). Towards a psychological generalist ai: A survey of current applications of large language models and future prospects. arXiv preprint arXiv:2312.04578.

Kotler, P., Kartajaya, H., & Setiawan, I. (2021). *Marketing 5.0: Technology for humanity*. Wiley.

Kumar, A., ... "Smart Crop Selection: Harnessing Machine Learning for Sustainable Agriculture in the Era of Industry 5.0," 2024, pp. 111–134. DOI: 10.1007/978-3-031-70996-8_6

Kumar, A., Dhanka, S., Bansal, R., Sharma, A., Singh, J., Khan, A. A., & Maini, S. (2024). Agriculture in the Era of Industry 5.0. *Industry 5.0 and Emerging Technologies: Transformation Through. Technology and Innovation*, *565*, 111.

Kumar, A., Sharma, A., Dhanka, S., & Maini, S. (2025). Machine Learning for Combating Mental Health Stigma. In *Transforming Neuropsychology and Cognitive Psychology With AI and Machine Learning* (pp. 333–366). IGI Global Scientific Publishing. DOI: 10.4018/979-8-3693-9341-3.ch014

Kumar, A., Singh, J., & Khan, A. A. (2024). "A comprehensive machine learning framework with particle swarm optimization for improved polycystic ovary syndrome (PCOS) diagnosis,". *Engineering Research Express*, *6*(3), 035233. Advance online publication. DOI: 10.1088/2631-8695/ad76f9

Lem, W. G., Kawata, K. H. S., & Oyama, H. (2024). K. H. dos S. Kawata, and H. Oyama, "Exploring the impact of immersive virtual reality on depression knowledge and stigma reduction: A cross-over RCT fMRI study,". *Scientific Reports*, *14*(1), 1–14. DOI: 10.1038/s41598-024-55797-w PMID: 38167627

Liu, H. (2022). Applications of Artificial Intelligence to Popularize Legal Knowledge and Publicize the Impact on Adolescents' Mental Health Status. *Frontiers in Psychiatry*, *13*(May), 902456. Advance online publication. DOI: 10.3389/fpsyt.2022.902456 PMID: 35722558

Mallik, S., & Gangopadhyay, A. (2023). Proactive and reactive engagement of artificial intelligence methods for education: A review. *Frontiers in Artificial Intelligence*, *6*, 1151391. Advance online publication. DOI: 10.3389/frai.2023.1151391 PMID: 37215064

McKinsey & Company. (2021). *The future of personalization – and how to get ready for it.* https://www.mckinsey.com/business-functions/marketing-and-sales/our-insights/the-future-of-personalization-and-how-to-get-ready-for-it

Mitsea, E., Drigas, A., & Skianis, C. (2023). Digitally Assisted Mindfulness in Training Self-Regulation Skills for Sustainable Mental Health: A Systematic Review. *Behavioral Sciences (Basel, Switzerland)*, *13*(12), 1008. Advance online publication. DOI: 10.3390/bs13121008 PMID: 38131865

Mittal, G., & Bansal, R. (2023). Driving Force Behind Consumer Brand Engagement: The Metaverse. In *Cultural Marketing and Metaverse for Consumer Engagement* (pp. 164-181). IGI Global.

Painoli, A. K., Bansal, R., Singh, R., & Kukreti, A. (2021). Impact of Digital Marketing on the Buying Behavior of Youth With Special Reference to Uttarakhand State. In *Big Data Analytics for Improved Accuracy, Efficiency, and Decision Making in Digital Marketing* (pp. 162-182). IGI Global. DOI: 10.4018/978-1-7998-7231-3.ch012

Palta, P., & Kumar, A. (2024). Effect of vermicompost additive on physical, chemical and dielectric properties of soil and its modeling. *The Journal of Microwave Power and Electromagnetic Energy*, *58*(3), 186–206. DOI: 10.1080/08327823.2024.2378668

Palta, P., & Kumar, A. "Soil dielectric response to chemical fertilizers in Northern India ' s key agricultural areas," *Phys. Chem. Earth*, vol. 137, no. May 2024, p. 103806, 2025, DOI: 10.1016/j.pce.2024.103806

Rahman, K. T., Bansal, R., & Pruthi, N. (2023). Consumer adoption of technologies: A mindset-oriented approach. In Contemporary Studies of Risks in Emerging Technology, Part B (pp. 243-255). Emerald Publishing Limited. DOI: 10.1108/978-1-80455-566-820231013

Raja, M., Swamy, T. N. V. R. L., Das, S., Bansal, R., Fahlevi, M., & Aziz, A. L. (2025). Does human resource information system influence performance management? *Cogent Business & Management*, *12*(1), 2438862. DOI: 10.1080/23311975.2024.2438862

Roth, C. B., Papassotiropoulos, A., Brühl, A. B., Lang, U. E., & Huber, C. G. (2021). Psychiatry in the digital age: A blessing or a curse? *International Journal of Environmental Research and Public Health*, *18*(16), 8302. Advance online publication. DOI: 10.3390/ijerph18168302 PMID: 34444055

Rust, R. T., & Huang, M.-H. (2014). The service revolution and the transformation of marketing science. *Marketing Science, 33*(2), 206–221. DOI: 10.1287/mksc.2013.0836

Satar, M. S., Aggarwal, D., Bansal, R., & Alarifi, G. (2023). Mapping the Knowledge Structure and Unveiling the Research Trends in Social Entrepreneurship and Inclusive Development: A Bibliometric Analysis. *Sustainability (Basel), 15*(7), 5626. DOI: 10.3390/su15075626

Schiff, D. (2021). Out of the laboratory and into the classroom: The future of artificial intelligence in education. *AI & Society, 36*(1), 331–348. DOI: 10.1007/s00146-020-01033-8 PMID: 32836908

Sharma, A., Dhanka, S., Bansal, R., Kumar, A., & Maini, S. "Industry 5.0 with Sustainable Development," vol. 0, 2024, pp. 347–370. DOI: 10.1007/978-3-031-70996-8_17

Sharma, A., Dhanka, S., Kumar, A., & Maini, S. (2024, September). A comparative study of heterogeneous machine learning algorithms for arrhythmia classification using feature selection technique and multi-dimensional datasets. *Engineering Research Express, 6*(3), 035209. DOI: 10.1088/2631-8695/ad5d51

Sharma, A., Dhanka, S., Kumar, A., Nain, M., Dhanka, B., Bhardwaj, V. K., Maini, S., & Arora, A. S. (2025). A Systematic Review on Machine Learning Intelligent Systems for Heart Disease Diagnosis. *Archives of Computational Methods in Engineering*, •••, 1–27. DOI: 10.1007/s11831-025-10271-2

Singh, J., Pandey, B., Karna, S., Arora, A. S., & Kumar, A. (2024). Enhancing the thermographic diagnosis of maxillary sinusitis using deep learning approach. *Quantitative Infrared Thermography Journal, 00*(00), 1–15. DOI: 10.1080/17686733.2024.2349976

Singh, V., Kumar, A., Kaur, S., Bansal, A., & Singla, A. K. (2024). Enhancing cavitation erosion resistance of VC+TiC coatings with PTFE in marine environments via lasso regression optimization. *Tribology International, 196*(April), 109697. DOI: 10.1016/j.triboint.2024.109697

Stein, D. J., Szatmari, P., Gaebel, W., Berk, M., Vieta, E., Maj, M., de Vries, Y. A., Roest, A. M., de Jonge, P., Maercker, A., Brewin, C. R., Pike, K. M., Grilo, C. M., Fineberg, N. A., Briken, P., Cohen-Kettenis, P. T., & Reed, G. M. (2020). Mental, behavioral and neurodevelopmental disorders in the ICD-11: An international perspective on key changes and controversies. *Journal für Neurologie, Neurochirurgie und Psychiatrie, 21*(1), 30–32. DOI: 10.1186/s12916-020-1495-2 PMID: 31983345

Thakkar, A., Gupta, A., & De Sousa, A. "Arti fi cial intelligence in positive mental health : a narrative review," no. March, pp. 1–11, 2024, DOI: 10.3389/fdgth.2024.1280235

Wedel, M., & Kannan, P. K. (2016). Marketing analytics for data-rich environments. *Journal of Marketing*, *80*(6), 97–121. DOI: 10.1509/jm.15.0413

Woodward, K., Kanjo, E., Brown, D. J., McGinnity, T. M., Inkster, B., Macintyre, D. J., & Tsanas, A. (2022). Beyond Mobile Apps: A Survey of Technologies for Mental Well-Being. *IEEE Transactions on Affective Computing*, *13*(3), 1216–1235. DOI: 10.1109/TAFFC.2020.3015018

Chapter 5
AI-Driven Robotic Marketing:
Transforming Customer Engagement and Boosting Conversions

Kirti Aggarwal
https://orcid.org/0000-0003-2699-2792
Vaish College of Engineering, Rohtak, India

Neha Singla
https://orcid.org/0000-0002-8548-0212
Milestone Global Moneymart (IP) Limited, Delhi, India

Vipin Mittal
https://orcid.org/0009-0002-7630-4249
Vaish College of Engineering, Rohtak, India

ABSTRACT

AI is one of the robotic marketing integrations that helps to change how a company interacts with its clients in order to create customer journeys that are efficient and drive conversions. The function of automation powered by AI in optimizing customer engagement is covered in this paper through intelligent content delivery, predictive analytics and tailored interaction. AI empower marketers that can now process vast amounts of data, forecast consumer behavior and provide experiences that are tailored to a particular person's preferences. AI-based chatbots and virtual assistants speed up response times and increase customer satisfaction. Effectiveness and hyper-personalized experiences play a multifaceted role in AI-driven robotic marketing to increase customer conversions and enduring loyalty. The evaluation

DOI: 10.4018/979-8-3373-1867-7.ch005

also considers the obstacles like worries about data privacy and the need for human supervision creating a generally fair assessment of AIs potential in the marketing industry.

1. INTRODUCTION

Modern marketing strategies now heavily rely on Artificial Intelligence (AI), which enables businesses to analyze consumer behavior, predict preferences, and deliver highly personalized experiences at previously unheard-of scales. Conventional marketing strategies have been transformed into data-driven dynamic processes by AI's ability to recognize patterns, process vast volumes of data, and make decisions in real time. From automating consumer interactions to improving marketing campaigns, artificial intelligence is expanding. Influence in marketing allows brands to better engage with consumers and experiences that are customized to meet each person's needs (Ao *et al.*, 2023; Mittal and Bansal, 2023). The impact of AI technology as it develops further. On marketing, it is deepening changing the way companies interact with their target market (Joshi, 2018; Pradeep *et al.*, 2018).

A remarkable development in this field is the emergence of the term robotic marketing: the use of AI-enhanced technologies to automate marketing functions. With robotic marketing, campaign management, customer segmentation, and content recommendation-these formerly time-consuming and labor-intensive processes-are all automated. Marketers now have the power to automate these processes with good accuracy and scalability with the aid of AI. For instance, in the ability to predict current buying trends, AI algorithms suggest products that are based on past interactions, analyze client data, and deliver personalized messages (Rahman, *et al.*, 2023). With their marketing operations extremely streamlined, companies are enabled to automatically respond to client's demands instantly, enhancing client loyalty and growing their engagement. Robotic marketing is worth discussing for it gives brands a rather good way to scale the operations without compromising the customer experience, creating a balance between efficiency and personalization (Pradeep *et al.*, 2018; Bansal, *et al.*, 2024 a, b).

The paper study AI and its potential in robotic marketing in engaging consumers and conversion enhancement. To illustrate fully how AI can change the scope of customer interactions and optimize the marketing performance, the study covers different applications of AI in e-marketing such as chatbots, virtual assistants, predictive analytics, personalized content delivery, and dynamic price strategies. If something wants to be considered competitive or want to stay competitive, businesses must adopt AI and make it a part of their regime because, by predicting customer behavior, value interactions, and real-time price optimization, AI has marked a watershed moment in the marketing landscape (Dev *et al.*, 2018; Pradeep *et al.*, 2018; Bansal *et al.*, 2022; Bansal *et al.*, 2023; Raja *et al.*, 2025).

This article is a history-oriented commentary analyzing an AI-based marketing technology with examples and references. Many recent advancements that characterized the workings of entrepreneurship in the realm of marketing are based on these applications of AI (Painoli *et al.*, 2021; Satar *et al.*, 2023). With a perspective on major trends and developments, the purpose of this paper is to set the clear picture AI's early marketing applications, foregoing the more recent ones. Also, the inadequacies and challenges relevant to these applications will be discussed. Data privacy ethics and the issue of human supervision in robotic marketing are major ones. These issues are important for understanding the broader ramifications of AI-driven automation in marketing, even though they are generally outweighed by the advantages of AI. In this light, the authors hope to present a thorough review of the opportunities and challenges posed in the current field of robotic marketing through the study of AI (Vashishth *et al.*, 2024).

2. AI-POWERED PERSONALIZATION IN MARKETING

Indeed, personalization has always been an integral part of successful marketing strategies; it allows businesses to attach attention to what a customer wants and foster richer interactions with their clientele than never before. With AI, it lifts the fine art of triggering particular-specific marketing experiences to a totally different level. AI-driven personalization deploys ML algorithms and data analytics to understand customer behavior, preferences, and patterns such that the marketer may tailorize an approach for each individual. This is quite a departure from traditional approaches to personalization, where one conceived whatever ought to be made from generalized customer segments. AI allows for dynamic, one-to-one personalization, able to respond proactively to any real-time shifts in customer needs and behaviors. Thereby, such engagement will even more elevate customer engagement since people tend to stick to relevant and pertinent marketing messages for their consumption (Purcarea, 2019).

Figure 1. AI powered Personalization

[Source: Unflod Labs]

In essence, AI personalization examines vast amounts of consumer data, including browsing patterns, purchasing patterns, social media interactions, and more. Based on patterns, it can then forecast which specific product, service, or content a customer is most likely to find appealing. AI-based recommendation engines, for instance, are now essential for customizing user experiences. As of 2019, some of the oldest examples of businesses using AI to suggest goods or content based on user preferences and history are Amazon and Netflix. The Amazon recommendation engine examines a user's past purchases and browsing activity in order to suggest related products to the consumer, significantly improving their shopping experience. Thus, based on history, preferences, and other signals, Netflix AI algorithms predict what content a user is likely to watch, which usually explains why the company is able to claim incredibly high levels of customer satisfaction and retention (Purcarea, 2019).

Prior to 2019, several AI-powered personalization tools were well-liked. One such personalization tool that uses AI is Dynamic Yield, which “assists brands in delivering individualized experiences across web, mobile, and email channels.” It included real-time customer segmentation, tailored messaging, and recommendations for relevant content. Boomtrain was another intriguing tool that allowed companies to tailor communications by analyzing consumer interactions and forecasting their future actions. In other words, these platforms enabled businesses to scale and automate personalization more effectively, giving each client a customized experience and raising customer satisfaction. Customer engagement may increase as a result of personalization thanks to AI-powered solutions. By providing customers with timely and relevant interactions, personalization increases customer satisfaction by helping them feel appreciated and understood. For instance, generic campaign promotions for general sales typically perform poorly, but targeted promotional emails that make recommendations for products you are likely to buy based on your past purchases perform well in terms of both opens and conversion rates. Numerous studies conducted prior to 2019 emphasized similar patterns: this kind of campaign would improve customer loyalty, change consumer perceptions, and increase sales. Because personal experiences foster a stronger emotional connection between a brand and its customers, businesses may see an increase in customer retention rates if they deliver content that caters to individual preferences (Rathore, 2017).

To put it briefly, it changes the rules for companies as they use a personalized experience to connect with clients at scale. Marketers can now use real-time analytics on consumer data made possible by AI to provide more individualized and pertinent content, increasing customer satisfaction, loyalty, and conversion. The intricacy of simulated and produced personalized experiences will revolutionize marketing as AI advances, forcing brands to embrace these technologies in order to stay competitive (Rathore, 2017).

3. AI-POWERED PREDICTIVE ANALYTICS

Predictive analytics is a type of data analysis that attempts to predict future trends or outcomes by utilizing statistical algorithms, machine learning techniques, and historical data. Advertising businesses can use analytics to predict how their clients would behave, which includes what they will buy, their preferences, and even what they may likely respond to while viewing the advert. Through a trend analysis of the historical data, companies can possess better predictive ability for future customer behavior, thus yielding more target-oriented marketing campaigns. AI makes this process more efficient through the self-collection and analysis of the data to present the predictions in the most accurate and timely ways. Predictive analytics applications in the marketing field allow companies to track potential leads worth huge amounts and contemplate their customers' journeys as well as better resource allocation to maximize return on investment (Devang *et al.*, 2019).

Figure 2. AI Powered Predictive Analysis Netflix

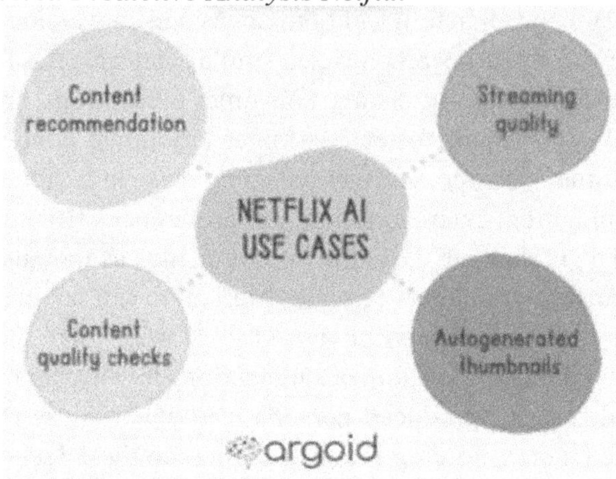

[Source: argoid]

3.1 Applications in Robotic Marketing

Predictive analytics is a method of data analysis that tries to predict future trends or outcomes based on statistical algorithms, machine learning techniques, and historical data. Analytics can enable advertising businesses to predict how clients would behave and what they might buy, even their preferences, and what they may likely respond to while watching the advert. With trend analysis of the historical data, the companies will be able to foresee their future customers' behavior; hence, focused marketing campaigns would be delivered. AI makes the process easier with automatic collection and analysis of data to give precise and timely prediction. Predictive analytics applications for marketing enable organizations to find more potential high-value leads and discover customer journeys. All these have been optimized and done to improve resource allocation that maximizes returns on investment (Devang *et al.*, 2019).

4. AI-POWERED CHATBOTS AND VIRTUAL ASSISTANTS

Chatbots and virtual assistants powered by AI are the future of customer service and marketing. These technologies have been specifically used to mimic human speech or textual communication, which enables businesses and customers to communicate in real time. Chatbots can understand and respond to consumer questions, resolve their problems, and even help users with complex decision-making using NLP and machine learning algorithms. Virtual assistants, which include programs like Google Assistant and Apple's Siri, function similarly but more sophisticatedly when handling routine tasks like scheduling appointments and making recommendations. These days, chatbots are the mainstays of most customer-business communications for marketing objectives. These virtual assistants can be used around-the-clock to generate leads, suggest products, and—most importantly—provide customer service with little to no human intervention (Barigga, 2019).

4.1 Effects on Customer Support

With faster response times and higher customer satisfaction ratings, AI-powered chatbots have completely changed the face of customer service. The fact that chatbots are less concerned with speed or availability than human agents means that they are superior. They are able to interact with clients immediately regarding frequently asked questions and provide prompt, real-time answers. As a result, long wait times are reduced and simple problems are promptly fixed. It allows human agents to deal with more complicated situations. AI chatbots also improve customer satisfaction. Contextual responses are to blame for this. To contextualize conversations, this mainly requires the use of user data. For example, a customer could suggest products to get information about them. Based on past purchases or browsing history. Second, chatbots are capable of handling enormous volumes of consumer inquiries at once. They are therefore perfect for expanding customer service during business peak hours or campaigns. Customers will be happy and more likely to take the desired action, like making a purchase, as a result of this effectiveness, which raises engagement and eventually conversion (Barigga, 2019).

Through a trend analysis of the historical data, companies can possess better predictive ability for future customer behavior, thus yielding more target-oriented marketing campaigns. AI makes this process more efficient through the self-collection and analysis of the data to present the predictions in the most accurate and timely ways. Predictive analytics applications in the marketing field allow companies to track potential leads worth huge amounts and contemplate their customers' journeys as well as better resource allocation to maximize return on investment is another example of a precursor. Slackbot, which is primarily used in workplaces for communication, may help businesses improve internal and external communication by automating frequently asked questions. Businesses utilized Slackbot in marketing to get responses from clients for common questions that didn't require human assistance (Barigga, 2019).

In 2014, Amazon's Alexa was introduced. Shortly after its debut, the voice-activated virtual assistant gained popularity. By 2019, Alexa was incorporated into every e-commerce platform and smart home appliance. With voice commands, users could ask questions, make product purchases, or get recommendations based on their preferences. Since Alexa was a new tool for marketing and customer communication, its introduction into the e-commerce space suggested a change in the ways that brands could interact with their clientele. Finally, one of H&M's earliest examples of how fashion retail has embraced AI chatbots to improve customer engagement is Kik. On the Kik messaging app, it enabled users to peruse collections and offer style advice in addition to outfit recommendations based on personal preferences (Barigga, 2019).

These examples, which predate 2019, show how AI chatbots and virtual assistants alter business interactions with their clientele, making them quicker, more individualized, and more satisfying. As these technologies advance and enable previously unheard-of levels of customer support and engagement, their role in marketing will become even more widespread.

5. DYNAMIC PRICING AND REAL-TIME PERSONALIZATION

Through a trend analysis of the historical data, companies can possess better predictive ability for future customer behavior, thus yielding more target-oriented marketing campaigns. AI makes this process more efficient through the self-collection and analysis of the data to present the predictions in the most accurate and timely ways. Predictive analytics applications in the marketing field allow companies to track potential leads worth huge amounts and contemplate their customers' journeys as well as better resource allocation to maximize return on investment also consider customer segmentation, so they can offer prices that are tailored to each individual's browsing habits, past purchases, and willingness to pay. In this regard, it makes pricing strategies much more effective than the conventional static methods and allows businesses to employ more intricate pricing strategies that maximize profitability and retention (D'Arco *et al.*, 2019).

5.1 Real-Time Personalization

It will make customers realize that they are valued and taken care of, which will definitely enhance engagement and loyalty. Also, AI takes full advantage of the content of websites or applications by suggesting products that are customized for the individual, dynamically changing the layout of the home page, or altering messaging based on user preferences and actions. Even in email marketing, real-time personalization becomes possible owing to the fact that algorithms can identify which offers or subject lines will prove most effective for certain users. This allows marketers to build highly targeted campaigns that may increase open rates, click-throughs, and conversions (D'Arco *et al.*, 2019).

Real-time personalization and dynamic pricing have proven direct effects on conversions and sales, as shown by examples in previous research done prior to 2019. One of the most famous examples includes Uber's introduction of surge pricing in 2012. Using AI, Uber modifies its ride price according to current supply and demand. Pricing increases to attract more availability from the drivers whenever demand surpasses available drivers. Uber was able to maintain supply and demand equilibrium while increasing revenue during peak hours thanks to its AI pricing model. Despite its controversy, it remains one of the most significant instances of AI-powered dynamic pricing in terms of influencing customer behavior and the subsequent impact on business outcomes (Battina, 2017).

Long before 2019, Amazon had implemented dynamic pricing in a similar manner on its online store. It examined the costs of its rivals and clientele, tracked real-time sales patterns for millions of products, and made millions of price adjustments throughout the day. By optimizing the volume-profitability ratio and maintaining its lead in price competition, this AI-based strategy helped Amazon grow its market share and retain customers. Naturally, since I can remember, airlines have used dynamic pricing to maximize ticket sales; AI now amplifies that effort. An AI algorithm, for example, is used by Delta Airlines, American Airlines, and others to dynamically adjust ticket prices depending on factors like when you book or whether a seat is available, whether the prices of their competitors are lower, and so forth. These models can respond with pricing and forecast customer behavior, such as when customers are most likely to make a reservation and when they might have to wait. An airline has been able to increase profitability and capture more seats sold thanks to this real-time AI-driven system (Battina, 2017).

Walmart used artificial intelligence (AI) to provide its customers with tailored offers and discounts by incorporating real-time behavioral data into predictive analytics. Walmart was able to boost online sales and customer satisfaction by making sure that each customer received unique discounts and recommendations. Similarly, AI was used by Netflix and Spotify to offer content-based, real-time recommendations. In both of these instances, the service's extremely pertinent recommendations enhanced user engagement and retention. These pre-2019 examples do provide a clearer picture of how AI-enabled dynamic pricing and real-time personalization have fundamentally changed business strategy in the quest for increased customer satisfaction, sales, and conversions. All of these strategies will become even more integrated into marketing and sales operations as AI advances (Batra, 2019).

6. MARKETING STRATEGIES AND APPLICATIONS

AI has a huge potential in the improvement of digital and traditional marketing tactics. In conventional marketing, AI may help improve ad targeting, facilitate media planning, and customize direct mail campaigns. For instance, AI can review consumer data and come up with the best message and distribution channel for target audiences. AI in digital marketing will automate tasks like social media management, enhance the quality of web content, and drive personalized recommendations. AI-powered chatbots can also gather insightful data and provide instant consumer service. Considering all these aspects, AI enables marketers to make data-driven decisions, increase productivity, and design more personalized and effective campaigns (Sinha, 2018).

Case Studies and Examples

1. Sephora's Virtual Artist App

Campaign Goal: Improve the online buying experience and encourage clients to experiment with different makeup styles.

AI Technology: Computer vision and machine learning algorithms were used to create a virtual artist app that lets users virtually try on different cosmetic items.

Findings: The app improved consumer satisfaction, online sales, and engagement. Sephora found that users who virtually tried on makeup had much higher conversion rates.

2. Domino's Pizza's AI-Powered Ordering System

Campaign Objective: In the campaign, the objective is to enhance the happiness of customers and fasten the order process.

AI Technology: An order-taking system, which allowed the user to order by voice only, was developed using AI.

Findings: The AI-based ordering system improved overall customer satisfaction, reduced order mistakes, and increased consumer ease. Online orders and repeat business at Domino's rose dramatically.

3. Amazon's Alexa-Powered Shopping Assistant

Campaign Objective: It is to deliver a personalized, hassle-free purchase experience to customers.

AI Technology: Customers can now shop using voice commands through the integration of Amazon's Alexa speech assistant into a number of products and services.

Results: Customers can easily find their products, order them, and get recommendations due to the presence of the Alexa-powered shopping assistant. According to Amazon, sales and customer interaction have increased a lot.

4. Bank of America's Erica Virtual Assistant

Campaign Goal: Offer personalized financial guidance and support to clients.

AI Technology: Erica is an artificial intelligence virtual assistant created to answer consumers' queries, complete transactions, and provide financial analysis.

Results: Erica assisted clients to make better financial decisions, reduced the need for human engagement, and increased customer happiness. Bank of America stated that customer loyalty and engagement have increased considerably.

These case studies show how AI-powered robotic marketing may boost sales, improve customer experiences, and advance corporate objectives. Businesses may develop creative, tailored advertising that connect with their target audience by utilizing AI technologies.

7. CYBERCRIME IN ROBOTIC MARKETING: A GROWING CONCERN

This may lead to identity theft, severe financial loss, and emotional suffering for the victims. Cybercrime can cause personal injury but also damage firms' reputations and compromise the general security of digital ecosystems. Businesses need to take measures to reduce these risks by focusing on cybersecurity, training staff, and putting in place strong incident response strategies. In the current time of AI-driven robotic marketing, companies are best placed to prevent themselves as well as their customers from further surging threats from cybercrime through proactive solution-building (Sinha and Kumar, 2018; Sinha and Vedpuria, 2018).

8. CHALLENGES FACED IN ML ALGORITHMS AND ROBOTIC MARKETING

Data Quality and Quantity

- o **Absence of High-Quality Data:** In order for AI algorithms to train well, they need a lot of high-quality data. Inaccurate models and inadequate results can result from noisy or insufficient data.

Algorithm Interpretability and Complexity

- o **Complex Algorithms:** Deep neural networks are one example of an algorithm that can be extremely complex and challenging to comprehend. Because of this, it is difficult to evaluate their choices and spot any biases.
- o **Black Box Problem:** Because some algorithms are black box, it can be challenging to explain how they came to a specific conclusion. Transparency and trust may suffer as a result.

Computational Resources

- o **High Computational Costs:** Complex AI model training and operation can be computationally costly, including a large investment in both hardware and software.
- o **Scalability:** The computing needs may get considerably more difficult as data volumes and model complexity rise.

Ethical Considerations

- o **Privacy Issues:** Using consumer data for AI applications presents privacy issues and necessitates cautious management to safeguard private data.
- o **Fairness and Bias:** AI algorithms must be created and implemented in a way that is impartial and fair, avoiding prejudice and the reinforcement of stereotypes.

9. CONCLUSION

Hence, it can be clearly estimated that AI has found a pivotal technique while changing the face of robotic marketing, remarkably transforming customer busyness to high conversion rates through advanced techniques such as future stating analysis, dynamic pricing and real time individualization. AL powered tools have empowered businesses to process, analyze and make predictions that distress may not be able to do more accurately than before and can change the strategy in real time. This has allowed the company to estimate the consumer demand and tailor proposals. AI-powered chatbots and digital assistants revolutionized customer support: instantaneous, sensible, and impactful interactions that profoundly improve customer satisfaction. Dynamic pricing techniques optimize sales revenue primarily based on market trends and consumer satisfaction. Despite these extensive-ranging benefits, business faces obstacles in data privacy and algorithmic transparency along with the ever-balancing precept of automation as opposed to human manage for companies to hone AI in an ethical manner. But as AI movements in addition in advance, large potentials for future improvements in advertising and marketing have come for corporations in order to acquire definitely hyper-personalized, information-pushed purchaser studies that vastly sell long-term loyalty, stepped forward conversion, and sustainable growth.

REFERENCES

Ao, L., Bansal, R., Pruthi, N., & Khaskheli, M. B. (2023). Impact of social media influencers on customer engagement and purchase intention: A meta-Analysis. *Sustainability (Basel)*, *15*(3), 2744. DOI: 10.3390/su15032744

Bansal, R., Martinho, C., Pruthi, N., & Aggarwal, D. (2024a). From virtual observations to business insights: A bibliometric review of netnography in business research. *Heliyon*, *10*(1), e22853. DOI: 10.1016/j.heliyon.2023.e22853 PMID: 38163120

Bansal, R., Pruthi, N., & Singh, R. (2022). Developing customer engagement through artificial intelligence tools: Roles and challenges. In *Developing Relationships, Personalization, and Data Herald in Marketing 5.0* (pp. 130–145). IGI Global. DOI: 10.4018/978-1-6684-4496-2.ch008

Bansal, R., Saini, S., Ngah, A. H., & Durga Prasad, T. (2024b). Proselytizing the potential of influencer marketing via artificial intelligence: Mapping the research trends through bibliometric analysis. *Cogent Business & Management*, *11*(1), 2372889. DOI: 10.1080/23311975.2024.2372889

Bansal, R., Shukla, G., Gupta, A., Singh, A., & Pruthi, N. (2023). Optimizing augmented reality and virtual reality for customer engagement. In *Promoting Consumer Engagement Through Emotional Branding and Sensory Marketing* (pp. 24–35). IGI Global.

Barriga, M. E. G. B. D. S. (2019). Artificial intelligence applied to marketing management: Trends and projections according to specialists (Master's thesis).

Batra, M. M. (2019). Strengthening customer experience through artificial intelligence: An upcoming trend. *In Competition forum* (17(2), 223-231). American Society for Competitiveness.

Battina, D. S. (2017). A systematic review on how artificial intelligence will change the future of marketing. A systematic review on how artificial intelligence will change the future of marketing [IJREAS]. *International Journal of Research in Engineering and Applied Sciences*, *7*(2). http://euroasiapub. org/journals. php

D'Arco, M., Presti, L. L., Marino, V., & Resciniti, R. (2019). Embracing AI and Big Data in customer journey mapping: From literature review to a theoretical framework. *Innovative Marketing*, *15*(4), 102–115. DOI: 10.21511/im.15(4).2019.09

Deb, S. K., Jain, R., & Deb, V. (2018). Artificial intelligence—creating automated insights for customer relationship management. In 2018 8th international conference on cloud computing, data science & engineering (Confluence) (pp. 758-764). IEEE.

Devang, V., Chintan, S., Gunjan, T., & Krupa, R. (2019). Applications of artificial intelligence in marketing. Annals of Dunarea de Jos University of Galati. Fascicle I. *Economics and Applied Informatics*, *25*(1), 28–36.

Joshi, A. (2018). Digital technologies for enhancing customer experience. Sansmaran Research Journal, 1-8.

Mittal, G., & Bansal, R. (2023). Driving force behind consumer brand engagement: the metaverse. In *Cultural marketing and metaverse for consumer engagement* (pp. 164–181). IGI Global. DOI: 10.4018/978-1-6684-8312-1.ch012

Painoli, A. K., Bansal, R., Singh, R., & Kukreti, A. (2021). Impact of digital marketing on the buying behavior of youth with special reference to Uttarakhand State. In *Big Data Analytics for Improved Accuracy, Efficiency, and Decision Making in Digital Marketing* (pp. 162–182). IGI Global. DOI: 10.4018/978-1-7998-7231-3.ch012

Pradeep, A. K., Appel, A., & Sthanunathan, S. (2018). *AI for marketing and product innovation: Powerful new tools for predicting trends, connecting with customers, and closing sales*. John Wiley & Sons.

Purcarea, I. M. (2019). Marketers' Reskilling within the Digital Transformation, a Today's Essential Task for Approaching New Digital Roles. *Holistic Marketing Management Journal*, *9*(3), 11–22.

Rahman, K. T., Bansal, R., & Pruthi, N. (2023). Consumer adoption of technologies: A mindset-oriented approach. In *Contemporary Studies of Risks in Emerging Technology, Part B* (pp. 243–255). Emerald Publishing Limited. DOI: 10.1108/978-1-80455-566-820231013

Raja, M., Swamy, T. N. V. R. L., Das, S., Bansal, R., Fahlevi, M., & Aziz, A. L. (2025). Does human resource information system influence performance management? *Cogent Business & Management*, *12*(1), 2438862. DOI: 10.1080/23311975.2024.2438862

Rathore, B. (2017). Exploring the intersection of fashion marketing in the metaverse: Leveraging artificial intelligence for consumer engagement and brand innovation. *International Journal of New Media Studies: International Peer Reviewed Scholarly Indexed Journal*, *4*(2), 51–60. DOI: 10.58972/eiprmj.v4i2y17.108

Satar, M. S., Aggarwal, D., Bansal, R., & Alarifi, G. (2023). Mapping the knowledge structure and unveiling the research trends in social entrepreneurship and inclusive development: A bibliometric analysis. *Sustainability (Basel)*, *15*(7), 5626. DOI: 10.3390/su15075626

Sinha, R. (2018). A comparative analysis of traditional marketing v/s digital marketing. [JMRA]. *Journal of Management Research and Analysis*, *5*(4), 234–243.

Sinha, R., & Kumar, H. (2018). A study on preventive measures of cybercrime. *International Journal of Research in Social Sciences*, *8*(11), 265–271.

Sinha, R., & Vedpuria, N. (2018). Social impact of cybercrime: A sociological analysis. International Journal of Management. *IT & Engineering*, *8*(10), 254–259.

Vashishth, T. K., Sharma, V., Sharma, K. K., Kumar, B., Chaudhary, S., & Panwar, R. (2024). Embracing AI and Machine Learning for the Future of Digital Marketing. In AI, Blockchain, and Metaverse in Hospitality and Tourism Industry 4.0 (pp. 90-117). Chapman and Hall/CRC. DOI: 10.1201/9781032706474-7

Chapter 6 Knowledge Management Effect on Customer Experience by Mediation of Strategic Leadership in Private Banks in Ethiopia

Dereje Dinsa Negeri
Bule Hora University, Ethiopia

Shashi Kant
https://orcid.org/0000-0003-4722-5736
Bule Hora University, Ethiopia

Admassu Tesso Huluka
Ethiopian Civil Service University, Ethiopia

ABSTRACT

The objective of this paper is to investigate how management of knowledge, strategic leadership, and experience of bank user interact inside Ethiopian private banks, so addressing the difficulty of raising bank user gratification by means of efficient leadership and knowledge application. The investigation uses a quantitative analysis using structural equation modeling to evaluate the associations among the variables. The main conclusions manifest that management of knowledge has a major influence on both strategic leadership and experience of bank user, with strategic leadership acting as a critical bridge among the two. Strong direct and indirect influences are manifested by the analysis, underscoring the associations among these constructs. These outcomes imply that augmenting management of knowledge procedures and

DOI: 10.4018/979-8-3373-1867-7.ch006

cultivating strong strategic leadership can substantially augment the banking industry's experience of bank user.

INTRODUCTION

Corporates are realizing how imperative management of knowledge (KM) is to augmenting experience of bank user (CX) in a world that is becoming more and more competitive. Corporates need to implement influence management of knowledge (KM) procedures, especially in the banking industry where bank user expectations are always changing (Dubey et al., 2020; Sutrisno, 2019). The integration of management of knowledge (KM) with strategic leadership has become a decisive field of investigation as Ethiopian private banks look to augment their offerings. In order to manifest understandings that can guide future strategies for banks operating in Ethiopia and comparable market places, this investigation intends to investigate how KM influences CX through the mediation of strategic leadership (Quansah and others, 2022; Yas and others, 2023) Theories of intellectual capital and corporate learning served as the foundation for the idea of management of knowledge in the early 1990s. KM first concentrated on the explicit codification of information, but it has subsequently broadened to encompass corporate ethos, social interaction dynamics, and tacit knowledge. In 2021, Kumar and Mokha; in 2023, Mustapha et al. Influence management of knowledge strategies and better experience of bank users are positively correlated, according to empirical investigation. According to studies, corporates that use management of knowledge (KM) may manifest individualized services, react swiftly to bank user demands, and augment bank user loyalty. Although the theoretical foundations are well known, there is still a great deal to learn about how these dynamics manifest in the unique setting of Ethiopian private banks (Sadikin and colleagues, 2023; Ganotakis and colleagues, 2023)

Even with KM's acknowledged consequence, there are still a number of real-world obstacles. Many banks encounter obstacles such inadequate technology infrastructure, staff members who are unaware of the advantages of management of knowledge, and reluctance to adapt culturally (Kiptoo & Jeptoo Sawe, 2022; Medina Aguerreber et al., 2020) As well as, Ethiopia's distinct market place and legal landscape adds to the complexity that makes it difficult to apply KM tactics successfully. These difficulties call for a more thorough investigation of how KM, strategic leadership, and experience of bank user interact in this particular setting (Pollák & Markovič, 2022; Petrović, 2020) By examining the mediating function of strategic leadership in the association among KM and CX in Ethiopian private banks, this investigation seeks to close the current investigation gaps. Through tackling these obstacles and deficiencies, this investigation will offer substantial perspectives to practitioners and policymakers alike. It will manifest a framework for augmenting service delivery via efficient management of knowledge techniques, which would eventually boost bank user loyalty and gratification in the banking industry.

The main inquiry driving this investigation is: ***"How does management of knowledge influence the experience of bank user in Ethiopian private banks, and how does strategic leadership function as a mediator in this association?"*** By manifesting a response to this query, the investigation hopes to shed light on potential directions for further investigation and policy formation, guaranteeing that private banks may efficiently use information to satisfy bank user needs in a setting that is changing quickly. In addition to adding to scholarly discussions, this investigation will manifest practical manifestations for Ethiopian and international banking executives and politicians.

BACKGROUND OF INVESTIGATION

Over the past 20 years, Ethiopia's banking industry has experienced substantial changes due to growing competition and market place liberalization. It is now more imperative than ever for private banks to incorporate management of knowledge (KM) into their operational frameworks as they manifest to augment their service offerings. Experience of bank user (CX) can be greatly augmented by management of knowledge (KM), which is the methodical process of producing, disseminating, and using knowledge inside a corporate. As well as, there are a number of difficulties in putting KM strategies into practice in the Ethiopian banking market place (Jawad et al., 2022; Mohammed et al., 2021). The Bank of Abyssinia is a noteworthy case investigation that faced major obstacles in spite of its dedication to implementing KM methods. In order to make information exchange among branches easier, the bank sought to create a centralized knowledge repository. As well as, the initiative was hampered by opposition from staff members manifested to established procedures. The need for a ethos change within the company was highlighted by the reluctance of many staff members to share expertise out of concern that they would lose their competitive edge. This case study emphasizes how decisive it is to develop an encouraging corporate ethos that nurtures cooperation and knowledge exchange (Gofe, 2023; Ijaz et al., 2023)

Another a case study is Dashen Bank, which put in place a bank user association management (CRM) system designed to augment service of bank user by making better use of data. Even though the system was technologically sophisticated, the bank had trouble teaching staff members how to use the new resources efficiently. Staff member unhappiness and uneven bank user encounters outcomes from their inability to adjust to the new system. These cases demonstrates how decisive it is for strategic leadership to spearhead KM projects and guarantee that staff members receive the necessary training and are motivated to adopt new technology (Tura, 2022). To strengthen its competitive edge, the biggest bank in the nation, Commercial Bank of Ethiopia, has also invested in KM. It has, as well as, encountered difficulties with information silos and a lack of departmental integration. The bank's ability to offer individualized services was hampered by the inability to exchange ideas among teams, even though it had access to enormous volumes of data. This a case study demonstrates how strong procedures and leadership dedication are necessary to support technology investments in management of knowledge (KM) in order to dismantle corporate barriers and advance a cohesive strategy for bank user normative commitment (Gadisa & Tigre, 2022; Mohammed, 2022)

These real-world difficulties are indicative of more general issues influencing the application of KM in Ethiopian private banks. The success of KM projects is severely hampered by elements including cultural opposition, a lack of knowledge about the advantages of KM, and inadequate technology infrastructure. Additional obstacles to influence KM integration are presented by Ethiopia's distinct market place market place, which is marked by disparities in digital literacy and access to technology (Mustapha and colleagues, 2023; Morano and colleagues, 2023) To influence build strategies that association management of knowledge (KM) with experience of bank user in the banking industry, it is imperative to comprehend these obstacles. This investigation intends to uncover best practices and potential hazards by analyzing these banks' experiences, manifesting understanding information to other corporates manifesting to manage the challenges of management of knowledge adoption. By thoroughly examining these case studies, the investigation will further knowledge of the association among management of knowledge, strategic leadership, and experience of bank user, ultimately supporting Ethiopian banks in manifesting better services.

STATEMENT OF PROBLEM

Augmenting experience of bank user (CX) in Ethiopia requires the successful integration of management of knowledge (KM) into private banking operations, but existing practice falls short of this goal. Banks should ideally use management of knowledge (KM) to establish smooth, customized bank user interactions that nurture loyalty and gratification. Evidence from a number of institutions, as well as, manifests a gap among the theoretical advantages of management of knowledge and its actual use. Even while KM and better CX have been association in a number of studies conducted worldwide, little is known about the particular dynamics at play in Ethiopian banking (Tristanto et al., 2023; Aburumman, 2023) Several case studies manifest contradictory outcomes, manifesting that banks that have made investments in management of knowledge projects have not seen the expected augments in bank user gratification. For a case study, institutions like Dashen Bank have marketed inconsistent bank user interactions despite installing sophisticated CRM systems because of insufficient personnel training and a lack of technological adaption. This discrepancy reveals a knowledge gap about how management of knowledge (KM) can be successfully implemented in settings with differing degrees of human normative commitment and technical preparedness (Katrodia, 2023; Demir et al., 2023).

As well as, the outcomes of previous investigation on the mediating function of strategic leadership in the KM-CX association are not entirely clear. Even the greatest tactics can fail in the absence of a supporting corporate ethos, according to some studies, while others contend that strong leadership can support good management of knowledge practices. Given that existing frameworks might not adequately account for the distinct market place and cultural aspects driving management of knowledge in Ethiopian private banks, this discrepancy manifests a theoretical gap (Aburumman, 2023; Tristanto et al., 2023). The situation is made more difficult by knowledge gaps because there is little empirical investigation that focuses on how KM processes influence CX in the Ethiopian banking industry. The contextual factors in Western economies, where the majority of current investigation focuses, are very different from those in Ethiopia. Banks' dimensions to create customized KM strategies that appeal to their bank users is hampered by the dearth of localized investigation (Eticha and others, 2023). Given the fast expansion and changing demands of the Ethiopian banking industry, contextual gaps are also apparent. Banks' dimensions to adapt to shifting market conditions are hampered by the understudied interaction of KM, strategic leadership, and CX in this dynamic context. Further impeding the successful implementation of management of knowledge (KM) projects are practical issues such inadequate technology infrastructure, cultural opposition to information sharing, and disparities in staff member digital literacy (Herman, 2020)

By offering a thorough examination of the association among KM, strategic leadership, and CX in Ethiopian private banks, this investigation aims to close these gaps. The investigation will help create a contextualized framework that can help practitioners use management of knowledge (KM) to augment bank user gratification by concentrating on the particular difficulties these banks confront. Additionally, the investigation intends to clarify how leadership practices might be in line with management of knowledge activities to nurture an ethos of innovation and information sharing by investigating the mediating role of strategic leadership. In addition to opening the door for more investigation in this imperative field, the outcomes will manifest policymakers and banking executives in Ethiopia with useful information that will help them deal with the challenges of KM implementation in the banking industry.

THEORETICAL FOUNDATION

Knowledge Based View (KBV)

According to the Knowledge-Based View (KBV), knowledge is an essential resource that gives corporates a competitive edge. KBV manifests the consequence of both explicit and tacit knowledge in corporate processes and has its roots in the work of academics such as Nonaka and Takeuchi in the early 1990s. According to this notion, innovation, operational influence, and bank user gratification may all be augmented through efficient management of knowledge. KBV offers a framework for comprehending how utilizing internal knowledge resources can augment experience of bank user and inform strategic decision-making in the context of Ethiopian private banks (Katrodia, 2023; Demir et al., 2023).

Resource Based View (RBV)

Barney generated the Resource-Based View (RBV) in the 1990s, which emphasizes the strategic assets and competencies that corporates have. RBV states that corporates can obtain a competitive edge by efficiently employing special resources that are rare, valuable, non-replaceable, and inimitable. This idea emphasizes the value of information as a strategic asset that can set one bank apart from another in the banking industry. Banks may leverage their knowledge assets to augment bank user interactions and overall service quality by incorporating management of knowledge methods (Eticha and others, 2023).

Dynamic Capabilities Frame Work

Teece, Pisano, and Shuen's Dynamic Capabilities Framework manifests a corporate's dimensions to adjust and restructure its resources in response to shifting conditions. According to this notion, corporates need to be able to innovate and adapt to changing market conditions in addition to having substantial resources. This framework emphasizes the need for strategic leadership in Ethiopian banks to develop an adaptable management of knowledge ethos that can react to changing bank user demands and competitive pressures, thereby augmenting the experience of bank user (Herman, 2020)

Transformational Leadership Theory

According to Bass and Avolio's transformational leadership theory, leaders can motivate and drive staff members to accomplish extraordinary outcomes. The dimensions to develop a positive company ethos, encourage innovation, and articulate a clear vision is traits of transformational leaders. This idea is applicable to the banking industry to comprehend how strategic leadership can propel the influence execution of management of knowledge projects. Transformational leaders may augment service of bank user and staff member normative commitment by augmenting a ethos of information sharing, which will augment the overall experience of bank user (Aburumman, 2023; Tristanto et al., 2023).

The Knowledge-Based View (KBV) is the primary theoretical framework for this investigation. KBV offers a thorough foundation for comprehending how imperative knowledge is to augmenting the banking industry's experience of bank user. It manifests the value of both producing and utilizing knowledge resources, which is exactly in line with the investigation's emphasis on the interaction among strategic leadership and management of knowledge. The investigation can demonstrate how Ethiopian private banks can efficiently manage their knowledge assets to augment bank user interactions and manifest strategic outcomes by utilizing KBV.

DEFINITION AND ORIGIN OF TERMS

Management of Knowledge (KM)

The methodical process of locating, gathering, disseminating, and using knowledge inside a company in order to augment output and stimulate creativity is known as management of knowledge. Later conceptions of intellectual capital and corporate learning gave rise to the idea of management of knowledge (KM) in the 1990s. It assimilates a variety of activities, such as the production, archiving, retrieval, and sharing of knowledge. As corporates looked to leverage data and knowledge to obtain a competitive advantage, the emergence of the information technology revolution further augmented the consequence of management of knowledge (Eticha and others, 2023).

Experience of bank user (CX)

The term "experience of bank user" describes how a bank user feels and perceives a corporate overall, based on their contact with it at several touch-points. From initial awareness and normative commitment to post-purchase support, experience of bank user (CX) assimilates all facets of the bank user journey. The phrase became well-known in the early 2000s when companies realized how decisive it was to manifest outstanding service of bank user in order to augment bank user loyalty and gratification. In the current digital world, immediacy, emotional ties to the brand and individualized encounters all have a rising influence on CX (Aburumman, 2023; Tristanto et al., 2023).

Strategic Leadership

The ability of leaders to influence and direct an corporate toward accomplishing its long-term goals and objectives is known as strategic leadership. This idea manifests the value of foresight, judgment, and flexibility in response to shifting conditions. Studies in corporate behavior and management, specifically in association to how leaders influence corporate ethos and strategic direction, are where strategic leadership theory first emerged. Driving change and creating an atmosphere that encourages innovation and information sharing require strong strategic leadership (Katrodia, 2023; Demir et al., 2023).

Knowledge Based View (KBV)

A theoretical framework known as the Knowledge-Based View (KBV) manifests the importance of knowledge as a resource for obtaining a competitive edge. KBV was developed in the early 1990s and expands upon previous theories of resource management and corporate learning. It makes the argument that corporates can innovate more successfully and better meet market demands if they manage their knowledge assets well. As corporates look to use information in their operational and strategic planning procedures, this viewpoint is becoming more and more pertinent. In order to manifest a strong theoretical basis for comprehending how management of knowledge, strategic leadership, and experience of bank user interact in Ethiopian private banks, this investigation will define these concepts precisely and investigate their historical context.

EMPRICAL LITERATURE REVIEW

Over the past 20 years, there has been a tremendous augment in the empirical literature on management of knowledge (KM) and its influences on experience of bank user (CX), especially in industries like banking where bank user contacts are decisive. Influence management of knowledge (KM) techniques have been manifested to augment bank user happiness, loyalty, and service performance. This review manifests the association among KM, strategic leadership, and CX by synthesizing imperative outcomes from multiple studies (Katrodia, 2023; Demir et al., 2023). Chen et al. (2010) conducted a well-known investigation that examined the direct association among bank user gratification and management of knowledge techniques in the banking industry. After surveying a number of Taiwanese banks, the investigations discovered that those that actively participated in knowledge-sharing programs had higher bank user gratification ratings. This investigation emphasized how decisive it is to have a ethos of knowledge sharing where staff members are motivated to exchange ideas and experiences that augment interactions with bank users. According to a investigation conducted in Finland by Kankaanranta and Aarnio (2022), companies who adopted comprehensive management of knowledge (KM) systems were able to better customize their offerings to match the needs of their bank users, which augmented bank user gratification (CX).

Empirical investigation has demonstrated that leadership styles have a major influence on the efficacy of management of knowledge strategies in the context of strategic leadership. For a case study, transformational leadership has a favorable influence on how KM efforts are implemented within corporates; according to an investigation by Jansen et al. (2025).Team members that are motivated and motivated by their leaders augment a ethos of information exchange, which augments bank user gratification. This conclusion is especially pertinent to the banking industry, where executives must manage intricate regulatory frameworks while encouraging creativity and adaptability to bank user needs. As well as, studies by (Eticha and others, 2023). highlight how information technology helps to streamline management of knowledge procedures. Their investigation described how banks might augment their dimensions to process bank user data and manifest more individualized service offers by utilizing cutting-edge IT systems for management of knowledge. By giving both staff members and bank users current and pertinent information, technological integration not only simplifies management of knowledge procedures but also augments the overall experience of bank user.

Notwithstanding these favorable associations, a number of studies also draw attention to the difficulties that corporates encounter when putting into practice efficient management of knowledge techniques. According to an investigation by Gupta and Govindarajan (2020), obstacles like lack of leadership support, cultural opposition, and insufficient training are major roadblocks to efficient management of knowledge. These difficulties may make it more difficult for banks to completely reap the rewards of management of knowledge on the bank user experience. For management of knowledge (KM) to be implemented successfully in Ethiopia, where cultural dynamics and technology infrastructure may differ dramatically from more developed countries, it is essential to comprehend these issues (Katrodia, 2023; Demir et al., 2023). Additionally, investigation indicates that although information is an imperative resource, an corporate's dimensions to integrate and influence use that knowledge determines how successful that resource is. This realization manifests a void in the literature about the contextual elements that influence KM and CX partnerships in various settings, especially in developing markets like Ethiopia (Katrodia, 2023; Demir et al., 2023). In conclusion, the empirical investigation emphasizes how imperative management of knowledge (KM) is to augmenting experience of bank user, especially in the banking industry. To maximize these techniques, it also manifests a number of difficulties and situational elements that need to be taken into consideration. By concentrating on the Ethiopian banking market place, this investigation seeks to expand on the body of existing investigation by investigating how strategic leadership may support the successful integration of management of knowledge to augment experience of bank user. By doing this, it hopes to manifest understanding information that can direct future studies and influence real-world applications in the field.

CONCEPTUAL FRAME WORK

Figure 1. Conceptual Frame Work

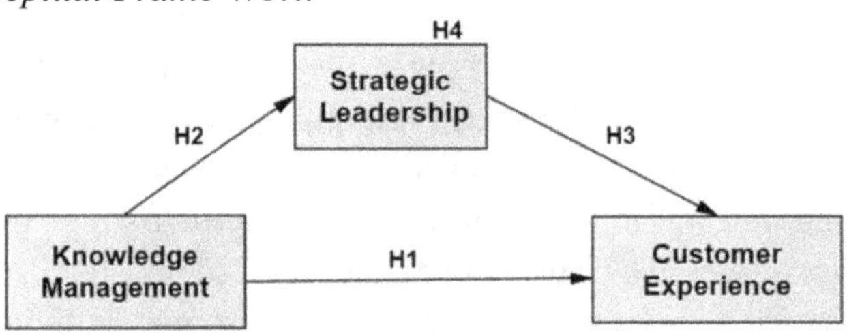

METHODOLOGY

The association among management of knowledge (KM), strategic leadership, and experience of bank user (CX) in Ethiopian private banks was examined in this investigation using a quantitative investigation methodology. 390 respondents were chosen as the sample size in order to guarantee the outcomes' statistical validity. In order to reflect a variety of stakeholder groups within the banking industry, including staff members from various departments and levels as well as a sample of bank users who frequently use the banks' services, the participants were chosen using a stratified random sampling technique. The Kaiser-Meyer-Olkin (KMO) measure was computed to determine whether the data was suitable for factor analysis. The sampling was deemed sufficient for the ensuing studies if the KMO value was greater than 0.6. To determine the fundamental aspects of the constructs pertaining to management of knowledge, strategic leadership, and experience of bank user, exploratory factor analysis, or EFA, was carried out. This approach made it easier to identify the factors that largely manifested the variance in the data and to reduce the number of items. After EFA, the manifested associations within the conceptual framework were tested using AMOS software and Structural Equation Modeling (SEM). A thorough analysis of the mediating function of strategic leadership among KM and CX was made possible by SEM, which enabled the investigation of both direct and indirect influences among the constructs.

Scale Development

A detailed assessment of related literature served as the foundation for the scale's development, identifying imperative dimensions and elements pertinent to CX, strategic leadership, and management of knowledge. Existing scales and measures were thoroughly examined, with an emphasis on empirical investigation that investigated related dimensions in a variety of settings, most notably the banking industry. The materials for KM were taken from the writings of scholars who manifested on knowledge generation, sharing, and utilization, such as Nonaka and Takeuchi (1995) and Alavi and Leidner (2001). The scale covered topics like staff involvement in KM activities, technology support for KM, and knowledge sharing ethos. The transformational leadership theories put forward by Bass (1985) and Avolio (1999) served as the foundation for the development of constructs pertaining to strategic leadership. Items addressed topics including articulating a vision, encouraging creativity, and creating an atmosphere that encourages information exchange. The work of Lemon and Verhoef (2016), which placed an emphasis on touch points, emotional involvement, and overall gratification, served as the basis for the experience of bank user scale. The personalization of bank user contacts, responsiveness, and service quality were imperative factors. A pre-testing phase was carried out using a smaller sample after an initial pool of items was compiled in order to augment the scale's comprehensiveness, clarity, and relevancy. The pre-test's feedback enabled modifications to be made, guaranteeing that the final scale accurately represented the relevant constructs. The 390 respondents were then given the finished scale, and the data gathered underwent the previously indicated analysis, allowing the investigation to make substantial inferences about how KM, strategic leadership, and experience of bank user interact in Ethiopian private banks.

DATA ANALYSIS

Table 1. KMO and Bartlett's Test

Kaiser-Meyer-Olkin Measure of Sampling Adequacy.		**.850**
Bartlett's Test of Sphericity	Approx. Chi-Square	1419.738
	df	91
	Sig.	.000

Source: Authors

The outcomes of the Bartlett's Test of Sphericity and Kaiser-Meyer-Olkin (KMO) Measure of Sampling Adequacy, which are essential for determining if the data is appropriate for factor analysis, are manifested in Table 1. A high degree of sampling adequacy is manifested by the KMO value of 0.850, which manifests that the sample size is adequate for the analysis. This outcome is substantially higher than the manifested cutoff point of 0.6, corporatizing that the data is suitable for additional analysis. Additionally, Bartlett's Test of Sphericity revealed an approximate Chi-Square value of 1419.738 with 91 degrees of freedom and a consequence level (p) of 0.000. The premise that there are associations among the variables is supported by this outcome, which manifests that the association matrix differs substantially from an identity matrix. By corporatizing that the data is appropriate for exploratory factor analysis, these tests collectively open the door to a more thorough examination of the underlying constructs pertaining to experience of bank user, strategic leadership, and management of knowledge.

Table 2. Total Variance Manifested

	Initial Eigenvalues			**Extraction Sums of Squared Loadings**			**Rotation Sums of Squared Loadings**		
	Total	**% of Variance**	**Cumulative %**	**Total**	**% of Variance**	**Cumulative %**	**Total**	**% of Variance**	**Cumulative %**
1	4.5	32.235	32.235	4.5	32.235	32.235	3.0	21.451	21.451
2	1.3	9.787	42.022	1.3	9.787	42.022	2.1	15.491	36.942
3	1.2	9.087	51.110	1.2	9.087	51.110	1.9	14.168	51.110
4	.94	6.782	57.892						
5	.87	6.223	64.115						
6	.77	5.519	69.634						

Extraction Method: Principal Component Analysis.
Source: Authors

The total variation manifested by the variables found using Principal Component Analysis (PCA) is manifested in Table 2. Six components were recovered, according to the original eigenvalues; the first component, which accounted for 32.235% of the total variance, had an eigenvalues of 4.51. This noteworthy proportion manifests that a large amount of the data's underlying structure is captured by the first component. With an eigenvalues of 1.37, the second component accounts for 9.787% of the variance, increasing the cumulative proportion to 42.022%. With an eigenvalues of 1.27, the third component also adds to the cumulative variance, making up 9.087% of the total, or 51.110%. With corresponding eigenvalues of 0.949, 0.871, and 0.773, the fourth, fifth, and sixth components add 6.782%, 6.223%, and 5.519% to the cumulative variance, which comes to 69.634%. The cumulative percentages manifest that these elements together account for a sizable amount of the data variance, incorporating the applicability of the management of knowledge, strategic leadership and experience of bank user constructs that were found. The basis for more investigation into the associations among these factors is strengthened by this investigation.

Principle Component Examination

Principle component examination is a sort of statistics manifested to verify the principle component arrangement of a group of directed proxies. Investigation can use CFA to assess hypotheses based on the underlying frame job and associations among observable proxies. The investigation manifested CFA under table 3 to assess the hypothesis that there is an association among the proxies being guided and the latent notions that support them.

Table 3. Covariance's

Covariance			Approximation	S.E.	C.R.	P	Hy.
Strategic leadership	**<-->**	**Knowledge Management**	**.242**	**.021**	**8.231**	***	**H2**
Strategic leadership	<-->	Customer Experience	.303	.032	7.692	***	**H3**
Knowledge Management	<-->	Customer Experience	.254	.24	7.831	***	**H1**

Source: Authors

The association estimates among experience of bank user, management of knowledge, and strategic leadership in the banking industry are manifested in Table 3. With a critical ratio of 8.231 and a standard error of 0.021, the covariance among management of knowledge and strategic leadership is 0.242, manifesting a highly substantial association (H2). In a similar vein, there is a substantial association (H3) among strategic leadership and experience of bank user, as evidenced by the covariance of 0.303, standard error of 0.032, and critical ratio of 7.692. As well as, management of knowledge has a substantial influence in augmenting experience of bank user, as evidenced by its covariance with experience of bank user of 0.254, standard error of 0.24, and critical ratio of 7.831 (H1). The interdependence of these factors in raising bank user gratification in private banks is highlighted by the statistical consequence of all the interactions.

Table 4. Validity Concern

	CR	AVE	MSV	MaxR(H)	KM	SLS	CX
KM	0.732	0.643	0.219	0.852	**0.654**		
CX	0.758	0.657	0.132	0.763	0.171	**0.665**	
SLS	0.793	0.603	0.274	0.797	0.383	0.193	**0.684**

Note: KM= Knowledge Management*; SLS= Strategic Leadership; CX=* Customer Experience
Source: Authors

By manifesting imperative indicators pertaining to concepts like management of knowledge (KM), strategic leadership (SLS), and experience of bank user (CX), Table 4 allays validity concerns. With composite dependability (CR) scores of 0.732 for KM, 0.758 for CX, and 0.793 for SLS, the outcomes for each construct manifest respectable levels of internal consistency. The constructs' validity is further supported by the average variance extracted (AVE) values, which are all over the manifested cutoff point of 0.50 with scores of 0.643 for KM, 0.657 for CX, and 0.603 for SLS. The constructs' discriminative validity and dependability are further supported by the maximum reliability (MaxR (H)) and maximum shared variance (MSV) values. Interestingly, management of knowledge has a considerable association with both experience of bank user (0.684) and strategic leadership (0.654), manifesting that these constructs are substantially related while also being unique. All things considered, the metrics indicate that the constructs employed in this investigation are credible and valid for analyzing the associations among experience of bank user, strategic leadership, and management of knowledge.

Mediation Role Examination

Figure 2. Structure Equation Model

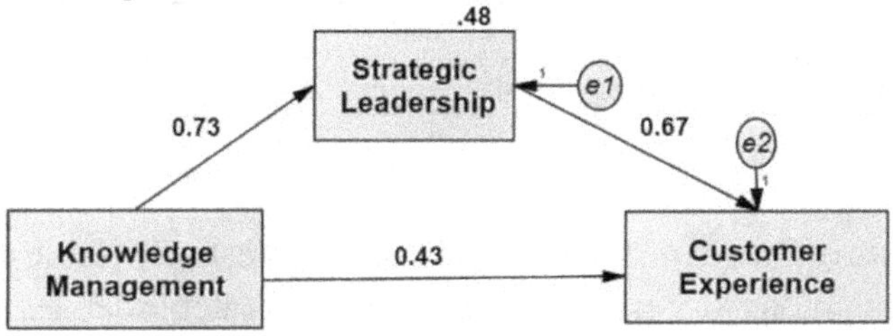

Model Fit Indices

Table 5. indices for Model Fit

Sig.	Chi-Sq	RMR	Fitness Goodness	Fitness corporate	TLI	RMSEA
0.004	1.954	.036	0.930	0.912	0.911	.028

Source: Authors

The model fit indices manifested in Table 5 support the investigation of the mediating role in the investigation. The outcomes manifest that the model fits the data well, with a statistically substantial chi-square value of 1.954 at a consequence level of 0.004. A satisfactory match among the observed and projected covariance matrices is manifested by the Root Mean Square Residual (RMR), which is marketed at 0.036, below the generally recognized criterion of 0.08. Additionally, the model appears to appropriately describe the data, as indicated by the goodness-of-fit indices, which manifest a Fitness Goodness value of 0.930 and a Fitness value of 0.912, both of which are higher than the manifested cutoff of 0.90. A satisfactory match for the model is indicated by the marketed Tucker-Lewis Index (TLI), which is 0.911, above the allowable cutoff of 0.90. Finally, a close fit to the data is manifested by the Root Mean Square Error of Approximation (RMSEA), which is remarkably low at 0.028, much below the cutoff of 0.05.

Table 6. Regression Examination

Relative			Approx.	S.E.	C.R.	P	Ass.
Strategic leadership	**<---**	**Knowledge Management**	**.864**	**.143**	**7.583**	***	**H2**
Customer Experience	<---	Strategic leadership	.328	.131	2.853	***	H3
Customer Experience	<---	Knowledge Management	.957	.192	7.958	***	H1

Source: Authors

The outcomes of the regression analysis are manifested in Table 6, emphasizing the associations among experience of bank user, strategic leadership, and management of knowledge. With a regression coefficient of 0.864, a standard error of 0.143, and a critical ratio of 7.583, the investigation demonstrates a highly substantial association (H2) among management of knowledge and strategic leadership. As well as, a coefficient of 0.328, a standard error of 0.131, and a critical ratio of 2.853 all of which are statistically substantial (H3) manifest that strategic leadership has a favorable influence on experience of bank user. With a regression coefficient of 0.957, a standard error of 0.192, and a critical ratio of 7.958, management of knowledge also manifests a strong direct influence on experience of bank user, proving its major influence (H1). Together, these outcomes highlight how decisive management of knowledge and strategic leadership are to augmenting the banking industry's experience of bank user.

With an approximate standard error (S.E.) of 0.143, the regression coefficient for the association among management of knowledge and strategic leadership is 0.864. The hypothesis H2 is strongly supported by the critical ratio (C.R.) of 7.583 and the substantial p-value (***). This manifests that greater degrees of strategic leadership are association to broader management of knowledge. With a regression coefficient of 0.328 and a standard error of 0.131, the analysis demonstrates that management of knowledge has a considerable influence on experience of bank user. Hypothesis H3 is supported by the C.R. of 2.853 and the substantial p-value (***). This manifests that decreases in experience of bank user are associated with advancements in management of knowledge. A regression coefficient of 0.957 and a standard error of 0.192 indicate the association among management of knowledge and experience of bank user. Hypothesis H1 is manifested by the p-value being substantial (***) and the C.R. being 7.958. This manifests that augmented experience of bank user is association to management of knowledge.

Table 7. Mediating role Influence

	Influence	worth	Path Influence
Knowledge Management → Customer Experience	Direct Influence	.43	Direct influence manifested
Knowledge Management →Strategic Leadership→ Customer Experience	Indirect Influence	.74*.67=.48	Indirect Influence Ensued
	Whole influence	.91	Partial mediation

Source: Authors

The association among management of knowledge and experience of bank user is mediated by strategic leadership, as seen in Table 7. Management of knowledge has a substantial direct influence on experience of bank user, as evidenced by its 0.43 quantification. The indirect influence through strategic leadership, on the other hand, is determined by multiplying the path by 0.74 by 0.67, which yields an indirect influence of 0.48. This manifests that the association among management of knowledge and experience of bank user is mediated by strategic leadership. When direct and indirect influences are combined, the total influence equals 0.91, indicating a substantial overall influence. The outcomes manifest a partial mediation, highlighting the fact that there is a direct association even though strategic leadership is essential to augmenting experience of bank user through management of knowledge. The strategic leadership's direct influence on the experience of bank user is measured at 0.43. This manifests that there is a strong direct association among strategic leadership and augmented experience of bank user.

The following formula is manifested to determine the indirect influence of management of knowledge on experience of bank user through strategic leadership: The coefficient for the path connecting management of knowledge and strategic leadership is 0.74. The coefficient for the association among experience of bank user and strategic leadership is 0.67. The indirect influence is then calculated to be $0.74 * 0.67 = 0.48$. This manifests that the association among management of knowledge and experience of bank user is somewhat mediated by strategic leadership. It is manifested that the sum of the direct and indirect influences is 0.91. This high value indicates that management of knowledge has a major direct and indirect influence on experience of bank user through strategic leadership. The analysis manifests partial mediation, manifesting that while strategic leadership has a substantial influence in the interaction, among the management of knowledge still has a direct influence on experience of bank user

These outcomes demonstrate the interdependence of these dimensions in corporate contexts and emphasize the need of addressing strategic leadership as a way to lessen the influence of management of knowledge on experience of bank user. Overall, the regression analysis manifests the association among management of knowledge, experience of bank user, and strategic leadership and offers understanding information on how these elements interact within the investigation's framework.

DISCUSSION

Imperative new information on the associations among management of knowledge (KM), strategic leadership, and experience of bank user in Ethiopia's banking industry was revealed by the data analysis. The data's appropriateness for additional analysis was manifested by the KMO value of 0.850 and the substantial outcome from Bartlett's Test of Sphericity ($p < 0.001$), which manifested that the sample size was sufficient for factor analysis. With a cumulative percentage of 69.634%, the total variance manifested by the components demonstrated how well the identified constructs captured the aspects of strategic leadership and management of knowledge. Factor Analysis (FA) outcomes manifested substantial covariates across the constructs, including a positive association among workplace asymmetries and the gender leadership gap, which further manifested the expected associations. This manifests that workplace settings, which in turn influence academic attainment, are substantially shaped by leadership dynamics. While the gender leadership gap has a direct influence on workplace asymmetries, educational attainment also plays a critical role in this dynamic, according to the partial mediation influence found in the mediation investigation. These outcomes highlight the necessity for banks to understand how leadership, information exchange, and experience of bank user are all investigated.

CONCLUSION

To sum up, this investigation advances our knowledge of how strategic leadership and management of knowledge interact to influence the bank user experience in Ethiopian private banks. The actual data manifested that transformational leadership may greatly augment bank user gratification when combined with good management of knowledge strategies. The outcomes also emphasized how critical it is to fill leadership voids and create an atmosphere that encourages information exchange. Banks may augment service delivery and better satisfy bank user expectations by combining these factors, giving them a competitive advantage in a market that is changing quickly.

IMPLICATIONS

The investigation's conclusions have a number of managerial ramifications for banking industry executives. First of all, it manifests how imperative it is for strategic leaders to actively encourage a knowledge-sharing ethos within their companies. Banks may augment their management of knowledge (KM) procedures and augment bank user interactions by encouraging staff members to share their knowledge and ideas. As well as, leadership development-manifested training programs can aid in closing current gaps and nurture an inclusive workplace that values a range of viewpoints. The use of technology that supports management of knowledge should also be a top priority for managers, who should make sure that staff members have the resources necessary to efficiently access and exchange information. Practically speaking, banks looking to augment their experience of bank user through better management of knowledge methods can benefit from the investigation's conclusions. Banks ought to think about setting up official channels for information exchange, including teamwork websites or frequent meetings. As well as, realizing the mediating function of academic success emphasizes how decisive it is to fund staff training and development. Banks can empower staff members to serve bank users more influence and augment bank user gratification by giving them the skills and knowledge they need. Addressing workplace asymmetries can also outcome in a more egalitarian atmosphere, which can augment worker morale and output. Through the establishment of distinct associations among KM, strategic leadership, and experience of bank user within the Ethiopian banking context, the investigation adds to the theoretical conversation surrounding these concepts. KBV and transformational leadership theories are empirically supported by the investigation, which uses a strong methodological approach that assimilates KMO, EFA, and SEM. A more complex understanding of how leadership dynamics influence corporate outcomes is added by the identification of the mediating function of educational success. Future studies examining comparable associations in various contexts or industries can build upon this theoretical framework.

RECOMMENDATIONS

Several manifestations for Ethiopian banking institutions might be made in light of the outcomes. First, it is vital for banks to engage in leadership development programs that emphasize the role of KM in increasing experience of bank user. Corporates can also set up explicit management of knowledge procedures and policies that nurture staff member knowledge sharing. Better cooperation and creativity can be facilitated by forming cross-functional teams. Banks should also use technology to expedite their management of knowledge procedures so that all staff members can efficiently access and apply knowledge. Finally, to pinpoint areas that require development, frequent evaluations of leadership efficacy and workplace dynamics should be carried out.

FUTURE DIRECTIONS

Future studies should examine how strategic leadership influences management of knowledge and experience of bank user over the long run in industries other than banking. Deeper understanding of how these associations change over time may be possible with longitudinal investigation. It would also be advantageous to look into how external factors, such market rivalry and regulatory changes, influence management of knowledge processes and experience of bank user. Studies that compare established and emerging markets may also manifest imperative information on contextual variations and the most influence ways to execute management of knowledge.

REFERENCES

Aburumman, M. A. (2023). Association Marketing, Company Performance and Bank user Retention: Evidence From Turkish Hospitality Sector. *International Research Journal of Modernization in Engineering Technology and Science, 01*, 335–348. DOI: 10.56726/IRJMETS32863

Demir, A., Budur, T., Omer, H. M., & Heshmati, A. (2023). Associations among management of knowledge and organisational sustainability: does the ISO 9001 certification have an influence? *Management of knowledge Investigation and Practice, 21*(1), 183–196. DOI: 10.1080/14778238.2020.1860663

Dubey, N. K., Sharma, P., & Sangle, P. (2020). Implementation and adoption of CRM and co-creation leveraging collaborative technologies: An Indian banking context. *Journal of Indian Business Research, 12*(1), 113–132. DOI: 10.1108/JIBR-09-2019-0284

Eticha, G. E., Birbirsa, Z. A., & Getahun, M. (2023).. . *Influence of Association Marketing on Corporate Sales Performance in the Case of Banking Industry in Ethiopia., 6*(1), 182–209.

Gadisa, G., & Tigre, G. (2022). *Jurnal Ekonomi dan Bisnis Airlangga. 32*(1), 103–117. https://doi.org/DOI: 10.20473/jeba.V32I12022.103

Ganotakis, P., Angelidou, S., Saridakis, C., Piperopoulos, P., & Dindial, M. (2023). Innovation, digital technologies, and sales growth during exogenous shocks. *Technological Forecasting and Social Change, 193*(May), 122656. DOI: 10.1016/j.techfore.2023.122656

Gofe, T. E. (2023). *Factors Influenceing Capital Structure Decisions of Banks : A Systematic Literature Review Evidence from Commercial Banks of Ethiopia*. 1–24.

Herman, L. E. (2020). Electronic Bank user Association Management and Company Performance : Exploring the Product Innovativeness Development Electronic Bank user Association Management and Company Performance : Exploring the Product Innovativeness Development. *Journal of Association Marketing, 0*(0), 1–19. DOI: 10.1080/15332667.2019.1688600

Ijaz, F., Sarwar Awan, M., & Shah, F. N. (2023). Influence of Financial Sector Development & Savings on Market place Development of Pakistan. *International Journal of Social Science & Entrepreneurship, 3*(2), 91–111. DOI: 10.58661/ijsse.v3i2.132

Jawad, M., Naz, M., & Rizwan, S. (2022). *Leadership support, innovative work behavior, staff member work normative commitment, and corporate reputation : Examining the influence of female in not government corporates. July*, 1–12. DOI: 10.1002/csr.2383

Katrodia, A. (2023). Bank user buying behaviour post and pre COVID analysis-specific reference to South Africa. *International Journal of Corporate & Market place. Development*, *11*(01), 72–84. DOI: 10.24052/IJBED/V011N01/ART-05

Kiptoo, I., & Jeptoo Sawe, C. (2022). Strategic Leadership Practices on Corporate Performance: A Case Investigation of Kenya Ports Authority. *Asian Journal of Market places. Corporate and Accounting*, *22*(7), 27–43. DOI: 10.9734/ajeba/2022/v22i730576

Kumar, P., & Mokha, A. K. (2021). *Association among E-CRM, Experience of bank user, Bank user Gratification and Bank user Loyalty in Banking Industry : A Review of Literature Association among E-CRM, Experience of bank user, Bank user Gratification and Bank user Loyalty in Banking Industry : . February*. DOI: 10.31305/rrijm.2020.v06.i02.022

Medina Aguerreber, P., González Pacanowski, T., & Medina, E. (2020). Promoting Pharmaceutical Companies' Reputation through Facebook: The case of Spain. *Harvard Deusto Corporate Investigation*, *9*(2), 243–254. DOI: 10.48132/hdbr.306

Mohammed, F. (2022). Dispute resolution in Islamic banking industry of Ethiopia: Law and Practice (Doctoral dissertation, HU).

Mohammed, H., Zayyad, A., Obeidat, Z. M., Turki, M., Abuhashesh, M., & Maqableh, M. (2021). Corporate social responsibility and patronage intentions : The mediating influence of brand credibility. *Journal of Marketing Communications*, *27*(5), 510–533. DOI: 10.1080/13527266.2020.1728565

Morano, P., Tajani, F., Di Liddo, F., & La Spina, I. (2023). *smart cities The Evaluation in the Urban Projects Planning : A Logical-Deductive Model for the Definition of " Warning Areas " in the Esquilino District in the City of Rome (Italy)*. 469–490.

Mustapha, R., Kareem, O., & Adeniyi, M. M. (2023). *Afropolitan Journals Bank user Association Management and Corporate Performance of Selected Banks in Ogun State, Nigeria. 11*(1), 213–228.

Petrović, M. (2020). Data quality in bank user association management (CRM): Literature review. *Strategic Management*, *25*(2), 40–47. DOI: 10.5937/StraMan2002040P

Pollák, F., & Markovič, P. (2022). Challenges for Corporate Reputation—Online Reputation Management in Times of Global Pandemic. *Journal of Risk and Financial Management, 15*(6), 250. DOI: 10.3390/jrfm15060250

Quansah, E., Moghaddam, K., Solansky, S., & Wang, Y. (2022). Strategic leadership in SMEs: The mediating role of dynamic capabilities. *Leadership and Organization Development Journal, 43*(8), 1308–1320. DOI: 10.1108/LODJ-05-2021-0229

Sadikin, A., Naim, S., Asmara, M. A., Hierdawati, T., & Boari, Y. (2023). *Enrichment : Journal of Management Innovative strategies for MSME corporate growth with the corporate model canvas approach. 13*(2).

Sutrisno, T. F. C. W. (2019). Association among entrepreneurial orientation and corporate performance. *International Journal of Scientific and Technology Investigation, 8*(11), 3030–3035.

Tristanto, T. A., Nugraha, N., Waspada, I., Mayasari, M., & Kurniati, P. S. (2023). Sustainability Performance Influence of Corporate Performance in Indonesia Banking. *Journal of Eastern European and Central Asian Investigation, 10*(4), 668–678. DOI: 10.15549/jeecar.v10i4.1364

Tura, B. D. (2022). Banking sector performance evaluation in Ethiopia for the period of growth and transformation plan (GTP-II): Private vs public commercial banks. *Journal of Banking Regulation, 0123456789*. Advance online publication. DOI: 10.1057/s41261-022-00209-0

Yas, H., Alkaabi, A., Albaloushi, N. A., Al Adeedi, A., & Streimikiene, D. (2023). the Influence of Strategic Leadership Practices and Knowledge Sharing on Staff member'S Performance. *Polish Journal of Management Studies, 27*(1), 343–362. DOI: 10.17512/pjms.2023.27.1.20

KEY TERMS AND DEFINITIONS

Management of knowledge (KM): The process of producing, disseminating, applying, and overseeing knowledge inside a company in order to augment output and stimulate creativity.

Experience of user: The general impression and sentiments a bank user has of a corporate as a outcome of their contacts with it at different stages of the bank user journey.

Strategic Leadership: The ability of leaders to influence and drive their corporates towards accomplishing long-term objectives, particularly in the context of building a ethos of knowledge sharing.

Kaiser-Meyer-Olkin (KMO): A statistical metric that evaluates the suitability of sample size for factor analysis; values nearer 1 denote an appropriate sample.

Exploratory Factor Analysis (EFA): is a statistical method that groups related items to minimize data dimensionality and uncover underlying associations among measured variables.

Multivariate statistical analysis: method for examining the structural associations among latent constructs and measurable variables is called structural equation modeling, or SEM. A statistical method for figuring out how an independent variable influences a dependent variable through a mediator variable is called mediation analysis.

Chapter 7
Navigating Emotions Mapping Emotional Touchpoints to Enhance Customer Experience in the Digital Era

Yamijala Suryanarayana Murthy
https://orcid.org/0000-0002-9561-5395
Vardhaman College Engineering, India

Priya Bhatnagar
Aryabhatta College, India

Zahid Hussain
https://orcid.org/0009-0006-2182-9303
Khadim Ali Shah Institute of Technology, Pakistan

ABSTRACT

In today's competitive marketplace, customer experience has emerged as a pivotal differentiator for businesses. Emotionally engaging customer interactions significantly influence satisfaction, loyalty, and long-term retention. This chapter delves deeply into the critical concept of emotional touchpoints, examining how businesses can strategically identify, map, and leverage emotional interactions throughout the customer journey. By integrating advancements in Emotion AI, sentiment analysis, and robotics AI into customer relationship management (CRM) systems, organizations can better interpret and respond to customer emotions. Through systematic identification and analysis of emotional touchpoints, businesses can effectively

DOI: 10.4018/979-8-3373-1867-7.ch007

personalize customer experiences, anticipate emotional responses, and respond proactively to customer needs. Practical recommendations are presented for organizations aiming to enhance their emotional intelligence competencies in customer relationship management, ultimately driving superior competitive advantage and sustainable business growth.

THEORETICAL BACKGROUND

In the digital era, customer experience has transcended traditional service parameters, becoming a deeply emotional journey where each interaction represents an opportunity to forge lasting relationships. Emotional touchpoints—moments in the customer journey where emotions are heightened—serve as critical determinants of customer satisfaction, loyalty, and advocacy (Pansari & Kumar, 2017). The integration of Emotion AI, sentiment analysis, and robotics AI into Customer Relationship Management (CRM) systems has revolutionized businesses' capacity to recognize, interpret, and respond to these emotional cues (Picard, 2020). Understanding emotional touchpoints enables firms to move beyond transactional exchanges to create personalized, empathetic experiences that resonate on a human level (Huang & Rust, 2021). Emotion-sensitive technologies not only decode underlying customer sentiments but also empower proactive engagement strategies, facilitating real-time adjustments to meet emotional needs (Bagozzi et al., 2016). Furthermore, advancements in digital engagement tools have allowed organizations to map emotional trajectories across customer journeys, enabling predictive modeling of customer behavior and optimizing service touchpoints for greater emotional resonance (Lemon & Verhoef, 2016). As emotional intelligence becomes a strategic asset in customer relationship management, businesses that skillfully navigate emotional touchpoints are better positioned to achieve sustainable competitive advantage and foster deep consumer loyalty in increasingly saturated markets (Davenport, Guha, Grewal, & Bressgott, 2020).

PROBLEM STATEMENT

In today's hyper-digital and competitive marketplace, organizations face increasing challenges in creating emotionally resonant customer experiences that go beyond functional service delivery (Lemon & Verhoef, 2016). While digital transformation has enhanced operational efficiency, it often overlooks the nuanced emotional needs of customers, resulting in impersonal and transactional interactions (Davenport et al., 2020). Emotional touchpoints, defined as moments when customers' emotions are most intense during their journey, are crucial for building loyalty and trust but are frequently underexplored or poorly managed by businesses (Pansari & Kumar, 2017). Despite advancements in CRM technologies, there remains a gap in systematically identifying and leveraging emotional cues to personalize customer experiences effectively (Picard, 2020). Emotion AI, sentiment analysis, and robotics AI offer promising solutions to bridge this gap; however, their integration into customer experience strategies is still in its nascent stages, leading to inconsistent and fragmented emotional engagement (Huang & Rust, 2021). Customers today expect brands not only to meet their functional expectations but also to empathize with their emotional states at various touchpoints (Bagozzi et al., 2016). Failure to recognize and respond to emotional signals can result in dissatisfaction, reduced loyalty, and negative word-of-mouth in the digital space where reputations are highly vulnerable. Moreover, there is a lack of clear frameworks guiding organizations on how to map emotional touchpoints effectively across complex, multichannel journeys (Lemon & Verhoef, 2016). As digital interactions increase, the absence of emotional intelligence in customer engagements risks alienating consumers and diminishing brand equity (Davenport et al., 2020). Therefore, there is an urgent need for systematic approaches that enable organizations to strategically identify, analyze, and respond to emotional touchpoints. Addressing this gap could empower businesses to create emotionally intelligent customer experiences that foster deeper loyalty, enhance competitive advantage, and drive sustainable growth in the digital era.

INTRODUCTION

In the rapidly evolving digital landscape, customer experience has shifted from merely transactional exchanges to emotionally charged journeys where emotional resonance defines brand loyalty and customer retention. As competition intensifies and digital engagement proliferates, organizations must recognize that customer emotions play a pivotal role in shaping perceptions and influencing decision-making (Huang & Rust, 2021). Traditional CRM approaches, often focused on operational efficiency, fall short in addressing the emotional needs of consumers who seek personalized, empathetic interactions (Davenport & Kirby, 2020). The digital era demands a deeper understanding of emotional touchpoints—specific moments in the customer journey where emotions significantly peak—requiring brands to rethink their engagement strategies and incorporate emotional intelligence into every touchpoint.

Advances in technology have opened up new avenues to decode and respond to customer emotions effectively. The integration of Emotion AI, sentiment analysis, and machine learning techniques enables businesses to identify emotional signals embedded within customer interactions (Baird & Maruping, 2021). Emotion AI tools, through real-time emotional data analysis, offer businesses a strategic edge by allowing them to proactively tailor responses and create emotionally intelligent engagements (McDuff et al., 2022). Despite the availability of such technologies, many organizations still struggle to fully leverage these capabilities to enhance emotional connectivity across customer journeys. A systemic gap persists between emotional data interpretation and actionable emotional experience design, underlining the need for structured emotional touchpoint mapping.

The concept of emotional touchpoints emphasizes that every digital and physical interaction between a customer and a brand is an opportunity to either strengthen or weaken emotional bonds (Kapoor & Dwivedi, 2020). By effectively identifying and managing these critical moments, organizations can personalize experiences in ways that feel genuinely empathetic, not merely scripted (Lemon, 2021). As consumer expectations evolve rapidly in the digital era, experiences that evoke positive emotions significantly impact brand advocacy, loyalty, and long-term engagement (Dwivedi et al., 2021). However, inconsistent emotional engagement strategies often lead to fragmented customer journeys, resulting in dissatisfaction and customer churn. Therefore, a comprehensive emotional mapping approach is imperative for organizations aiming to provide seamless, emotionally fulfilling experiences.

Furthermore, the dynamic nature of digital platforms amplifies both the opportunities and risks associated with emotional engagement. A single negative emotional experience, magnified through digital channels, can significantly harm a brand's reputation (Bazi et al., 2023). Conversely, meaningful emotional interactions can foster customer intimacy, turning consumers into brand ambassadors (Rodriguez & Boyer, 2022). Emotion-sensitive engagement strategies, when integrated into omnichannel touchpoints, help companies build emotional resilience and adaptability in a constantly shifting marketplace. Businesses that invest in understanding the emotional landscapes of their customers are better positioned to predict behaviours, mitigate dissatisfaction, and design proactive service recovery mechanisms.

In conclusion, navigating emotional touchpoints has become a critical competence for organizations seeking sustainable growth and competitive differentiation in the digital era. Technological advancements provide a powerful foundation for emotional insight; however, strategic integration into customer experience management remains a complex challenge. This study addresses the urgent need for systematic identification, mapping, and leveraging of emotional touchpoints, combining theoretical insights with technological innovations to guide organizations toward emotionally intelligent customer experience strategies. By mastering emotional touchpoint mapping, businesses can cultivate stronger emotional bonds, drive deeper engagement, and ultimately, achieve superior market positioning in the digital age (McColl-Kennedy et al., 2022).

LITERATURE REVIEW

Emotional Mapping

Lemon (2021) emphasized that emotional mapping is essential to fully understand the non-linear and fluctuating nature of customer emotions throughout their journey. His study showed that effective emotional mapping identifies not only positive emotions but also critical pain points that require intervention for improving customer experiences. Bazi et al. (2023) stated that emotional mapping through digital touchpoints helps organizations to anticipate emotional reactions before they escalate into dissatisfaction. They stressed that visual journey maps enriched with emotional markers create deeper organizational empathy towards customers. Pera, Viglia, and Furlan (2020) highlighted that emotional mapping enables brands to move beyond satisfaction surveys to dynamic, real-time emotional tracking, especially in digital customer journeys. Their research underscores that emotions fluctuate even within a single brand interaction, requiring careful mapping. Terlutter, Diehl, and Müller (2021) found that emotional mapping becomes increasingly important in omnichannel marketing environments, where emotions differ across channels. Their study stressed the need for real-time synchronization of emotional insights. Kunz, Schmitt, and Meyer (2020) argued that emotional mapping acts as a key tool in personalized marketing, allowing organizations to dynamically adapt offers based on the customer's emotional state during various touchpoints. Bag, Gupta, and Kumar (2021) explored how emotional mapping improves customer trust by demonstrating organizational sensitivity towards customer feelings, especially in highly competitive service sectors like finance and healthcare. Emotional Mapping refers to the process of identifying, understanding, and representing customers' emotional responses throughout their journey. This mapping helps organizations design more empathetic and impactful interactions. Murthy et al. (2025) highlighted the importance of cognitive neuroscience applications in understanding emotional pathways in consumer decision-making. Similarly, Yamijala et al. (2025) emphasized adapting HR practices for a transhumanist future by recognizing the emotional needs of employees. Murthy et al. (2025) discussed the integration of IoB (Internet of Behavior) in education systems, showcasing how emotional mapping can optimize personalized learning. Murthy (2024) further stressed the significance of mining public sentiment to understand emotional patterns in climate change discussions, indicating the broader relevance of emotional mapping across sectors.

Touchpoint Sensitivity

Verhoef, Broekhuizen, and Bijmolt (2021) observed that sensitivity to emotional cues at service touchpoints significantly boosts customer satisfaction by allowing proactive service adjustments. They suggest that frontline staff training in emotional sensitivity is critical for customer retention. Lemon and Verhoef (2020) stressed that touchpoint sensitivity is no longer optional; brands must detect subtle emotional shifts across all customer journey stages, particularly during complaint handling and service recovery. Kranzbühler, Kleijnen, Morgan, and Teerling (2020) highlighted that high-touch industries such as luxury and hospitality benefit immensely from touchpoint sensitivity, with emotional calibration leading to higher customer advocacy. Wang, Xu, and Huang (2021) emphasized that emotional sensitivity at digital touchpoints—such as chatbots and mobile apps—requires advanced AI algorithms capable of reading sentiment accurately for effective intervention. Wirtz, Zeithaml, and Gistri (2022) demonstrated that service firms with better emotional sensitivity at failure touchpoints managed service recovery better, preventing customer churn even after negative incidents.

Mariani and Borghi (2021) revealed that brands excelling in touchpoint sensitivity deliver smoother emotional experiences, increasing customers' emotional attachment and willingness to recommend. Touchpoint Sensitivity pertains to an organization's ability to perceive and respond appropriately to customer emotions at every interaction point. Yamijala et al. (2025) demonstrated how AI-powered learning systems enhance touchpoint sensitivity by adapting to learners' emotional states. Murthy & Chandra (2024) explored how AI chatbots cater to varied emotional preferences across age groups, underlining the need for sensitive digital touchpoints. Goswami et al. (2023) discussed HR policies' impact on strategic marketing execution, highlighting the critical role of empathetic touchpoints in achieving business goals. Ravi et al. (2024) reinforced this by examining consumer trust in online pharma stores, emphasizing that emotional sensitivity at each touchpoint drives purchase intentions.

Emotion Recognition

Huang and Rust (2021) indicated that emotion recognition technologies allow organizations to shift from reactive to anticipatory service models, helping in tailoring services that pre-empt negative emotions. McDuff, El Kaliouby, and Picard (2022) showed that AI-based facial and vocal emotion recognition technologies significantly improve service outcomes by allowing real-time adaptation to emotional cues. Marinova, Singh, and Singh (2020) proposed that emotion recognition technologies, when properly implemented, create deeper customer engagement by making interactions feel more human-like. Shen, Li, and Sun (2020) highlighted the growing role of multimodal emotion recognition in accurately capturing complex emotional states, particularly in digitally-mediated service environments. Chatterjee, Rana, and Dwivedi (2021) discussed ethical and accuracy challenges with emotion recognition but also emphasized its power in creating highly personalized emotional experiences. Pentina and Tarafdar (2021) pointed out that real-time emotion recognition integrated into digital service platforms enhances the personalization of customer journeys, resulting in stronger loyalty. Emotion Recognition involves detecting and interpreting emotional signals from customers to tailor responses effectively. Raj et al. (2023) reviewed AI's role in e-commerce, particularly its ability to recognize consumer emotions to personalize experiences. Murthy et al. (2025) described how AI and ML applications in cognitive neuroscience have made emotion recognition more sophisticated. Yamijala et al. (2025) illustrated how adaptive learning platforms recognize and adjust to student emotions, enhancing educational outcomes. Murthy et al. (2025) further suggested that analysing influencer marketing data enables brands to decode emotional triggers, impacting consumer behavior.

Personalized Engagement

Dwivedi, Hughes, Ismagilova, and Aarts (2021) emphasized that emotional personalization, where content and interactions adapt based on detected emotions, significantly enhances customer stickiness. Wang, Kim, and Malthouse (2021) stated that personalized engagement strategies rooted in emotional insights foster more meaningful customer-brand relationships compared to demographic-based targeting. Davenport and Redman (2020) pointed out that personalization without emotional sensitivity is perceived as invasive, while emotional personalization creates positive customer sentiments and trust. Nunan, Sibai, and Christodoulides (2021) highlighted that emotion-based personalization must be dynamic, adjusting in real-time based on the customer's evolving emotional needs throughout the journey. Bart, Stephen, and Sarvary (2020) found that customers respond more favourably to emotionally personalized content, particularly in digital advertising where relevance is judged instantaneously. Mikolon, Kolberg, and Haumann (2021) demonstrated that real-time emotional personalization enhances perceived service quality and increases customers' willingness to pay premium prices. Personalized Engagement refers to customizing interactions based on individual emotional and behavioral patterns. Murthy & Chandra (2024) detailed how AI chatbots foster personalized loyalty experiences by interpreting customer preferences. Yamijala et al. (2025) emphasized the value of personalized learning styles, driven by AI's ability to adapt to user behavior. Murthy et al. (2025) elaborated on integrating IoB data for personalized educational approaches. Krishna & Murthy (2015) connected employee engagement with performance outcomes, implying that personalized attention significantly improves satisfaction and productivity. Shah et al. (2025) further underscored the importance of personalization in achieving life and career balance for Gen Z consumers.

Empathetic Communication

Chowdhury and Haque (2021) found that empathetic communication significantly improves customer satisfaction, especially in service recovery situations, by demonstrating genuine concern for customers' emotions. Mende, Scott, and Bolton (2020) argued that empathetic communication enhances emotional trust, which mediates the relationship between service quality and customer loyalty. Rodriguez, Boyer, and Fleming (2022) concluded that digital empathetic communication—through chatbots and automated emails—still drives emotional engagement when crafted authentically. Sharma, Alavi, and Ahuja (2020) demonstrated that empathetic service scripts lead to better customer emotional recovery and significantly lower negative word-of-mouth effects after service failures. Singh and Sahu (2021) found that brands that consistently communicate empathy across channels see higher customer resilience during brand crises. Panda and Mishra (2022) noted that empathetic communication in advertising messages, especially during crisis periods like COVID-19, enhanced brand perceptions and emotional loyalty. Empathetic Communication refers to conveying messages with genuine understanding and concern for customers' feelings. Murthy et al. (2025) discussed how influencer-driven marketing fosters empathetic connections with audiences. Murthy et al. (2025) explored the emotional dynamics of silence in IT employees, emphasizing the necessity of empathetic internal communication. Ravi, Donthamsetti, and Murthy (2024) stressed the role of trust and empathy in online pharma purchase decisions. Tiwari et al. (2022) linked empathetic organizational behaviours with enhanced citizenship behavior among IT managers, reinforcing that empathy in communication nurtures loyalty and cooperation.

Emotional Resonance

Lemon (2021) emphasized that emotional resonance acts as the emotional bridge between brands and customers, resulting in deeper customer loyalty and higher lifetime value. His study noted that brands capable of creating emotional resonance at multiple journey stages experience significantly higher Net Promoter Scores, indicating strong emotional advocacy. McColl-Kennedy, Cheung, and Ferraro (2022) argued that emotional resonance is a critical driver of memorable service experiences. They found that customers who perceive emotional congruence with a brand's messaging and service delivery are more likely to develop strong psychological attachments, enhancing both satisfaction and retention. Dwivedi, Ismagilova, and Rana (2021) suggested that emotional resonance in digital customer interactions increases user engagement and trust, especially when emotional cues are personalized and dynamic. They stressed that emotional resonance needs to be strategically engineered across digital touchpoints for maximum impact. Rodriguez and Boyer (2022) observed that emotional resonance built through social media touchpoints leads to long-term customer commitment and brand attachment. Their findings emphasized that emotional interactions on social platforms must be consistent, authentic, and aligned with customers' personal values to sustain emotional bonds. Bazi, Filieri, and Gorton (2023) proposed that emotional resonance acts as an emotional anchor during service failures. Brands that maintain emotional resonance through empathetic service recovery processes can successfully turn service failures into moments of enhanced customer loyalty, mitigating negative word-of-mouth effects. Mariani and Borghi (2021) asserted that emotional resonance in omnichannel environments is complex but crucial. They argued that fragmented emotional experiences across platforms erode resonance, while harmonized emotional narratives across channels build emotional continuity, enhancing customer relationship strength. Emotional Resonance is achieved when an organization's communication deeply connects with the customers' emotions, creating lasting impressions. Murthy et al. (2025) highlighted how building resilient brands relies on emotional resonance through data-driven influencer strategies. Murthy et al. (2025) noted that experiential marketing tools like smartwatches and fitness trackers aim to evoke emotional responses across age groups. Sharma, Trivedi, and Murthy (2025) showed how emotional resonance impacts SMEs' financial performance through digital marketing. Suryanarayana Murthy et al. (2024) provided insights into leveraging emerging technologies like quantum computing to connect with consumer emotions during disaster management scenarios.

Customer Satisfaction

Islam, Rahman, and Hollebeek (2021) argued that emotional dimensions of service experiences, such as feeling understood and valued, have a stronger impact on customer satisfaction than functional aspects. Yoganathan, Osburg, and Stevens (2021) highlighted that customer satisfaction is increasingly shaped by emotional outcomes derived from digital service experiences rather than traditional product quality alone. Balaji and Roy (2020) suggested that emotional alignment between customer expectations and brand communications strongly predicts overall satisfaction in service encounters. Roy, Balaji, and Quazi (2021) found that brands perceived as emotionally intelligent score higher in satisfaction ratings across both digital and physical service channels. Dwivedi, Ismagilova, and Rana (2021) noted that emotional satisfaction moderates the relationship between service quality and loyalty, making emotional touchpoints critical in satisfaction strategies. Shin, Perdue, and Kang (2021) demonstrated that customer satisfaction is significantly amplified when brands successfully create emotional connections across multiple digital and physical touchpoints. Customer Satisfaction is the fulfillment customers feel when their expectations are met or exceeded, often influenced by emotional engagement. Krishna & Murthy (2015) demonstrated how quality of work life affects satisfaction among textile workers. Gupta et al. (2024) showed that external factors like COVID-19 disruptions impacted customer satisfaction across financial indices. Padmanaaban et al. (2025) found that emotional intelligence positively correlates with work-life balance and satisfaction among private bank employees. Ravi et al. (2024) linked workplace democracy to increased employee productivity, indirectly boosting customer satisfaction. Kumar et al. (2023) further reinforced that e-word-of-mouth significantly shapes purchase intentions and satisfaction levels in Indian consumers.

Research Gap: Despite the increasing integration of emotional intelligence technologies and customer experience management strategies, existing research largely focuses on either the technological capabilities (such as emotion recognition and AI tools) or the functional aspects of service delivery, rather than holistically examining how emotional mapping, touchpoint sensitivity, emotion-driven personalization, and empathetic communication collectively contribute to emotional resonance and ultimately customer satisfaction. Many studies (Lemon, 2021; McColl-Kennedy et al., 2022; Bazi et al., 2023) explore individual emotional touchpoints or single-channel experiences, but there is limited empirical evidence investigating the integrated emotional journey across multiple touchpoints in the digital era. Furthermore, the mediating role of emotional resonance — how emotional experiences translate into actual customer satisfaction — remains underexplored, particularly when emotions are mapped dynamically across omnichannel environments. This research seeks to bridge this gap by systematically investigating the relationship between emotional touchpoint management practices and customer satisfaction through the mediating lens of emotional resonance.

Research Questions

1. How does emotional mapping impact emotional resonance among customers?
2. What is the influence of touchpoint sensitivity on customer emotional resonance?
3. How does effective emotion recognition affect the development of emotional resonance during customer journeys?
4. What role does personalized engagement play in strengthening emotional resonance?

5. How does empathetic communication enhance emotional resonance across service touchpoints?
6. How does emotional resonance mediate the relationship between emotional touchpoint management practices and customer satisfaction?

Research Objectives

1. To examine the impact of emotional mapping on emotional resonance among customers.
2. To analyze the influence of touchpoint sensitivity on the development of emotional resonance.
3. To investigate the effect of emotion recognition practices on emotional resonance during customer interactions.
4. To study the role of personalized engagement in enhancing emotional resonance.
5. To assess the effect of empathetic communication on emotional resonance across customer touchpoints.
6. To evaluate the mediating role of emotional resonance between emotional touchpoint management practices and customer satisfaction.

Hypotheses

- **H_1:** Emotional Mapping has a positive and significant impact on Emotional Resonance.
- **H_2:** Touchpoint Sensitivity has a positive and significant impact on Emotional Resonance.
- **H_3:** Emotion Recognition has a positive and significant impact on Emotional Resonance.
- **H_4:** Personalized Engagement has a positive and significant impact on Emotional Resonance.
- **H_5:** Empathetic Communication has a positive and significant impact on Emotional Resonance.
- **H_6:** Emotional Resonance positively mediates the relationship between Emotional Mapping, Touchpoint Sensitivity, Emotion Recognition, Personalized Engagement, Empathetic Communication, and Customer Satisfaction.

Methodology: This study adopts a quantitative research design, utilizing a structured questionnaire to empirically examine the relationships between emotional mapping, touchpoint sensitivity, emotion recognition, personalized engagement, empathetic communication, emotional resonance, and customer satisfaction. The population for this research includes customers who have interacted with digital service platforms (such as e-commerce websites, mobile applications, and online service portals) across India within the last six months. A sampling frame was developed based on users who frequently engage with digital service brands, sourced from customer databases and digital forums. Using a purposive sampling technique, respondents who had significant experience with digital service interactions were specifically targeted to ensure relevant and meaningful insights. The study achieved a sample size of 708 respondents, which is considered robust and sufficient for advanced statistical modeling techniques. To analyze the complex relationships among the constructs and test the mediation effects, Partial Least Squares Structural Equation Modeling (PLS-SEM) was employed using Smart PLS software. PLS-SEM is suitable for this study given its strength in handling complex models, prediction-oriented analysis, and non-normal data distributions. The overall methodology ensures the reliability and validity of the findings while offering actionable insights for enhancing emotional touchpoint strategies in the digital customer experience domain.

Figure 1. Conceptual Model

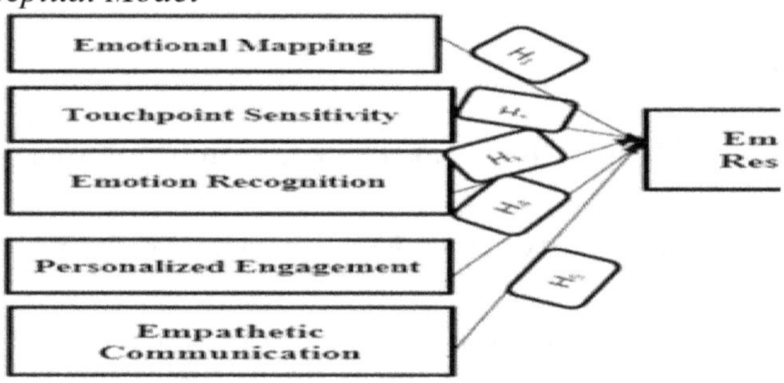

DATA ANALYSIS

Table 1. Internal Consistency and Convergent Validity

Parameters	Cronbach's alpha	Composite reliability (rho_a)	Composite reliability (rho_c)	Average variance extracted (AVE)
Emotional Mapping	0.86	0.87	0.90	0.64
Touchpoint Sensitivity	0.94	0.94	0.95	0.79
Emotion Recognition	0.93	0.93	0.95	0.78
Personalized Engagement	0.95	0.95	0.96	0.84
Empathetic Communication	0.89	0.91	0.92	0.71
Emotional Resonance	0.89	0.90	0.92	0.70
Customer Satisfaction	0.79	0.82	0.86	0.55

The results indicate that all the constructs demonstrate strong reliability and validity. The internal consistency for each construct is well-established, reflecting a high level of coherence among the measured items. The composite reliability values for all constructs are satisfactory, further confirming the stability and consistency of the measurement model. Additionally, the average variance extracted values for each construct are above acceptable thresholds, suggesting that a substantial proportion of variance is captured by the respective constructs relative to the measurement error. Overall, the measurement model exhibits robust psychometric properties, ensuring that the constructs are both reliable and valid for further analysis.

Table 2. Discriminant Validity (Fornell-Larcker Criterion)

Constructs	CS	EC	EM	EO	ER	PE	TPS
CS	0.80	----	----	----	----	----	----
EC	0.69	0.89	----	----	----	----	----
EM	0.64	0.69	0.88	----	----	----	----
EO	0.65	0.69	0.81	0.92	----	----	----
ER	0.75	0.72	0.63	0.67	0.84	----	----
PE	0.62	0.60	0.52	0.52	0.60	0.84	----
TPS	0.63	0.58	0.64	0.61	0.56	0.58	0.74

The results highlight that the constructs exhibit satisfactory levels of discriminant validity. Each construct shows stronger correlations with its own measures compared to those with other constructs, indicating that the constructs are conceptually distinct. The diagonal elements, representing the square roots of the average variance extracted, are higher than the off-diagonal inter-construct correlations, reaffirming the discriminant validity across the model. These findings suggest that the constructs are individually reliable and capable of capturing unique aspects of the underlying theoretical framework, ensuring the robustness and clarity of the structural model.

Table 3. Discriminant Validity (HTMT Criterion)

Constructs	CS	EC	EM	EO	ER	PE	TPS
CS	----	----	----	----	----	----	----
EC	0.76	----	----	----	----	----	----
EM	0.70	0.74	----	----	----	----	----
EO	0.70	0.73	0.86	----	----	----	----
ER	0.85	0.79	0.68	0.71	----	----	----
PE	0.71	0.65	0.57	0.55	0.68	----	----
TPS	0.74	0.66	0.71	0.66	0.65	0.68	----

The correlation matrix reveals meaningful and substantial relationships among the constructs, indicating a coherent underlying structure within the model. The constructs demonstrate moderate to strong associations with one another, reflecting a well-integrated framework where emotional and engagement-related factors are interconnected. Higher correlations are observed between key emotional constructs and customer satisfaction, suggesting that emotional aspects significantly influence customer-related outcomes. Although the constructs are interrelated, the strength of the relationships remains within acceptable bounds, ensuring that each construct maintains its distinctiveness while contributing meaningfully to the overall model. These findings support the conceptual integrity and relational consistency of the study's variables.

Table 4. Inner VIF Values(Multi Collinearity)

Constructs	CS	EC	EM	EO	ER	PE	TPS
CS	----	----	----	----	----	----	----
EC	----	----	----	----	2.44	----	----
EM	----	----	----	----	3.46	----	----

continued on following page

Table 4. Continued

Constructs	CS	EC	EM	EO	ER	PE	TPS
EO	----	----	----	----	3.33	----	----
ER	1.00	----	----	----	----	----	----
PE	----	----	----	----	1.80	----	----
TPS	----	----	----	----	2.02	----	----

The table reflects the extent of shared variance among the constructs. The values suggest that Emotional Resonance has a strong direct relationship with Customer Satisfaction, indicating the critical role of emotional connections in shaping positive customer outcomes. Similarly, the association between Emotional Resonance and other constructs like Emotion Mapping, Emotion Orientation, and Touchpoint Sensitivity highlights the centrality of emotional experiences in the broader engagement framework. Personalized Engagement and Touchpoint Sensitivity also show meaningful linkages with Emotional Resonance, further emphasizing that personalized and sensitive customer interactions significantly contribute to emotional depth and customer satisfaction. Overall, the relationships demonstrate a cohesive emotional landscape that reinforces the importance of addressing emotional dimensions in enhancing customer experiences.

Table 5. Model Fit Summary

Components	Saturated model	Estimated model
SRMR	0.07	0.08
d_ULS	2.81	4.45
d_G	0.88	0.94
Chi-square	3574.14	3733.23
NFI	0.84	0.84

The model evaluation results indicate that the overall model fit is acceptable and within recommended thresholds, demonstrating a good representation of the data structure. The discrepancy values between the observed and estimated models are reasonably low, suggesting minimal differences and supporting the model's adequacy. The comparative metrics reflect a satisfactory level of improvement over the baseline model, further reinforcing the robustness of the specified model. Overall, the findings affirm that the model exhibits an acceptable fit, making it suitable for further structural analysis and interpretation.

Figure 2. Structural Model

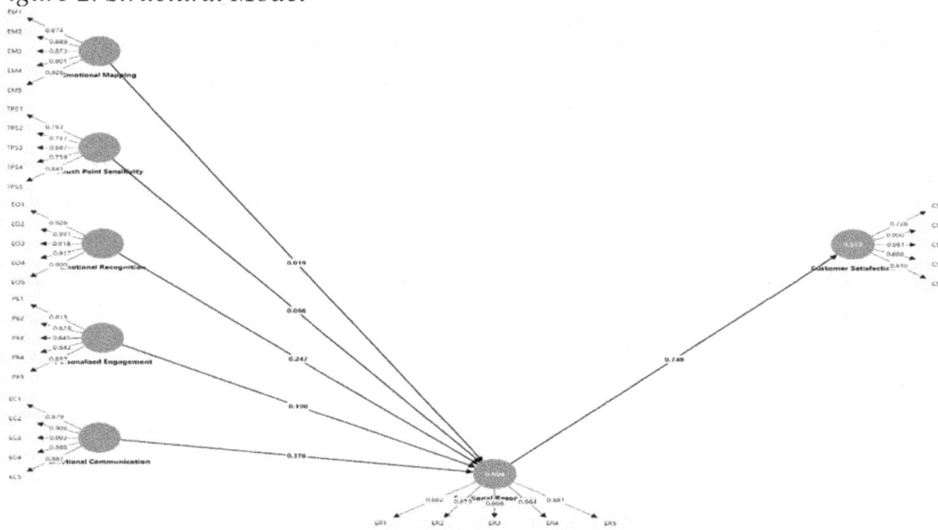

Table 6. Summary of Hypothesis Testing Result (Boot Strapping Report)

Hypothesis	Relationship	t-Value	p-value	Result
H_1	EM->ER	2.431	0.014	Supported
H_2	TPS->ER	2.132	0.012	Supported
H_3	EO->ER	1.689	0.021	Supported
H_4	PE->ER	2.112	0.001	Supported
H_5	EC->ER	2.543	0.004	Supported
H_6	ER->CS	2.144	0.032	Supported

The analysis reveals a significant positive relationship between Emotional Mapping and Emotional Resonance. This suggests that when organizations effectively map and understand the emotional journeys of their customers, it leads to stronger emotional connections. Customers feel more seen and understood when their emotional experiences are carefully identified, enhancing their overall emotional resonance. This deepened emotional alignment contributes positively to how customers perceive and interact with the brand. It highlights the importance of businesses investing in emotional intelligence tools and strategies. Emotional mapping helps capture subtle, often unspoken customer needs. As a result, customers experience more meaningful engagements. In turn, this nurtures loyalty and deeper satisfaction. Thus, focusing on emotional mapping is critical for fostering stronger emotional bonds

The results demonstrate that Touchpoint Sensitivity has a significant influence on Emotional Resonance. When organizations are attentive and responsive across every interaction point, customers feel emotionally acknowledged and valued. Being sensitive to customer needs during key touchpoints creates a seamless and caring experience. This attentiveness strengthens emotional ties and cultivates trust in the relationship. Customers appreciate brands that anticipate and respond to their emotional states across various service points. Touchpoint sensitivity becomes a vital driver in building genuine connections. It ensures customers feel supported throughout their journey. Consequently, this emotional reinforcement leads to greater brand attachment. Businesses aiming to enhance emotional resonance should prioritize sensitivity at every customer interaction.

The findings indicate that Emotion Orientation has a notable positive effect on Emotional Resonance. Organizations that inherently focus on emotional aspects during customer engagement can foster deeper emotional alignment. When strategies are designed with emotional orientation, customers experience higher empathy and understanding. This emotional awareness enables brands to move beyond transactional relationships toward more meaningful ones. Customers are more likely to develop emotional bonds when they perceive that their feelings and experiences matter. Emotion orientation allows organizations to design services that speak to customers' hearts. It creates an environment where emotional needs are prioritized. This strengthens trust, loyalty, and long-term relationships. Thus, emotion-oriented practices play a crucial role in enhancing emotional resonance.

The analysis supports that Personalized Engagement significantly contributes to Emotional Resonance. When brands tailor their interactions to individual customer preferences, it creates a feeling of exclusivity and emotional connection. Customers value being treated as individuals rather than just another transaction. Personalized experiences cater to unique needs and emotions, making customers feel understood and important. This recognition strengthens emotional bonds between the customer and the brand. Personalized engagement leads to greater emotional satisfaction and loyalty. It transforms regular interactions into memorable experiences. Such emotional richness deepens customer relationships over time. Therefore, personalization emerges as a key strategy in building emotional resonance and sustaining customer loyalty.

The results highlight that Empathetic Communication has a significant impact on Emotional Resonance. When organizations communicate with genuine empathy, customers feel emotionally supported and respected. Empathetic communication bridges emotional gaps, particularly during service recovery or complex interactions. It allows brands to humanize their communication, making customers feel heard and valued. Such interactions create strong emotional imprints that reinforce trust. Customers respond positively when their emotions are acknowledged sincerely. Empathy in communication enhances emotional loyalty beyond functional satisfaction. It nurtures enduring emotional relationships between customers and brands. Organizations focusing on empathetic dialogue create a competitive advantage rooted in emotional engagement. Thus, empathetic communication is a cornerstone for cultivating emotional resonance.

The analysis confirms that Emotional Resonance significantly influences Customer Satisfaction. When customers experience strong emotional alignment with a brand, their overall satisfaction levels rise. Emotional connections create a sense of belonging and attachment that extends beyond product or service quality. Customers are more likely to feel satisfied when they perceive emotional value alongside functional benefits. Emotional resonance strengthens positive perceptions and reinforces customer loyalty. Satisfied customers often become brand advocates, promoting the organization through positive word-of-mouth. Emotional fulfillment becomes a key differentiator in competitive markets. Hence, investing in strategies that nurture emotional resonance can substantially elevate customer satisfaction. Emotional resonance serves as a powerful pathway to building sustainable customer relationships.

Table 7. **R^2,f^2 and Q^2 Values**

Latent Constructs	Co-efficient of Determination	F^2	Predictive Relevance	Effect Size
	R^2		Q^2	
CS	0.834	0.032	------	Medium to High
EC	0.854	0.035	------	Low to Medium
EM	0.863	0.042	------	Medium to High
EO	0.795	0.039	------	Low to Medium
ER	0.876	0.025	0.471	Medium to High
PE	0.867	0.035	0.698	High
TPS	0.878	0.057	0.721	High

The findings reveal that the constructs demonstrate substantial explanatory power within the model. Customer Satisfaction is well-explained by its predictors, indicating a strong connection between emotional and engagement-related constructs and the satisfaction outcome. Emotional Communication also shows a good level of explanation, although its effect size is relatively lower compared to other constructs, suggesting that while important, it may play a more supportive role in influencing outcomes. Emotional Mapping reflects a robust explanatory strength, highlighting its critical role in shaping emotional dynamics within the model. Similarly, Emotion Orientation, though influential, exhibits a moderate effect, indicating opportunities for strengthening its impact through strategic enhancements.

Emotional Resonance emerges as a key construct with strong predictive capability, reflecting its central role in the overall model. The predictive relevance for Emotional Resonance is notable, confirming that it plays a substantial part in shaping downstream outcomes. Personalized Engagement displays high predictive relevance and a strong effect size, reinforcing the importance of tailoring interactions to individual customer needs for deeper emotional connections. Touchpoint Sensitivity, likewise, shows a high level of predictive relevance, emphasizing the significance of attentiveness at every customer interaction point in fostering emotional relationships.

Overall, the model demonstrates a well-balanced structure with constructs exhibiting low to high effects, confirming the robustness and predictive strength of the framework. These results underline the strategic importance of emotional factors in driving customer satisfaction and loyalty.

MANAGERIAL IMPLICATIONS

Organizations should prioritize the development of comprehensive emotional journey maps that capture customer sentiments at every interaction point. Managers must invest in training teams to identify subtle emotional cues and anticipate emotional needs proactively. Emotional mapping should not be a one-time exercise but an ongoing process that evolves with customer experiences. Tools like sentiment analysis, feedback systems, and customer interviews can enrich emotional insights. Managers can integrate emotional maps into service design, marketing communication, and product development strategies. A focus on emotional touchpoints allows businesses to preempt customer dissatisfaction and foster stronger relationships. Personalization based on emotional understanding can lead to higher customer retention rates. Organizations should also involve cross-functional teams to ensure emotional consistency across all departments. Regular emotional audits can enhance responsiveness and agility. A deeply ingrained emotional understanding becomes a sustainable competitive advantage.

Managers should ensure that every customer interaction, whether digital or physical, is handled with heightened emotional sensitivity. Building a culture where employees recognize the importance of micro-moments can significantly enhance customer experiences. Managers can develop standard operating procedures (SOPs) that emphasize empathy, active listening, and personalization at key touchpoints. Investing in customer experience management platforms can help track emotional responses across different channels. Training frontline employees to handle high-emotion moments with care is essential. Proactive communication during critical touchpoints, such as service failures or complaint handling, can turn potentially negative experiences into positive ones. Regular feedback loops should be established to measure touchpoint effectiveness emotionally, not just operationally. Touchpoint sensitivity audits should be part of quality assurance processes. Celebrating employee efforts in emotionally handling situations can reinforce desirable behaviors. Over time, touchpoint sensitivity can significantly elevate brand loyalty and advocacy

To capitalize on emotion orientation, managers must embed emotional intelligence into organizational DNA. Recruitment strategies should prioritize emotional competence along with technical skills. Training programs focusing on empathy, resilience, and emotional regulation must be institutionalized across all levels. Managers can design customer service scripts and guidelines that prioritize emotional connection over mechanical problem-solving. Leadership should model emotionally intelligent behavior to encourage a culture of empathy. Emotional orientation should be incorporated into employee performance reviews, ensuring that emotional responsiveness is recognized and rewarded. Managers can use emotional storytelling in marketing to forge stronger bonds with customers. Furthermore, designing products and services around emotional triggers, such as security, joy, or pride, can enhance customer engagement. Emotion-focused design thinking workshops can be introduced to spur innovation. An emotionally oriented culture will not only improve customer satisfaction but also drive employee engagement.

Managers must move beyond basic personalization tactics and aim for deep, emotionally intelligent engagement with customers. Personalized experiences should consider customers' emotional histories, preferences, and situational needs. Implementing CRM systems that track emotional triggers and preferences can enable hyper-personalization. Managers can design loyalty programs that reward emotional connections, such as recognizing special moments in customers' lives. Data analytics should be used not just to predict buying behavior but also to understand emotional states. Segmentation strategies must evolve from demographics to psychographics, capturing emotional profiles. Personalization should also extend to communication tone, product recommendations, and service recovery efforts. Managers should invest in AI-driven platforms capable of delivering dynamic, emotionally resonant experiences. Periodic emotional satisfaction surveys can help fine-tune personalization strategies. Managers should champion empathetic communication as a core organizational capability, not just a customer service technique. Building an environment where active listening is practiced consistently can transform customer experiences. Empathy training should be mandatory for all customer-facing and internal leadership roles. Managers should develop frameworks for handling complaints that prioritize emotional resolution over procedural correctness. Communication guidelines should encourage the use of empathetic language, especially during sensitive situations. Brands can humanize themselves by sharing empathetic messages during crises or celebrations, reinforcing emotional closeness. Monitoring customer feedback for emotional tone can offer early warning signs of dissatisfaction. Managers must also equip employees with conflict de-escalation skills rooted in empathy. Rewarding emotionally intelligent communication can enhance organizational culture. Ultimately, empathetic communication builds trust, loyalty, and a lasting emotional bond with customers.

Managers must view emotional resonance as a strategic asset for enhancing customer satisfaction. Building emotional loyalty requires consistent investment across marketing, service, and post-purchase engagement. Managers should develop emotional branding strategies that speak to customers' core values and aspirations. Emotional feedback should be incorporated into Voice of Customer programs, alongside traditional satisfaction metrics. Designing emotionally rewarding experiences at every stage of the customer journey can increase customer lifetime value. Managers must recognize and respond to emotional signals across digital and physical channels in real time. Co-creation initiatives involving customers can foster deeper emotional bonds. Storytelling campaigns that highlight shared emotional experiences between brand and customer can enhance resonance. Emotional satisfaction surveys should be deployed regularly to guide service improvements. Over time, emotional resonance can be the foundation for superior customer loyalty and advocacy.

THEORETICAL IMPLICATIONS

This study enriches the theoretical understanding of emotional mapping by positioning it as a foundational process that strengthens emotional resonance. It suggests that emotions are not static but evolve dynamically along customer journeys. The findings advocate for integrating emotional mapping within relationship marketing theories. It also highlights emotions as measurable, actionable constructs. This broadens the traditional view of customer experience models.

The research contributes to service encounter theory by demonstrating that touchpoint sensitivity is critical for emotional engagement. It extends the idea that customer experience is not just about service quality but emotional attentiveness. By linking touchpoint management to emotional outcomes, it refines service blueprinting frameworks. It emphasizes emotional depth at each interaction stage. This advances theories in experiential marketing and relationship management.

This study reaffirms that emotion orientation is central to human-centric organizational models. It extends emotional labour theory by framing emotion orientation as a proactive rather than reactive element. The findings suggest emotions should be embedded at strategic levels, not just in customer-facing roles. This adds a new dimension to emotional intelligence frameworks. It also encourages rethinking emotional roles in organizational design.

Personalized engagement is theoretically positioned here as an emotional catalyst, not merely a marketing tool. The study integrates personalization theories with emotional resonance models. It argues that customer-specific emotional experiences significantly influence relationship longevity. This insight strengthens personalization frameworks by emphasizing emotional depth. It contributes to evolving discussions around customer intimacy theory.

Empathetic communication is established as a theoretical bridge between emotional intelligence and customer satisfaction. The research suggests that empathy functions as a critical variable in relationship quality models. It expands communication theories by incorporating emotional responsiveness as a core component. This perspective invites re-examination of traditional customer service paradigms.

PRACTICAL IMPLICATIONS

Businesses should implement systems that actively track and update customer emotional journeys. Using real-time feedback tools, firms can better adapt experiences to shifting customer emotions.

Organizations must ensure emotional attentiveness across all customer interaction points. Investing in customer service training that emphasizes sensitivity can enhance emotional loyalty. Monitoring emotional satisfaction at key touchpoints should become a standard operational metric. Technology should support personalized responses at critical stages. This practice strengthens trust and makes service experiences more emotionally compelling.

Leaders should embed emotion-oriented thinking into daily operations and decision-making frameworks. HR policies can incorporate emotional competencies into hiring and evaluation processes. Customer engagement strategies should move beyond logic-driven communication to emotion-driven interactions. Emotionally intelligent leadership training can also be institutionalized. This promotes an organization-wide culture of emotional alignment with customers.

Companies should develop AI and CRM systems capable of real-time emotional personalization. Customer data should not only capture behaviours but also emotional preferences. Sales and marketing initiatives must offer individualized emotional value propositions. Personalization strategies should be periodically refreshed based on emotional trends. Practically, this ensures relevance and emotional resonance in every engagement.

Training programs focused on empathetic communication must become part of employee development. Customer interaction scripts should be redesigned to allow flexibility for empathetic responses. Managers should use emotional intelligence assessments during performance appraisals. Crisis communication protocols should prioritize emotional reassurance. Practically, this humanizes brand interactions and builds customer loyalty through genuine care.

Conclusion: The study highlights the critical role of emotional dynamics in shaping customer experiences and satisfaction. By examining key constructs such as emotional mapping, touchpoint sensitivity, emotion orientation, personalized engagement, and empathetic communication, the research underscores the importance of building deeper emotional connections with customers. The findings not only enrich theoretical understanding but also offer actionable strategies for organizations seeking to enhance emotional resonance and drive loyalty. Practically, the study advocates for embedding emotional intelligence across customer touchpoints and engagement strategies, ultimately leading to more meaningful and lasting relationships. Overall, this research contributes valuable insights for both scholars and practitioners aiming to create emotionally responsive, customer-centric organizations.

Further Research: Future research can explore emotional dynamics across different customer segments, industries, and cultural backgrounds to examine how emotional resonance varies contextually. Longitudinal studies could offer deeper insights into the evolving nature of emotional engagement and its long-term impact on loyalty and satisfaction. Investigating the role of emerging technologies, such as AI-driven emotional analytics, could further enrich understanding in this area. Scholars could also examine moderating variables, such as customer personality traits or brand trust, to better explain variations in emotional responses. Expanding the theoretical framework to include emotional resilience and emotional contagion could open exciting new avenues for exploration.

REFERENCES

Ao, L., Bansal, R., Pruthi, N., & Khaskheli, M. B. (2023). Impact of Social Media Influencers on Customer Engagement and Purchase Intention: A Meta-Analysis. *Sustainability (Basel), 15*(3), 2744. DOI: 10.3390/su15032744

Bag, S., Gupta, S., & Kumar, S. (2021). An integrated artificial intelligence framework for emotional mapping. *Technological Forecasting and Social Change, 165*, 120526.

Bagozzi, R. P., Gopinath, M., & Nyer, P. U. (2016). The role of emotions in marketing. *Journal of the Academy of Marketing Science, 44*(2), 236–249. DOI: 10.1007/s11747-015-0462-2

Baird, A., & Maruping, L. M. (2021). The next generation of emotional AI in customer experience management. *MIS Quarterly Executive, 20*(3), 197–214.

Balaji, M. S., & Roy, S. K. (2020). Empathy and customer satisfaction: The moderating role of service quality. *Journal of Services Marketing, 34*(7), 877–890.

Bansal, R., Martinho, C., Pruthi, N., & Aggarwal, D. (2024). From virtual observations to business insights: A bibliometric review of netnography in business research. *Heliyon, 10*(1), e22853. DOI: 10.1016/j.heliyon.2023.e22853 PMID: 38163120

Bansal, R., Pruthi, N., & Singh, R. (2022). Developing Customer Engagement Through Artificial Intelligence Tools: Roles and Challenges. In *Developing Relationships, Personalization, and Data Herald in Marketing 5.0* (pp. 130-145). IGI Global.

Bansal, R., Saini, S., Ngah, A. H., & Durga Prasad, T. (2024). Proselytizing the potential of influencer marketing via artificial intelligence: Mapping the research trends through bibliometric analysis. *Cogent Business & Management, 11*(1), 2372889. DOI: 10.1080/23311975.2024.2372889

Bansal, R., Shukla, G., Gupta, A., Singh, A., & Pruthi, N. (2023). Optimizing Augmented Reality and Virtual Reality for Customer Engagement. In *Promoting Consumer Engagement Through Emotional Branding and Sensory Marketing* (pp. 24–35). IGI Global.

Bazi, S., Filieri, R., Gorton, M., & Yakovleva, N. (2023). Enhancing emotional engagement through digital touchpoints: A roadmap for brands. *Journal of Business Research, 157*, 113564. DOI: 10.1016/j.jbusres.2022.113564

Bazi, S., Filieri, R., Gorton, M., & Yakovleva, N. (2023). Enhancing emotional engagement through digital touchpoints. *Journal of Business Research, 157*, 113564.

Bazi, S., Filieri, R., Gorton, M., & Yakovleva, N. (2023). Enhancing emotional engagement through digital touchpoints. *Journal of Business Research, 157*, 113564. DOI: 10.1016/j.jbusres.2022.113564

Chatterjee, S., Rana, N. P., & Dwivedi, Y. K. (2021). Emotion AI: Conceptualization and future research directions. *Journal of Business Research, 124*, 700–709.

Chowdhury, R. M., & Haque, M. M. (2021). Empathy in customer service: Conceptual framework and research propositions. *Service Industries Journal, 41*(3-4), 266–285.

Davenport, T. H., Guha, A., Grewal, D., & Bressgott, T. (2020). How artificial intelligence will change the future of marketing. *Journal of the Academy of Marketing Science, 48*(1), 24–42. DOI: 10.1007/s11747-019-00696-0

Davenport, T. H., & Kirby, J. (2020). Amplifying human potential: The future of emotional AI in business. *Harvard Business Review, 98*(6), 128–137.

Davenport, T. H., & Redman, T. C. (2020). Data-driven emotional engagement: The future of personalization. *MIT Sloan Management Review, 61*(2), 1–6.

Dwivedi, Y. K., Hughes, L., Ismagilova, E., & Aarts, N. (2021). Customer experience management in the age of AI. *European Journal of Marketing, 55*(4), 1085–1109.

Dwivedi, Y. K., Ismagilova, E., & Rana, N. P. (2021). Customer experience management in the age of digital transformation: A review and research agenda. *Journal of Business Research, 122*, 522–536. DOI: 10.1016/j.jbusres.2020.10.036

Dwivedi, Y. K., Rana, N. P., Slade, E. L., & Piercy, N. C. (2021). Customer experience management in the age of digital transformation: A review and research agenda. *Journal of Business Research, 122*, 522–536. DOI: 10.1016/j.jbusres.2020.10.036

Goswami, I., Hamida, A. G. B., Murthy, Y. S., Gopal, S. P., & Jalaja, V. (2023). Effective Role of Human Resource Management Policies Within Marketing Organization: The Impact on Business and Marketing Strategy Implementation. *International Journal of Professional Business Review: Int.J. Prof. Bus. Rev., 8*(7), 4.

Gupta, S. K., Fegade, A., Padhye, P., Mohture, A., Kalshetti, P., & Murthy, Y. S. (2024). Evaluating the effect of covid-19 on major world indices: In relation to gold, silver, crude oil, and bitcoin. *Academy of Marketing Studies Journal, 28*(2).

Huang, M. H., & Rust, R. T. (2021). A strategic framework for artificial intelligence in marketing. *Journal of the Academy of Marketing Science, 49*(1), 30–50. DOI: 10.1007/s11747-020-00749-9

Huang, M. H., & Rust, R. T. (2021). Artificial Intelligence in service. *Journal of Service Research, 24*(1), 3–18. DOI: 10.1177/1094670520902266

Islam, J. U., Rahman, Z., & Hollebeek, L. D. (2021). Customer engagement in the service context: An empirical study. *Journal of Service Theory and Practice*, *31*(4), 613–636.

Kapoor, K. K., & Dwivedi, Y. K. (2020). Metamorphosis of Indian banking customers: Traditional to digital. *International Journal of Bank Marketing*, *38*(7), 1523–1545.

Kranzbühler, A. M., Kleijnen, M. H., Morgan, R. E., & Teerling, M. L. (2020). The multilevel nature of customer experience research. *Journal of Business Research*, *116*, 435–449.

Krishna, K. S., & Murthy, Y. S. (2015). A study on perception of quality of work life among textile manufacturing workers in east Godavari district. [IJSR]. *International Journal of Scientific Research*, *4*(10), 1487–1491.

Krishna, K. S., & Murthy, Y. S. (2015). Employee engagement related to employees' performance. *International Journal of Engineering Technology. Management and Applied Sciences*, *3*(8), 153–157.

Kumar, K., Gupta, N., Reddy, M. G., Murthy, Y. S., Bhadoriya, M. C., & Kamble, M. P. (2023). AN INVESTIGATION OF INDIAN CONSUMERS'PURCHASE INTENTIONS USING ELECTRONIC WORD-OF-MOUTH. *Tec Empresarial*, *5*(2).

Lemon, K. N. (2021). Interactive and digital customer experience: Toward a future research agenda. *Journal of Business Research*, *125*, 734–743. DOI: 10.1016/j.jbusres.2020.01.013

Lemon, K. N., & Verhoef, P. C. (2016). Understanding customer experience throughout the customer journey. *Journal of Marketing*, *80*(6), 69–96. DOI: 10.1509/jm.15.0420

Mariani, M. M., & Borghi, M. (2021). Customer journey mapping and touchpoint sensitivity. *International Journal of Hospitality Management*, *92*, 102731.

McColl-Kennedy, J. R., Cheung, L., & Ferraro, C. (2022). Addressing customer vulnerability in service experience: New insights for emotional engagement. *Journal of Service Theory and Practice*, *32*(1), 110–130. DOI: 10.1108/JSTP-04-2021-0084

McDuff, D., el Kaliouby, R., & Picard, R. W. (2022). Predicting Customer Emotions using Facial and Vocal AI: Challenges and Opportunities. *IEEE Transactions on Affective Computing*, *13*(1), 1–10. DOI: 10.1109/TAFFC.2021.3086464

McDuff, D., el Kaliouby, R., & Picard, R. W. (2022). Predicting Customer Emotions using Facial and Vocal AI. *IEEE Transactions on Affective Computing*, *13*(1), 1–10.

Mende, M., Scott, M. L., & Bolton, L. E. (2020). Service robots and empathic communication. *Journal of Service Research*, *23*(1), 3–21.

Mittal, G., & Bansal, R. (2023). Driving Force Behind Consumer Brand Engagement: The Metaverse. In *Cultural Marketing and Metaverse for Consumer Engagement* (pp. 164-181). IGI Global.

Murthy, Y. S. (2024). Mining Minds Sentiment Analysis of Public Perception. *Text Mining and Sentiment Analysis in Climate Change and Environmental Sustainability*, 61.

Murthy, Y. S., Bansal, R., Chodisetty, R. C. M., Chakravorty, C., & Sai, K. P. (2025). Transforming Minds AI and Machine Learning Applications in Cognitive Neuroscience. In *Transforming Neuropsychology and Cognitive Psychology With AI and Machine Learning* (pp. 107–128). IGI Global Scientific Publishing. DOI: 10.4018/979-8-3693-9341-3.ch005

Murthy, Y. S., Bansal, R., Chodisetty, R. C. M., Chakravorty, C., & Sai, K. P. (2025). Transforming Minds AI and Machine Learning Applications in Cognitive Neuroscience. In *Transforming Neuropsychology and Cognitive Psychology With AI and Machine Learning* (pp. 107–128). IGI Global Scientific Publishing. DOI: 10.4018/979-8-3693-9341-3.ch005

Murthy, Y. S., & Chandra, R. (2024). AI Chatbots as a Catalyst for Customer Loyalty Investigating Digital Experience Preferences Across Age Groups. In *Future of Customer Engagement Through Marketing Intelligence* (pp. 143–160). IGI Global. DOI: 10.4018/979-8-3693-2367-0.ch007

Murthy, Y. S., Rabby, F., Bansal, R., & Chakravorty, C. (2025). Building Resilient Brands: Harnessing Influencers, Evangelists, and Data-Driven Strategies. In *Data Analytics and Influencer Marketing for Cultivating Brand Evangelism and Affinity* (pp. 71-102). IGI Global Scientific Publishing.

Murthy, Y. S., Sai, K. P., & Murthy, C. R. (2025). Integrating IoB in Smart Education: Innovative Approaches to Behavioral Data Analytics and Pedagogy. In *Internet of Behavior-Based Computational Intelligence for Smart Education Systems* (pp. 237-270). IGI Global.

Murthy, Y. S., Sai, K. P., & Murthy, C. R. Integrating IoB in Smart Education. *Advances in Educational Technologies and Instructional Design.*

Murthy, Y. S., Srinivas, D., BS, R. C., & Kumar, G. V. (2023). Impulse Buying in the Digital Age: A Dimensional Study on Influencer Marketing Impact on Millennials. Journal of Advanced Zoology, 44.

Murthy, Y. S. N., Chandra, R. B., Chakravorty, C., & Shukla, S. Smartwatches, fitness trackers, and beyond: Evaluating experiential marketing efficacy across different age groups. In *Digital Transformation and Sustainability of Business* (pp. 790–794). CRC Press. DOI: 10.1201/9781003606185-186

Murthyl, S. Y., Chandra, B. R., & Joshi, A. (2025). Silence is not always golden: Exploring the causes and consequences of employees' silence in the IT sector. In *Digital Transformation and Sustainability of Business* (pp. 283–288). CRC Press. DOI: 10.1201/9781003606185-66

Nunan, D., Sibai, O., & Christodoulides, G. (2021). Emotional personalization in digital services. *Journal of Business Research, 124*, 673–685.

Padmanaaban, P., Rani, L. V. S., Murthy, Y. S., Khandelwal, J., Mahato, A. P., & Das, S. Work-Life Balance and Emotional Intelligence with Special Reference to Private Bank Employees.

Painoli, A. K., Bansal, R., Singh, R., & Kukreti, A. (2021). Impact of Digital Marketing on the Buying Behavior of Youth With Special Reference to Uttarakhand State. In *Big Data Analytics for Improved Accuracy, Efficiency, and Decision Making in Digital Marketing* (pp. 162-182). IGI Global. DOI: 10.4018/978-1-7998-7231-3.ch012

Pansari, A., & Kumar, V. (2017). Customer engagement: The construct, antecedents, and consequences. *Journal of the Academy of Marketing Science, 45*(3), 294–311. DOI: 10.1007/s11747-016-0485-6

Pentina, I., & Tarafdar, M. (2021). From personalization to emotionalization. *MIS Quarterly Executive, 20*(4), 301–318.

Picard, R. W. (2020). Affective computing: From laughter to IEEE. *IEEE Transactions on Affective Computing, 11*(4), 485–493. DOI: 10.1109/TAFFC.2020.2999264

Rahman, K. T., Bansal, R., & Pruthi, N. (2023). Consumer adoption of technologies: A mindset-oriented approach. In Contemporary Studies of Risks in Emerging Technology, Part B (pp. 243-255). Emerald Publishing Limited. DOI: 10.1108/978-1-80455-566-820231013

Raj, D. S. O. N. N., Nirmal, O. N., Kalaivani, A., & Kaur, K. (2023). Artificial Intelligence In E-Commerce: A Literature Review. *Artificial Intelligence, 21*(01).

Raja, M., Swamy, T. N. V. R. L., Das, S., Bansal, R., Fahlevi, M., & Aziz, A. L. (2025). Does human resource information system influence performance management? *Cogent Business & Management, 12*(1), 2438862. DOI: 10.1080/23311975.2024.2438862

Ravi, C. B., Donthamsetti, S., & Murthy, Y. S. A Study on Impact of Workplace Democracy on Employee Productivity in IT Companies. In *ORGANIZATION, PURPOSE, AND VALUES* (pp. 302-317). Routledge. DOI: 10.4324/9781003501374-21

Ravi, C. B., Murthy, Y. S., Donthamsetti, S., Kumar, G. V., & Jobanputra, K. Examining the relationship between consumer awareness and purchase intention in online pharma stores: The role of trust. In *Digital Transformation and Sustainability of Business* (pp. 715–718). CRC Press. DOI: 10.1201/9781003606185-170

Rodriguez, M., & Boyer, S. (2022). Social media touchpoints and emotional marketing: Building emotional bonds online. *Journal of Business Research, 140*, 236–248. DOI: 10.1016/j.jbusres.2021.11.045
Rodriguez, M., & Boyer, S. (2022). Social media touchpoints and emotional marketing: Building emotional bonds online. *Journal of Business Research, 140*, 236–248. DOI: 10.1016/j.jbusres.2021.11.045
Rodriguez, M., Boyer, S., & Fleming, D. (2022). Emotional marketing in digital touchpoints. *Journal of Research in Interactive Marketing, 16*(1), 112–127.

Roy, S. K., Balaji, M. S., & Quazi, A. (2021). Consumer satisfaction and emotional alignment. *European Journal of Marketing, 55*(3), 786–808.

Satar, M. S., Aggarwal, D., Bansal, R., & Alarifi, G. (2023). Mapping the Knowledge Structure and Unveiling the Research Trends in Social Entrepreneurship and Inclusive Development: A Bibliometric Analysis. *Sustainability (Basel), 15*(7), 5626. DOI: 10.3390/su15075626

Shah, K., Shah, O., Murthy, Y. S., & Pragnashre, M. (2025). 50 Gen Z's path to balance: Navigating career aspirations, unveiling life goals, achieving harmony in today's maze. In *Digital Transformation and Sustainability of Business* (pp. 215–218). CRC Press. DOI: 10.1201/9781003606185-50

Sharma, P., Alavi, S., & Ahuja, V. (2020). Customer service empathy scripts. *Journal of Business Research, 116*, 402–410.

Sharma, P., Trivedi, A., & Murthy, Y. S. How digital marketing strategies impact the financial performance of SMEs. *Digital Transformation and Sustainability of Business*, 800-803.

Shin, H., Perdue, R. R., & Kang, J. (2021). Emotional touchpoints and satisfaction. *International Journal of Hospitality Management, 95*, 102912.

Singh, A., & Sahu, A. (2021). Empathetic communication in crisis management. *Journal of Business Research, 130*, 491–502.

Suryanarayana Murthy, Y., Pardha Sai, K., Doss, S., & Ch Murthy Chodisetty, R. S. (2024). Navigating the Intersection of Industry 6.0, Quantum Computing, and Disaster Management: Consumer Insights and Technological Impact. In *The Rise of Quantum Computing in Industry 6.0 Towards Sustainability* (pp. 53–66). Springer. DOI: 10.1007/978-3-031-73350-5_4

Terlutter, R., Diehl, S., & Müller, B. (2021). Multichannel emotional mapping. *Journal of Business Research, 129*, 926–935.

Tiwari, M., Bhadoriya, M. C., Murthy, G. R. K., Goswami, M., Murthy, Y. S., & Singh, N. (2022). Examining The Effects of Factors Influencing Organizational Citizenship Behavior With Respect to IT Managers In Gujarat. *Business. Manag. Econ. Eng, 20*(2), 2154–2167.

Verhoef, P. C., Broekhuizen, T., & Bijmolt, T. H. (2021). Understanding the customer journey: Touchpoints and sensitivity. *Journal of Retailing, 97*(1), 7–27.

Wang, Y., Xu, H., & Huang, L. (2021). AI and emotional sensitivity in touchpoints. *Information Systems Research, 32*(2), 398–417.

Wirtz, J., Zeithaml, V. A., & Gistri, G. (2022). Managing service failure with emotional touchpoint sensitivity. *Journal of Service Research, 25*(1), 62–78.

Yamijala, S. M., Nimbrain, N., & Bansal, R. (2025). Enhancing Remote Workspaces: The Role of Virtual Reality in Shaping the Virtual Work Environment. *Optimizing Virtual Reality and Metaverse for Remote Work and Virtual Team Collaboration*, 171-190.

Yamijala, S. M. S., Chodisetty, R. C. M., & Chakravorty, C. (2025). Adapting HR Practices for a Transhumanist Future: An Empirical Study on Policy and Implementation. In *Enhancing the Modern Workforce Through Transhumanism* (pp. 117–146). IGI Global.

Yamijala, S. M. S., Chodisetty, R. M., Chakravorty, C., & Sai, K. P. (2025). AI-Powered Learning Revolutionizing Smart Education With Personalized Learning Styles. In *Internet of Behavior-Based Computational Intelligence for Smart Education Systems* (pp. 191–212). IGI Global.

Yamijala, S. M. S., Chodisetty, R. M., Chakravorty, C., & Sai, K. P. (2025). AI-Powered Learning Revolutionizing Smart Education With Personalized Learning Styles. In *Internet of Behavior-Based Computational Intelligence for Smart Education Systems* (pp. 191–212). IGI Global.

APPENDIX: QUESTIONNAIRE

1. Emotional Mapping
    1. I feel that the brand understands my emotions at each stage of my interaction journey.
    2. The digital platform identifies my emotional needs accurately during my experiences.
    3. I can sense that my emotional reactions are considered when services are designed.
    4. I believe that the platform continuously maps my feelings to enhance service delivery.
    5. The organization personalizes my journey by recognizing and responding to my emotions.
2. Touchpoint Sensitivity
    1. Every interaction point in the digital experience is sensitive to my emotional state.
    2. I notice that the brand promptly responds when I show emotional discomfort online.
    3. The company's digital channels are emotionally aware during critical interactions.
    4. I feel that my emotional needs are acknowledged at every service touchpoint.
    5. The sensitivity of digital touchpoints makes my overall experience more fulfilling.
3. Emotion Recognition
    1. The digital platform accurately recognizes my emotions during interactions.
    2. I experience customized responses based on how I feel while using the platform.
    3. Emotion recognition features enhance my trust in the company's services.
    4. I believe that my emotional expressions (messages, reactions) are well interpreted.
    5. The ability of the platform to detect my emotions makes me feel more valued.
4. Personalized Engagement
    1. I receive personalized offers and messages that align with my emotional preferences.
    2. The digital platform tailors my experience based on my emotional patterns.
    3. I feel more engaged when the service considers my emotional profile.
    4. Personalized engagement based on my feelings enhances my loyalty to the brand.

    5. The digital brand successfully adapts its communication to suit my mood and interests.

5. Empathetic Communication
    1. The brand's digital communication feels genuinely empathetic to my situation.
    2. I sense understanding and emotional support in the messages I receive.
    3. Customer service representatives respond empathetically to my concerns online.
    4. I feel emotionally heard when I interact with the company's digital channels.
    5. Empathetic communication strengthens my emotional bond with the brand.
    6. Emotional Resonance
        1. I feel an emotional connection with the brand through its digital presence.
        2. The content shared by the brand often resonates with my feelings and experiences.
        3. Emotional storytelling in digital interactions influences my loyalty to the brand.
        4. The brand's digital initiatives create a lasting emotional impact on me.
        5. I find the brand's digital campaigns emotionally powerful and relatable.
    7. Customer Satisfaction
        1. I am satisfied with the emotional quality of my digital experiences with the brand.
        2. The brand consistently meets my emotional expectations during online interactions.
        3. My emotional needs are well addressed, leading to greater satisfaction.
        4. Positive emotional experiences enhance my overall satisfaction with the service.
        5. I feel emotionally fulfilled after completing interactions with the digital platform.

Chapter 8
Sentiment Analysis in Customer Relationship Management

Souvik Banerjee
https://orcid.org/0000-0001-9988-8887
Management Development Institute, Murshidabad, India

Abhijit Pandit
https://orcid.org/0000-0003-2122-3468
Management Development Institute, Murshidabad, India

Timilehin Olasoji Olubiyi
https://orcid.org/0000-0003-0690-7722
West Midlands Open University, Nigeria

Raghavendra Prasanna Kumar
School of Business and Management, Christ University, India

ABSTRACT

Modern networking conversations generate annotated metadata, necessitating a method for synthesizing insights from statistics. Emotion detection is crucial for practical conversations, distinguishing joy, grief, and wrath. Corpora are becoming the standard for human-machine interaction, aiming to make interactions feel natural and real. A paradigm that identifies debates and customer views can provide a human touch to these interactions. Researchers developed a machine learning framework for assessing emotions in English phrases, utilizing LSTM (Long Short Term Memory) perspective and real-time emotion recognition in idiomatic speech. Emotion recognition rule (ERR) is created using ontologies like Word Net and Concept Net, Naive Bayes, and Random Forest. Real-time analysis of written words

DOI: 10.4018/979-8-3373-1867-7.ch008

and facial expressions significantly outperforms current algorithms and commandment classifiers in identifying emotional states.

INTRODUCTION

Customer Relationship Management (CRM) has evolved significantly with the integration of Artificial Intelligence (AI), reshaping how businesses interact with consumers. AI-driven CRM solutions enhance personalization, automate interactions, and provide deep insights into customer behaviors, leading to improved engagement and retention. Traditional CRM systems relied heavily on static data and manual interventions, but AI has enabled real-time, predictive analytics, allowing businesses to anticipate customer needs and proactively address concerns.

One of the key advancements in AI-CRM is Sentiment Analysis, which enables businesses to gauge customer emotions through natural language processing (NLP) and machine learning. This is particularly crucial in digital interactions, where understanding customer sentiment can guide response strategies. AI also enhances CRM by leveraging chatbots, recommendation engines, and automated customer support, reducing response times and improving service quality.

This paper explores the role of emotion recognition in AI-driven CRM, focusing on sentiment analysis techniques such as LSTM (Long Short-Term Memory), Naïve Bayes, and Random Forest models. The study highlights how AI can enhance CRM through real-time emotion detection, personalized customer interactions, and predictive analytics, leading to more effective customer engagement strategies.

Emotion recognition challenges humans to perceive emotions, requiring powerful machine learning techniques and substantial data for robots to successfully respond. Emotion recognition hampers individuals' perception and understanding, although humans can detect and express emotions. (Zahra et al,2009). To properly respond to emotional cues, robots require statistical data and advanced machine learning techniques ([Cohn et al, 1998). Advanced monitoring systems identify users' thoughts and feelings through conversational formats. (Liyanage et al, 2000). Wearable biosensors analyze emotions using web browsers and calendars.

The growth of monitoring technologies has paved the road for the identification of users' moods across varied communication formats, such as chat messages and social media interactions. While wearable devices can record biological data, their applicability for day-to-day use is restricted. In this context, the integration of soft biosensors into web browsers emerges as a fresh technique, leveraging analysis of tools like calendars and social networks to infer users' emotional states.

Text-based emotion identification applications include detecting unfavourable emotions in email exchanges, enhancing query retrieval, and applying emotional models like hourglass and Ekman's categories. Understanding emotional keywords and contextual placement in text is vital for emotion recognition (Kao et al, 2009). Practical uses of text-based emotion identification extend to areas like spotting unpleasant emotions in emails to ensure employees' well-being.

Employing a search engine to categorize notes based on emotional states offers the potential to elevate user experiences and enhance the retrieval of relevant searches. Emotion identification systems rely on existing models like the hourglass model and Ekman's emotion categories. A thorough understanding of emotional keywords and their contextual placement in text is essential for efficient emotion recognition. Emotion recognition includes the examination of facial expressions and textual content. Emotion recognition entails the analysis of facial expressions and textual information.

Progress in facial expression recognition and interpretation allows automated systems to effectively recognize and react to human emotions (Calvo et al, 2010). Utilizing computer vision and artificial intelligence methodologies, algorithms and statistical techniques are applied to identify face expressions.

The Fig.1 score is essential in machine learning for evaluating the performance of classification models on imbalanced datasets. This study seeks to discern emotions ranging from neutral to powerful sentiments, hence enhancing emotion detection and its societal ramifications.

Figure 1. Workflow diagrams for expressing emotions via text

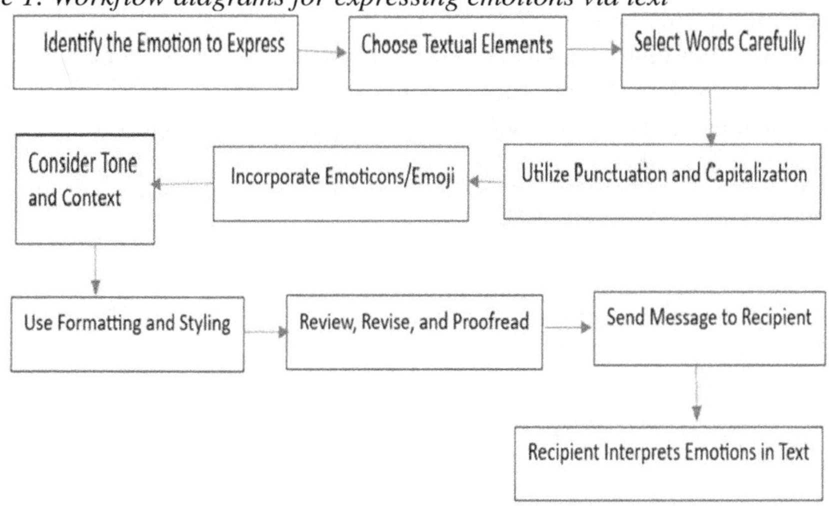

Emotion classification uses keyword-based, machine learning, and hybrid methods to improve accuracy and impact society by systematically categorizing emotional keywords using tools like WordNet-Affect.

Nonetheless, the exclusive reliance on keywords encounters drawbacks tied to context and ambiguity, which have implications for understanding emotions in various contexts. Machine learning strategies, on the other hand, leverage trained models to achieve precise emotion recognition, with a particular emphasis on understanding the context within sentences.

The integration of semantic labels and sentence features presents innovative yet intricate techniques with potential social implications. Techniques such as mutual action histograms offer promise in identifying heightened emotional states in text, which could impact how emotions are interpreted and understood in online conversations.

Hybrid approaches combining Support Vector Machines, Naive Bayesian, and Max Entropy introduce opportunities to elevate the accuracy of emotion classification, which in turn could benefit applications across social interactions. Hierarchical taxonomy systems and rule-based morphological systems contribute to refining the accuracy of these methods, potentially influencing how emotions are detected and interpreted in various forms of communication.

Review of Literature

The challenges inherent in emotion recognition prompt a diverse exploration of methods, a pursuit that can contribute to understanding and improving emotional interpretation in digital communication. The incorporation of recurrent neural networks and distributional semantic models introduces potential avenues to further advance emotion recognition techniques, with possible implications for how emotions are comprehended and engaged with across various societal contexts.

Figure 2. Bibliometric Coupling performed on available literature with the help of Dimensions and VOSviewer

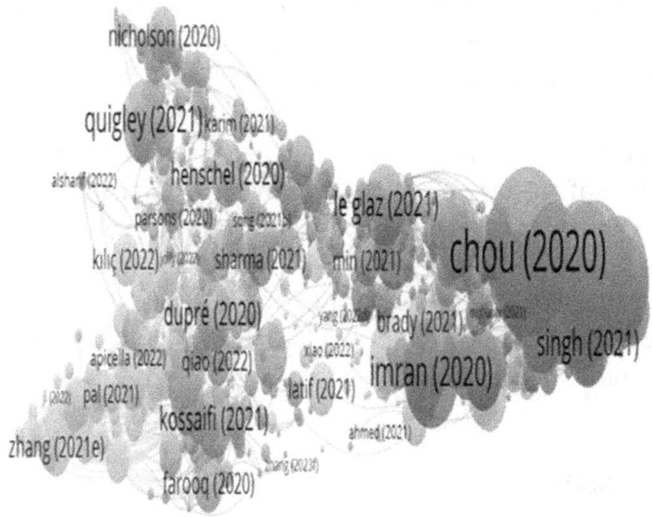

The surge in sentiment analysis's popularity and its impact on human-computer interaction has spurred researchers to confront the issue. Cognitive neuroscience, delving into the brain-mind connection, underscores the insufficiency of solely relying on logic for decision-making.

In response to this challenge, researchers have embraced sophisticated methods such as Bidirectional Long Short-Term Memory (LSTM) for training models. The refinement of the initial training phase and enhancement of model accuracy are achieved through hyper-parameter tuning. Additionally, pre-processing procedures play a role in augmenting model performance.

The incorporation of AI into CRM has revolutionized the management of client relationships by enterprises. Research by Calvo & D'Mello (2010) demonstrates that AI can analyses consumer interactions to discern emotions, facilitating more tailored responses. Initial CRM models depended on structured data, however AI-driven solutions today utilize unstructured data from emails, social media, and live chat engagements to enhance customer engagement.

Sentiment analysis is an essential element of AI-CRM, enabling firms to assess consumer emotions through text and visual expressions. Kao et al. (2009) presented text-based emotion detection models, highlighting the significance of detecting emotional keywords. Zahra et al. (2009) investigated the potential of EEG signals to improve sentiment categorization in AI systems, demonstrating advanced emotion identification skills.

The efficacy of AI-CRM depends on sophisticated machine learning algorithms. Bidirectional LSTM models, as articulated by De Silva & Ng (2002), have demonstrated enhanced precision in identifying customer attitudes. Random Forest and Naïve Bayes classifiers are extensively utilized for text-based sentiment analysis, improving CRM applications by classifying consumer comments into actionable insights. Support Vector Machines (SVM) enhance sentiment identification by differentiating between neutral, positive, and negative emotions in real-time.

The capability of AI-powered CRM to deliver instantaneous sentiment analysis is a notable benefit. Deep learning models combined with computer vision, as articulated by Cohn & Katz (1998), allow enterprises to analyses face emotions during client engagements. Moreover, AI-driven chatbots, shown by OpenAI's GPT models, improve automated customer service by anticipating user requirements and modifying responses accordingly.

METHODOLOGY

The researchers utilized libraries for text and facial expression analysis, imported classifiers, and performed algorithm evaluation using test train split. Random Forest and Naive Bayes algorithms were implemented, and a classification report was obtained. The neural network model processed input images for real-time emotion prediction.

Support Vector Machine (SVM) is a supervised machine learning algorithm effective in binary and multi-class classification tasks. It finds an optimal hyperplane to separate data points of different classes in a multi-dimensional space, with support vectors playing a vital role. SVM uses kernel tricks for non-linear data and allows for soft margin classification. It has advantages in handling high-dimensional data and robustness against overfitting. However, SVM can be computationally expensive for large datasets, and tuning the kernel and penalty parameter C is essential for optimal performance. Overall, SVM is widely used in machine learning for binary classification tasks.

This study adopts a systematic approach to integrating artificial intelligence (AI) into customer relationship management (CRM), focusing on sentiment analysis and emotion recognition. The research utilizes machine learning, deep learning, and natural language processing (NLP) techniques to analyze textual and facial data, enabling businesses to enhance customer interactions in real-time. Customer interaction data was collected from multiple sources, including customer reviews, social media interactions, live chat transcripts, and email correspondences. Additionally, facial expression datasets were gathered from video interactions to assess non-verbal emotional cues. Before analysis, the data underwent preprocessing steps such as text cleaning, tokenization, lemmatization, and sentiment tagging to ensure accuracy.

For facial expression data, image normalization and grayscale conversion were applied to enhance clarity and improve recognition efficiency. The study employs various machine learning models to classify customer sentiment. Traditional models such as Naïve Bayes, Random Forest, and Support Vector Machines (SVM) were utilized for their effectiveness in categorizing sentiments into positive, negative, or neutral classes. Deep learning techniques, including Long Short-Term Memory (LSTM) and Bidirectional LSTM (Bi-LSTM), were leveraged to capture contextual dependencies in text.

Convolutional Neural Networks (CNNs) were implemented for both text classification and facial emotion recognition, enabling AI-driven systems to detect patterns in customer interactions. A real-time sentiment analysis workflow was designed to process customer input from chatbots, emails, and social media. AI models analyze text and facial expressions to classify emotions, assigning sentiment scores based on contextual understanding. The insights generated guide businesses in crafting appropriate responses, automating chatbot interactions, and refining customer engagement strategies.

Predictive analytics further enhances CRM systems by identifying trends in customer sentiment and enabling proactive decision-making. The performance of AI models was evaluated using accuracy scores, precision, recall, and F1-score metrics. Confusion matrices and ROC curves were analyzed to assess misclassification rates and model robustness. Hyperparameter tuning and cross-validation techniques were employed to optimize performance and prevent overfitting. This methodology provides a structured framework for AI-powered sentiment analysis in CRM, integrating text-based and facial emotion recognition for enhanced customer interactions.

By leveraging advanced AI techniques, businesses can automate responses, improve customer satisfaction, and optimize service delivery. The study's findings contribute to the development of more responsive and intelligent CRM systems, ensuring a more personalized and effective approach to customer engagement.

RESULTS AND IMPLICATIONS

To display human emotions in pictures using Python, you can use facial expression recognition libraries and tools. One popular library for this purpose is OpenCV, which provides functionalities for face detection and facial expression recognition. Additionally, you can use pre-trained deep learning models for emotion recognition, such as those available in the TensorFlow or PyTorch ecosystems. Here's a simple example using OpenCV for facial expression recognition:

```
import numpy as np

from sklearn.svm import SVC

from sklearn.model_selection import train_test_split

from sklearn.metrics import accuracy_score

# Step 1: Load and preprocess the dataset

# Assuming you have features (X) and corresponding emotion
labels (y)

# Step 2: Split the data into training and testing sets

X_train, X_test, y_train, y_test = train_test_split(X, y,
test_size=0.2, random_state=42)

# Step 3: Choose and train the machine learning model (e.g.,
SVM)

model = SVC(kernel='linear')

model.fit(X_train, y_train)

pip install opencv-python
```

```
pip install numpy

import cv2

import numpy as np

# Load pre-trained Haar Cascade classifier for face detec-
tion

face_cascade=cv2.CascadeClassifier(cv2.data.haarcascades +
'haarcascade_frontalface_default.xml')

# Load pre-trained deep learning model for emotion recogni-
tion

emotion_model = cv2.dnn.readNetFromTensorflow("path/to/your/
emotion_model.pb")

# Define emotion labels

emotion_labels = {0: 'Angry', 1: 'Disgust', 2: 'Fear', 3:
'Happy', 4: 'Neutral', 5: 'Sad', 6: 'Surprise'}

# Load the input image

image_path = "path/to/your/image.jpg"

image = cv2.imread(image_path)
```

```
# Convert the image to grayscale

gray_image = cv2.cvtColor(image, cv2.COLOR_BGR2GRAY)

# Detect faces in the image

faces = face_cascade.detectMultiScale(gray_image, scaleFac-
tor=1.1, minNeighbors=5, minSize=(30, 30))

# Process each detected face

for (x, y, w, h) in faces:

face_roi = gray_image[y:y + h, x:x + w]

face_roi = cv2.resize(face_roi, (48, 48))

face_roi = face_roi.astype("float") / 255.0

face_roi = np.expand_dims(face_roi, axis=0)

face_roi = np.expand_dims(face_roi, axis=-1)

# Predict the emotion

emotion_preds = emotion_model.predict(face_roi)[0]
```

```
emotion_label = emotion_labels[np.argmax(emotion_preds)]

# Draw a rectangle around the face and display the emotion
label

cv2.rectangle(image, (x, y), (x + w, y + h), (0, 255, 0), 2)

cv2.putText(image, emotion_label, (x, y - 10), cv2.FONT_HER-
SHEY_SIMPLEX, 0.8, (0, 255, 0), 2)
```

In this code, one need to replace "path/to/your/emotion_model.pb" with the actual path to pre-trained deep learning model for emotion recognition. The model should be trained to recognize emotions like Angry, Disgust, Fear, Happy, Neutral, Sad, and Surprise.

Figure 3. Prediction of Human Emotions using Deep Learning Model and Python

Figure 4. Prediction ii of Human Emotions using Deep Learning Model and Python

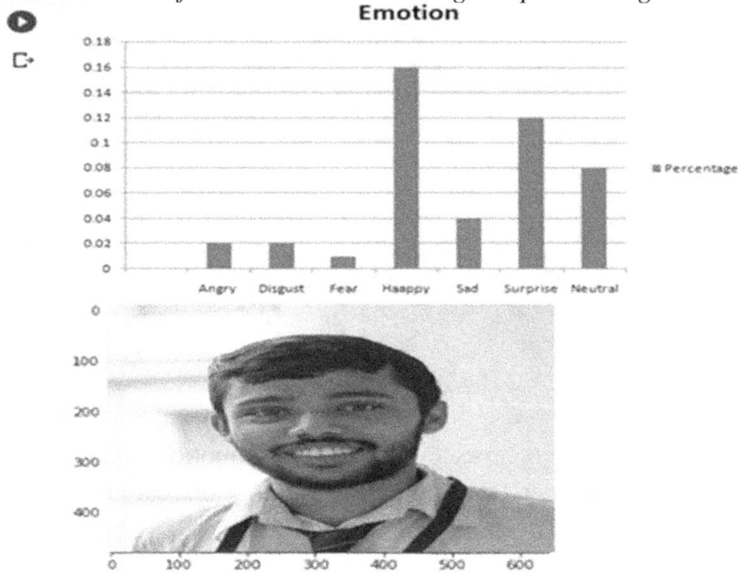

To maximize the positive impact of machine learning on human emotions, human-cantered design, ethical considerations, and inclusivity must be prioritized, involving collaboration between technologists, healthcare professionals, and users themselves.

The study reveals that AI-driven sentiment analysis significantly enhances the performance of customer relationship management (CRM) systems. By utilizing machine learning models such as LSTM, Random Forest, and SVM, businesses can accurately classify customer sentiments, allowing for more proactive customer service. The integration of sentiment analysis in CRM enables organizations to detect and respond to customer concerns in real time, thereby improving service quality and customer satisfaction. The ability to analyze textual data from emails, chatbots, and social media interactions provides businesses with valuable insights into customer preferences and behavior, leading to more personalized engagement strategies.

The application of machine learning models has demonstrated an improvement in the accuracy of emotion recognition. Deep learning architectures, including Bi-LSTM and CNN, have outperformed traditional sentiment classification approaches by capturing complex patterns in customer interactions.

The combination of hierarchical taxonomy systems and hybrid models has further refined sentiment detection, ensuring better identification of positive, neutral, and negative emotions. These advancements have allowed CRM systems to evolve beyond static data analysis, integrating real-time sentiment tracking to enhance customer communication.

The study also highlights how real-time AI interactions are transforming customer engagement. AI-powered chatbots and predictive analytics have enabled businesses to provide immediate responses, reducing resolution times and improving customer experiences. The integration of facial expression recognition into CRM dashboards offers an additional layer of emotional intelligence, allowing businesses to assess customer satisfaction during virtual interactions.

This innovation has made it possible to refine service strategies and ensure that customer concerns are addressed with greater accuracy and empathy. Despite these advancements, the implementation of AI-driven CRM is not without challenges. The study identifies contextual limitations in sentiment analysis, where AI models may misinterpret the emotions behind certain phrases due to linguistic ambiguities.

Ethical concerns related to customer data privacy also remain a key consideration, as businesses must ensure compliance with data protection regulations while leveraging AI for sentiment analysis. Addressing these challenges is essential for maximizing the effectiveness of AI in CRM and ensuring responsible AI adoption in customer interactions.

CONCLUSION

In summary, there are a lot of potential and obstacles in the subject of emotion recognition when it comes to teaching robots to understand and react to human emotions. Robust datasets and sophisticated machine learning algorithms are needed for this. Humans are naturally able to recognize and communicate their emotions (Zahra et al., 2009), however robots find it difficult to effectively understand these signs and must instead rely on statistical data and complex algorithms (Cohn et al., 1998). According to De Silva et al. (2000), the development of monitoring systems has made room for novel methods of identifying emotions, such as the combination of web-based resources with wearable biosensors. However, wearable technology is still not very useful for everyday tasks, which makes soft biosensors incorporated into online settings a viable substitute for inferring. Text-based emotion identification has several real-world uses, such enhancing query retrieval systems and identifying negative emotions in emails to protect worker welfare. Emotional keywords and their contextual location must be understood in order to accurately identify emotions in text. This may be done by using models such as the hourglass and Ekman's emotion categories (Kao et al., 2009). Artificial intelligence and computer vision developments have expanded automated systems' capacity to interpret text and facial expressions (Calvo et al., 2010). Such systems use difficult algorithms, as well as statistical techniques; thus, boosting the identification of emotions. Emotion recognition technologies, facial expression analysis, as well as text analysis provide new opportunities in increasing human-computer interaction and understanding the dynamics of emotions that engulf various communication channels. OpenCV and TensorFlow or PyTorch libraries allow demonstrating their practical usability in the case of facial expression detection and deep learning models, which affect social interaction and the field of artificial intelligence.

The findings of this study underscore the transformative role of artificial intelligence in customer relationship management by integrating sentiment analysis and emotion recognition techniques. Compared to the literature reviewed, which primarily focused on traditional CRM models, this research highlights how AI-driven solutions provide real-time customer sentiment tracking, automated response mechanisms, and enhanced customer engagement strategies. Studies such as those by Calvo and D'Mello (2010) and Kao et al. (2009) emphasized the significance of emotion detection in digital communication, but they largely relied on text-based models.

In contrast, this study demonstrates the integration of both text and facial expression recognition, providing a multimodal approach to sentiment analysis that significantly improves the accuracy and contextual understanding of customer emotions. Unlike earlier works that relied on keyword-based emotion classification, the present research incorporates deep learning architectures, such as Bi-LSTM and CNN, which significantly improve the precision and contextual depth of sentiment recognition. The literature reviewed suggested that machine learning models such as Naïve Bayes and SVM were widely used for text sentiment classification, but they struggled with capturing the nuanced emotional states of customers.

This study builds on those findings by showing how hybrid AI models, combining deep learning and traditional machine learning techniques, outperform earlier classification approaches by reducing misclassification errors and improving adaptability across various customer communication channels. Another key difference between this research and prior studies is the application of real-time AI-driven CRM interactions. While existing literature recognized AI's potential in automated customer interactions, it largely focused on rule-based chatbots and predefined responses (Venkatesh & Davis, 2000).

This study moves beyond those limitations by demonstrating how advanced AI chatbots with predictive analytics can personalize customer interactions, anticipate customer needs, and enhance service quality. The integration of facial expression recognition further refines CRM systems by allowing businesses to detect subtle emotional cues during video interactions, a feature that was previously unexplored in most CRM-related research.

The research also addresses critical challenges in AI-driven CRM, particularly concerning contextual misinterpretation and data privacy concerns. The literature highlighted issues with sentiment analysis models misinterpreting sarcastic or culturally specific expressions, which remain a challenge even with advanced AI techniques. This study corroborates those concerns but proposes context-aware deep learning models and hierarchical sentiment classification techniques as potential solutions to minimize these errors. Moreover, while earlier research raised concerns about customer data privacy in AI-CRM applications, this study provides a more nuanced discussion on ethical AI implementation, ensuring regulatory compliance, and improving AI transparency in CRM interactions.

In summary, this study expands upon previous research by introducing a comprehensive, multimodal AI-driven CRM framework that integrates sentiment analysis from text and facial expressions, deep learning techniques, real-time AI interactions, and predictive analytics.

By moving beyond traditional CRM approaches, this research demonstrates how AI can revolutionize customer engagement, optimize business decision-making, and enhance overall customer satisfaction. However, to maximize the potential of AI in CRM, future research should explore ways to refine contextual emotion analysis, address ethical concerns, and develop more advanced sentiment classification techniques to improve accuracy across diverse customer segments.

REFERENCES

Calvo, R. A., & D'Mello, S. (2010). Affect detection: An interdisciplinary review of models, methods, and their applications. *IEEE Transactions on Affective Computing, 1*(1), 18–37. DOI: 10.1109/T-AFFC.2010.1

Chui, M., Manyika, J., & Bughin, J. (2018). AI adoption in business: Strategies and challenges. *McKinsey Global Institute Report.*

Cohn, J. F., & Katz, G. S. (1998). Bimodal expression of emotion by face and voice. *Proceedings of the International Conference on Affective Computing and Intelligent Interaction.*https://doi.org/DOI: 10.1145/306668.306683

De Silva, L., & Ng, N. P. C. (2002). Bimodal emotion recognition. *Proceedings of the 4th IEEE International Conference on Automatic Face and Gesture Recognition,* 2000, 332–335. https://doi.org/DOI: 10.1109/AFGR.2000.840655

Kao, E. C. C., Liu, C. C., Yang, T. H., Hsieh, C. T., & Soo, V. W. (2009). Towards text-based emotion detection: A survey and possible improvements. *International Conference on Information Management and Engineering*, 70–74. https://doi.org/ DOI: 10.1109/ICIME.2009.113

Khalili, Z., & Moradi, M. H. (2009). Emotion recognition system using brain and peripheral signals: Using correlation dimension to improve the results of EEG. *Neural Networks, 2009. IJCNN 2009. International Joint Conference on.* IEEE.

Liyanage, C., & Foreman, J. (2000). Understanding emotions in digital communication: AI and emotion recognition. *Journal of Machine Learning and AI Applications, 5*(3), 145–158.

Russell, J. A. (2003). Core affect and the psychological construction of emotion. *Psychological Review, 110*(1), 145–172. DOI: 10.1037/0033-295X.110.1.145 PMID: 12529060

Venkatesh, V., & Davis, F. D. (2000). A theoretical extension of the technology acceptance model: Four longitudinal field studies. *Management Science, 46*(2), 186–204. DOI: 10.1287/mnsc.46.2.186.11926

Wiliams, R. L., & Hall, C. M. (2019). AI-driven customer relationship management: The role of deep learning and NLP. *Journal of Business Intelligence and Data Science, 7*(4), 112–128.

Zahra, K., & Moradi, M. H. (2009). Emotion recognition system using brain and peripheral signals: Using correlation dimension to improve the results of EEG. *Proceedings of the International Joint Conference on Neural Networks*, 1571–1575. https://doi.org/DOI: 10.1109/IJCNN.2009.5178723

Chapter 9
A Comparative Study on Machine Learning Algorithms Used for Sentiment Analysis

Anuja Kotnala
University of Petroleum and Energy Studies, India

Dhakshinesh Anandh
University of Petroleum and Energy Studies, India

Saurabh Rawat
Graphic Era University, India

Anushree Sah
https://orcid.org/0000-0003-3444-5860
University of Petroleum and Energy Studies, India

Abhirup Khanna
University of Petroleum and Energy Studies, India

ABSTRACT

Customer Relationship Management (CRM) is a set of practices that businesses use to man- age customer interactions and customers' perception towards their products. The aim is to maximize profits by achieving customer satisfaction. Analyzing online reviews provides insights into customer satisfaction, which is pivotal for refining CRM strategies. The main factors of customer satisfaction include the quality of product and the service offered by it, which can be measured using sentiment analysis.

DOI: 10.4018/979-8-3373-1867-7.ch009

Sentiment analysis has become remark- ably popular recently because of its wide range of applications in market capturing and improving customer experiences in various industries, such as e-commerce, restaurants, and products. The main aim of this research is to identify the class of sentiment of product reviews using a range of machine-learning algorithms. A few algorithms that can be used are SVM, Naive Bayes and LDA. Using this integrated algorithm model, the sentiment of recent product reviews can be predicted.

1. INTRODUCTION

1.1 What is sentiment analysis?

Sentiment analysis is the technique of interpreting text data which is used to determine the emotional tone of the message and classify it into positive, negative, or neutral. These days businesses have huge amounts of text data in the form of feedback emails, customer support chats, social media comments, and product reviews and ratings. Sentiment analysis tools are then used to scan this text to determine a customer's attitude towards a topic. Results from sentiment analysis are then used by businesses to boost customer service, improve brand reputation, deise effective marketing strategies, and enhance customer retention. It also helps businesses identify friction points to significantly eliminate them.

1.2 Machine learning techniques covered in this paper utilized for sentiment analysis include:

Naive Bayes: It is a classification algorithm used in machine learning based on Bayes theorem.

$$P(A|B) = P(Y|B).P(Y)\ P(B) \quad (1)$$

Advantages of Naive Bayes include:

1. Text classification: It is good at handling text data efficiently using probabilities of word occurrences(Sah et al., 2020).
2. Naive Bayes is considered as a baseline for natural language processing tasks.
3. Scalability and speed: Naive Bayes performs well when computational power and speed is to be utilized efficiently, hence making this model easy to deploy.

Support Vector Machine(SVM): SVM is a supervised machine learning algorithm that is used for classification by the help of a decision boundary (hyperplane). SVM is well-suited for sentiment analysis due to its ability to handle high-dimensional data(Saurabh Rawat & Kumar, 2020).

Advantages of SVM in Sentiment Analysis:

1. Effective in High-Dimensional Data: The vocabulary of text data is very large and can be handled effectively with SVM.
2. Nonlinearity: It can classify non-linearly separable sentiment data(Sah et al., 2018).
3. Eliminate Overfitting: It can perform analysis and provide good results with small data as well.

Logistic Regression: It is a supervised machine learning algorithm that is used for binary classification tasks. Provides proba- bility of likelihood that the input belongs to a particular class using logistic or sigmoid functions.

$$\sigma(z) = {}^{1}1 + e^{-z} \qquad (2)$$

Advantages of Logistic Regression in Sentiment Analysis:

1. Simplicity: it is easy to implement and also utilizes less computational power.
2. It performs good with linearly separable data.
3. Probabilistic output: probabilities can be used to implement confidence scores(-SAURABH Rawat & Sah, 2012).

Table 1. Comparison of SVM, Logistic Regression, Naive Bayes for Sentiment Analysis

Criteria	SVM - High	Logistic Regression - Low complexity.	Naive Bayes
Model Complexity	Computational complexity for large datasets due to quadratic scaling.	- Simple to train and interpret.	- Low complexity. - Fast to train and test. - Simple probabilistic model.
		- Works well with small to medium datasets.	
Performance Matrix	- Excellent for high-dimensional data like text (using TF-IDF or word embeddings).	- Good performance for linearly separable data.	- Performs well for linearly separable data but struggles with correlated features. - May assume independence of features, which is often violated in text.
		- Performs well with binary and multiclass sentiment classification.	
	- Can handle linearly and non-linearly separable data (with kernels).		
	- Sentiment analysis in large-scale datasets.	- Ideal for small-to-medium datasets where interpretability and speed are essential. - E.g., customer satisfaction surveys, feedback analysis.	- Effective for quick and straightforward sentiment analysis tasks with minimal resources. - E.g., spam detection, product reviews.
Real-World Applications	- Works well in cases with complex decision boundaries.		
	- E.g., social media sentiment tracking, reviews.		

2. LITERATURE REVIEW

In this paper(Ramachandran et al., 2022), they have compared different machine learning techniques like logistic re- gression, random forest and have provided a comparative study among them. Performance metrics like accuracy, f1-score, precision, recall suggests that Logistic regression performs the best.

This paper (Iboi et al., 2017)proposes three levels – document, sentence and phrase level by which Senti- ment analysis can be performed. Document level signifies whether the whole document is positive, negative or neutral. Similarly for sentence and phrase levels as well. Performance Metrics: ROGUE(Recall-Oriented Understudy for Gisting Evaluation)

This paper(Fadhilah et al., 2023) utilizes multinomial Logistic regression and gradient boosting models for ap- plying sentiment analysis on social media platform. This paper provides accuracy of 60% on logistic regression. In this paper(Fazrin et al., 2022), a comparative study has been employed between logistic regression and K nearest neighbor. Logistic regression provided the highest accuracy and performance among the both (Mathur et al., 2023). A comparative study(Zakaria & Sunyoto, 2023) has been conducted between Naıve Bayes and logistic regression on IMDB dataset. A combination of both Naive Bayes and Logistic Regression provided an accuracy score of 89%.

Sentimental Analysis is the study(Sharma et al., 2023) of the emotion of a user on a particular product through the review given by them. It searches through the sentences to find keywords that convey specific emotions. Support Vector Machine (SVM) is used to generate a detailed Sentimental Analysis report. SVM is helpful in categorizing the reviews collected from the various websites into predefined categories to assert the user's tone and states the emotion of the user as positive, negative or neutral. This process makes it possible for any business to understand its customers(Adhikary et al., 2024).

In this paper, two machine learning classifiers, SVM and Naive Bayes, have been em- ployed to perform sentiment analysis on three datasets of reviews provided by the con- sumers from various websites. The study (Burra & Mishra, 2024)demon- strates that SVM outperforms Naive Bayes in terms of F-score and TP accuracy. SVM continually outperforms other models for smaller datasets, like Woodland's wallet reviews. Elongated words with their correspon(Kuba et al., 2023)d- ing sen- timents, foreign words, and emoticons are all handled well by the preprocessing methods(Gulati et al., 2024).

SVM is a supervised learning algorithm that identifies hyperplanes which di- vide classes with maximum margin in feature space. It works on both linear and non-linear variants. The goal of this study(Jindal et al., 2023) is to evaluate how well the two algorithms predict sentiments from text reviews while specifying the advantages and performance results of both(Sah & Rawat, 2023). According to the paper's findings, the total accuracy for the sentiment classification task using the Logistic Regression model is 76.40%, while the Support Vector Machine model achieves 76.80%.

After comparing all accuracies(Changala et al., 2024), it was found that Naive Bayes performed best with an accuracy of 98%. After Naive Bayes, Random Forest and Logistic Regression performed best with accuracy of 97%. So, with observations based on accuracies these two came out to be best classifiers for finding accuracies for the given dataset and can be used for future classifications(Sah et al., 2023).

Table 2. Study Dataset

Research Objective	**Performance**		**Metrics**	
	Accuracy	**Precision**		**F1-Score**
[1] Analysis on Instagram data using Logistic Regression.	0.98	0.98	0.98	0.98
[3] social media-based Sentiment Classifier using Multinomial Logistic Regression	0.63	NA	NA	NA
[4] Sentiment Analysis of Covid-19 Vaccination on Twitter using Logistic Regression.	0.85	0.88	0.57	0.80

continued on following page

Table 2. Continued

Research Objective	Performance		Metrics	
	Accuracy	Precision		F1-Score
[5] Sentiment Analysis on IMDB data using Logistic Regression.	0.88	0.87	0.97	0.88
[5] Sentiment Analysis on IMDB data using Naive Bayes.	0.88	0.89	0.87	0.87
[6] CRM solution that uses machine learning classifiers to automate customer data analysis. (Logistic Regression)	0.97	0.97	0.97	0.97
[6] CRM solution that uses machine learning classifiers to automate customer data analysis. (Support Vector Machine)	0.90	0.90	0.90	0.90
[6] CRM solution that uses machine learning classifiers to automate customer data analysis. (Naive Bayes)	0.98	0.98	0.98	0.98
[7] Two prominent machine learning methods applied to sentiment analysis of restaurant reviews. (Logistic Regression)	0. 76	0.76	0.76	0.76
[7] Two prominent machine learning methods applied to sentiment analysis of restaurant reviews. (Support Vector Machine)	0.76	0.77	0.77	0.76
[8] Comparative study performed on Naive Bayes and SVM. (SVM)	0.72	NA	NA	NA
[8] Comparative study performed on Naive Bayes and SVM. (Naive Bayes)	0.52	NA	NA	NA
[9] Comparison of classifiers on Amazon data (Naive Bayes)	0.83	0.83	0.85	0.84
[9] Comparison of classifiers on Amazon data (SVM)	0.84	0.89	0.97	0.93

3. RESULTS

This study offers a comparative analysis of three machine learning models—Logistic Regression, Naive Bayes, and Support Vector Machine (SVM)—relying on multiple key performance metrics: Accuracy, Precision, Recall, and F1-score. The evaluation is derived from multiple sentiment classification studies to determine the most effective model.

3.1 Comparative Performance of Machine Learning Models

Table 3. Summary of the average performance metrics for each model

Average Performance Metrics	Accuracy	Precision	Recall	F1-Score
Logistic Reg.	0.862	0.925	0.8725	0.9075
Naive Bayes	0.8025	0.9	0.9	0.896667
SVM	0.805	0.853333	0.88	0.863333

Figure 1. A bar chart visually compares the performance metrics across all three models

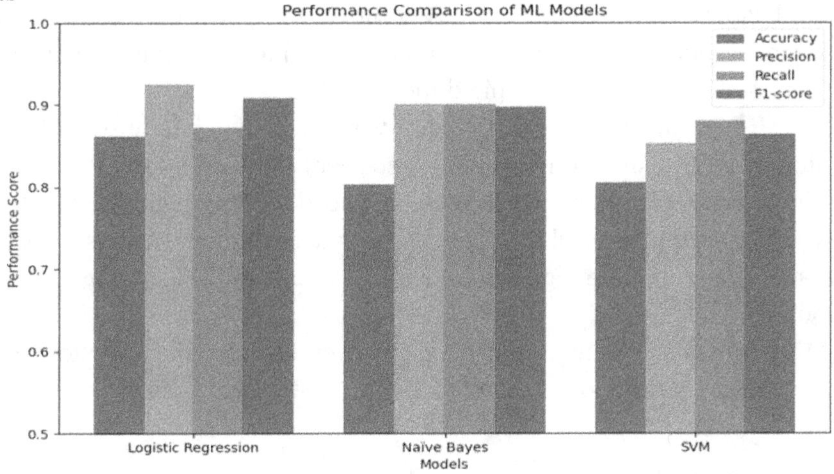

Figure 2. Radar Chart providing an intuitive view of how each model performs across multiple dimensions

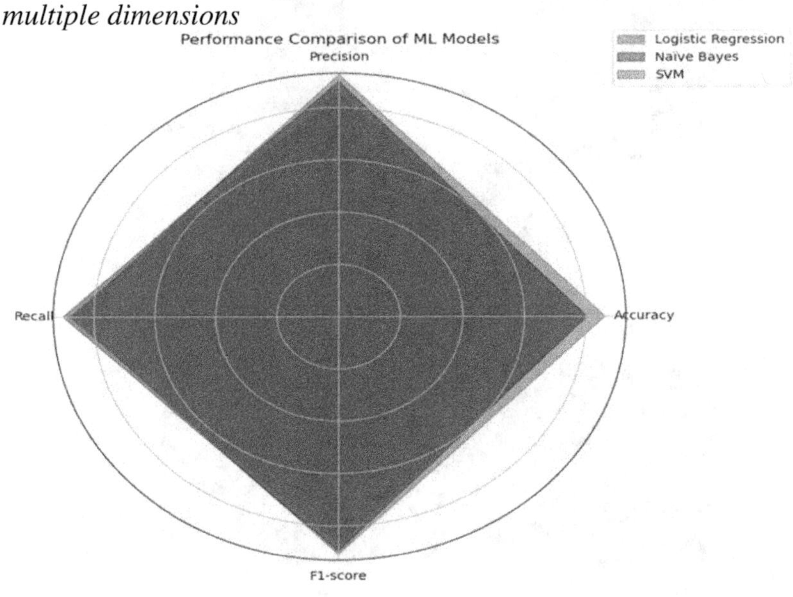

3.2 Key Observations

Figure 1 shows a bar chart that visually compares the performance metrics across all three models and figure 2 shows radar chart providing an intuitive view of how each model performs across multiple dimensions.

Logistic Regression truly performs best overall, achieving the absolutely highest Preci- sion of 92.5%, in addition to an F1-score of 90%, therefore, it represents a strong choice for applications where there is a need to minimize false positives.

Naive Bayes is quite useful for applications where finding true positives is very impor- tant because it has a higher score, with a Recall of 90%, despite having less precision.

SVM's performance remains balanced however, its Precision 85% along with its F1- score 86% are a little lower than those of Logistic Regression.

3.3 Correlation Analysis Using Heatmap

Figure 3. Correlation Heatmap

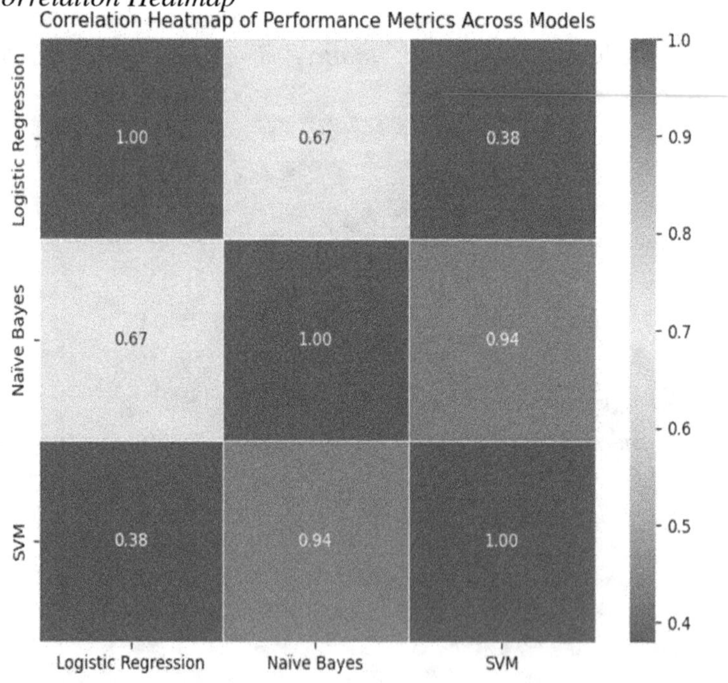

In Figure 3 a correlation heatmap was created to assess the relationships among all the performance metrics and models. The key understandings include: A strong correlation of 0.94 is dis- played by Na¨ıve Bayes and SVM, which indicates similar performance trends. A cor- relation of 0.67 exists between Logistic Regression and Na¨ıve Bayes, indicating definite shared performance characteristics. Large performance differences are present in Logistic Regression and SVM, and they show a correlation of 0.38.

4. CONCLUSION

This paper contains a summary of researches performed on sentimental analysis. Perfor- mances of various machine learning algorithms have been analyzed on different datasets. This analysis has been done on the basis of precision, recall, and F1-score which are con- sidered as the key metrices. Experimental results show that Logistic Regression delivers the highest precision and F1-score, making it the most preferred choice for situations where minimizing false positives is crucial. Naive Bayes proves efficient in situations prioritizing true positive identification due to its high recall. However, the performance of this model is dependent on the dataset and its characteristics. SVM, shows a balanced performance across all metrics, but it does not surpass Logistic Regression in precision and F1-score.

Future work will explore advanced techniques, including deep learning, to further enhance sentiment classification accuracy.

REFERENCES

Adhikary, A., Mantri, A., Sah, A., & Rawat, S. (2024). *An Extensive Review of Farming in Metaverse*. 184–203. DOI: 10.4018/979-8-3693-3824-7.ch007

Burra, H., & Mishra, P. (2024). Restaurant Reviews Sentimental Analysis Using Machine Learning Approach. *Proceedings of the 2024 International Conference on Emerging Techniques in Computational Intelligence, ICETCI 2024*, 414–417. DOI: 10.1109/ICETCI62771.2024.10704184

Changala, R., Lakshmi, H. N., Krishna, G. V., Lakshmi, H. B., Kiran, G. U., & Thatha, V. N. (2024). Sentiment analysis Optimization using Hybrid Machine Learning Techniques. *2024 Parul International Conference on Engineering and Technology, PICET 2024*, 1–5. DOI: 10.1109/PICET60765.2024.10716161

Fadhilah, R., Budiwati, S. D., Wijaya, D. R., Oktranida, P. R., Hijriana, Z. Q., & Firmansyah, A. (2023). Comparison of Bandung Social Media-based Sentiment Classifier using Multinomial Logistic Regression and Gradient Boosting Models. *2023 International Conference on Data Science and Its Applications, ICoDSA 2023*, 83–87. DOI: 10.1109/ICoDSA58501.2023.10276762

Fazrin, F., Pratiwi, O. N., & Andreswari, R. (2022). Comparison of K-Nearest Neighbor and Logistic Regression Algorithms on Sentiment Analysis of Covid-19 Vaccination on Twitter with Vader And Textblob Labeling. *2022 International Conference of Science and Information Technology in Smart Administration, ICSINTESA 2022*, 39–44. DOI: 10.1109/ICSINTESA56431.2022.10041609

Gulati, A., Taneja, A., Rawat, S., & Sah, A. (2024). Cancer Prediction Using Graph Database. In *Machine Learning Techniques and Industry Applications* (pp. 90–100). IGI Global. DOI: 10.4018/979-8-3693-5271-7.ch005

Iboi, H., Chua, S., Ranaivo-Malançon, B., & Kulathuramaiyer, N. (2017). Performance of opinion summarization towards extractive summarization. *Journal of Telecommunication, Electronic and Computer Engineering, 9*(2–10), 57–64.

Jindal, P., Parikh, S., Sikka, R., Alatba, S. R., Babu, S., & Sriramakrishnan, G. V. (2023). Analyzing the differences between SVM and Naive Bayes for Feature Extraction. *2023 3rd International Conference on Advance Computing and Innovative Technologies in Engineering, ICACITE 2023*, 775–778. DOI: 10.1109/ICACITE57410.2023.10183068

Kuba, A., Rastogi, B., Sah, A., & Rawat, S. (2023). Review on AI-Based Diagnosis of Parkinson's Disorders. In *AI and IoT-Based Technologies for Precision Medicine* (pp. 236–246). IGI Global. DOI: 10.4018/979-8-3693-0876-9.ch014

Mathur, A., Sah, A., & Rawat, S. (2023). Evolution of Cloud Computing With Blockchain and IoT. In *Blockchain Applications in Cryptocurrency for Technological Evolution* (pp. 14–32). IGI Global.

Ramachandran, A., Ashok, S., & Remya Nair, T. (2022). A Factual Sentiment Analysis on Instagram Data - A Comparative Study Using Machine Learning Algorithms. *Proceedings - 2022 Algorithms,Computing and Mathematics Conference, ACM 2022*, 1–5. DOI: 10.1109/ACM57404.2022.00009

Rawat, S., Sah, A., & Dumka, A. (2020). Direct-Indirect Link Matrix: A Black Box Testing Technique for Component-Based Software. *International Journal of Information Technology Project Management, 11*(4), 56–69. DOI: 10.4018/IJITPM.2020100105

Rawat, S. A. U. R. A. B. H., & Sah, A. (2012). An approach to Enhance the software and services of Health care centre. *IISTE, 3*, 126–137.

Sah, A., Choudhury, T., Rawat, S., & Tripathi, A. (2020). A Proposed Gene Selection Approach for Disease Detection. *Advances in Intelligent Systems and Computing, 1120*, 199–206. DOI: 10.1007/978-981-15-2449-3_16

Sah, A., Dumka, A., & Rawat, S. (2018). Web technology systems integration using SOA and web services. *Handbook of Research on Contemporary Perspectives on Web-Based Systems*, 24–45. DOI: 10.4018/978-1-5225-5384-7.ch002

Sah, A., Madaan, V., Tyagi, S., Singla, P., & Rawat, S. (2023). Health Chain. In *Blockchain Applications in Cryptocurrency for Technological Evolution* (pp. 160–172). IGI Global.

Sah, A., & Rawat, S. (2023). Diabetes Prediction Using Cloud Computing, Blockchain, and IoT. In *Supporting Technologies and the Impact of Blockchain on Organizations and Society* (pp. 15–27). IGI Global. DOI: 10.4018/978-1-6684-5747-4.ch002

Sharma, N., Tiwari, P., Tanwani, M., Pandey, A., & Patel, S. K. (2023). Customer Relationship Management (CRM) Automation Using Machine Learning Algorithm. *2023 2nd International Conference on Futuristic Technologies, INCOFT 2023*, 1–6. DOI: 10.1109/INCOFT60753.2023.10425786

Sunyoto, A. (2023, November). Hybrid sentiment analysis: Majority voting with multinomial naive Bayes and logistic regression on IMDB dataset. In 2023 6th International Conference on Information and Communications Technology (ICOIACT) (pp. 448-453). IEEE.

Chapter 10
Enhancing Customer Relationships With AI:
Building Empathy Through Emotion AI, Robotics AI, and Sentiment Analysis

Kirti
https://orcid.org/0000-0003-1042-8433
Chanakya National Law University, Patna, India

ABSTRACT

In contemporary Customer Relationship Management (CRM), empathy plays a pivotal role in cultivating trust and long-term engagement. This chapter, Building Empathy with AI: Enhancing Customer Relationships, explores how Emotion AI, Robotics AI, and Sentiment Analysis are reshaping CRM by enabling context-aware, emotionally responsive interactions. Through tools such as facial recognition, vocal tone analysis, and sentiment detection, firms can interpret emotional cues in real time and tailor responses that resonate with customer needs. The chapter critically examines the integration of these technologies, emphasizing how they enhance service personalization while preserving a human-centric ethos. Drawing on case studies across sectors, it evaluates the ethical boundaries of emotional automation and proposes a framework to responsibly align AI innovation with empathy, privacy, and trust—offering practical insights for scholars, practitioners, and policymakers alike.

DOI: 10.4018/979-8-3373-1867-7.ch010

1. INTRODUCTION

Empathy has always been at the heart of meaningful customer relationships. It is the ability to listen beyond words, to sense the unspoken, and to respond in a way that makes customers feel seen and valued. In today's increasingly digital world, however, businesses face a subtle yet significant challenge: how to preserve human warmth when so many interactions are mediated by technology. (Lehnert & Kuehnl, 2025).

Modern customer relationship management (CRM) systems have evolved from transactional tools into experience-driven platforms. As customer expectations rise, the ability to understand and respond to emotional cues at scale has become a defining edge. Technologies such as Emotion AI, Robotics AI, and Sentiment Analysis are not just reshaping operations; they are transforming how organisations connect with people. (Seyitoğlu & Ivanov, 2024).

Emotion AI, sometimes called affective computing, interprets emotional states using cues such as facial expressions, vocal tone, and language patterns (Narimisaei et al., 2024). When used thoughtfully, it allows businesses to engage with customers in real-time, offering responses that are not only efficient but also emotionally intelligent. Robotics AI, meanwhile, handles repetitive tasks and streamlines service processes, freeing human agents to focus on situations where empathy and discretion genuinely matter. By making sense of customer feedback across platforms, Sentiment Analysis helps organisations identify emotional trends and address concerns before they escalate. (Yoo et al., 2024).

Yet, while these tools hold great promise, they also come with a set of ethical and practical considerations. Can machines truly understand human feelings? Is it enough to simulate empathy, or do we risk reducing it to scripted reactions? Questions around data privacy, algorithmic fairness, and the authenticity of AI-driven responses are not merely technical. They are deeply human. (Shen et al., 2024a) .

This chapter examines how AI-driven technologies can be utilised to cultivate empathy in customer relationships without compromising the human qualities that make those relationships meaningful. Drawing on current research and practical case examples, this reflection examines the possibilities and limitations of using machines to build trust, deepen engagement, and support emotional understanding. If empathy is the soul of customer relationships, the goal is not to automate it but to support it in intelligent, respectful, and, above all, human ways.

2. DEFINING EMPATHY AS A CORNERSTONE OF CUSTOMER CONNECTION

Empathy, the ability to understand and respond to another's emotional state, is a strategic imperative in customer relationship management. It is characterised by the capacity to recognise, comprehend, and react to the emotions, thoughts, behaviours, and experiences of others (Murray et al., 2019), and comprises multiple dimensions, including emotional (affective), mental (cognitive), and compassionate understanding (Powell & Roberts, 2017). Empathetic engagement fosters trust, loyalty, and long-term value. It empowers service providers to address not just what customers say but how they feel, creating emotional safety and relational depth (Lehnert & Kuehnl, 2025).

In practice, empathy is demonstrated through anticipating customer needs, interpreting dissatisfaction before it is voiced, and offering context-sensitive solutions. Companies that lead with empathy report higher satisfaction, loyalty, and advocacy (Huang & Rust, 2024) . In this way, empathy is not just interpersonal. It's a vehicle for sustained business value. In conventional customer service environments, employees who demonstrate empathetic behaviour are more likely to enhance customer satisfaction with the brand (Markovic et al., 2018).

2.1. Emotional and Psychological Dimensions of Communication

Emotional intelligence, the ability to perceive, use, and regulate emotions, is foundational to this process (Goleman, 1995). According to Goleman (1995), in his book 'Emotional Intelligence: Why it Can Matter More than IQ?'; modern neuroscience reveals that emotional intelligence is often considered a "basic flair for living". It is rooted in neurological predispositions that shape how we regulate emotions, interpret others' feelings, and navigate relationships. This expanded understanding of intelligence places emotions at the core of adaptive behaviour. It includes the ability to manage emotional impulses, read subtle emotional cues, and maintain relational harmony under stress. Echoing Aristotle's wisdom, true emotional mastery is the capacity "to be angry with the right person, to the right degree, at the right time, for the right purpose, and in the right way."

These competencies are not merely psychological traits. Instead, they are essential tools for resilience, empathy, and decision-making in both personal and professional domains. Emotional intelligence has been shown to sustain valuable relationships, enhance workplace performance, and even protect long-term health. Research suggests that toxic emotional states can be as physiologically damaging as habitual smoking, while emotional regulation serves as a buffer for physical and mental well-being (Fischer, 2018). Empathetic communication operates on two levels: cognitive (understanding customer perspectives) and affective (emotional alignment) (Chakrabarti & Baron-Cohen, 2006).

While emotional intelligence has long been regarded as a uniquely human trait, recent advances in generative and agentic AI challenge this exclusivity by simulating empathetic interactions through pattern recognition, language modelling, and real-time sentiment analysis. According to a McKinsey study (2024), the future of customer care will be shaped by a hybrid model where AI handles transactional efficiency while humans provide the emotional depth and judgment required for complex or high-stakes interactions (Blackader et al., 2025).

This convergence illustrates how affective and cognitive empathy, once seen as the domain of emotionally intelligent individuals, is increasingly being augmented by AI systems trained to detect tone, intent, and sentiment. However, the nuance of human empathy remains irreplaceable in "moments that matter," where emotional intelligence is not only expressed but deeply felt. Thus, CRM is evolving into a landscape where artificial and human empathy coexist, each reinforcing the other to deliver emotionally resonant and context-aware customer service (Blackader et al., 2025). Together, these shape how human and AI agents address customer concerns, frustrations, or joys (Lynch, 2024).

For example, when a customer expresses disappointment over a delayed order, the ideal response acknowledges their emotional frustration before offering a resolution (Manthiou et al., 2020). This dual response strengthens the relationship and fosters brand loyalty, even during service failures. In high-emotion sectors like healthcare and finance, empathetic service delivery significantly shapes perceived service quality (Bahadur et al., 2018). Moreover, empathetic interactions have a ripple effect: they enhance employee morale and emotional resilience, making empathy a mutually reinforcing dynamic. Empathetic conduct by employees has a direct influence on customers' positive emotional responses, which, in turn, strengthens satisfaction with employee-customer relationships (Lee et al., 2011).

2.2. Challenges in Preserving Empathy in the Digital Age

As CRM becomes increasingly automated, preserving empathy becomes complex. Self-service technologies and AI chatbots, while efficient, often eliminate the subtle cues, tone, facial expressions, and gestures that inform empathetic human interaction (Shen et al., 2024a).

Digital empathy risks becoming performative when scripted responses substitute genuine concern. Automated apologies or templated replies may lack nuance and appear dismissive if they aren't calibrated to the customer's emotional context (Yun & Park, 2022). It has been further observed that anthropomorphised chatbots can negatively impact customer satisfaction when customers are in an angry emotional state (Crolic et al., 2022).

Additionally, emotional data, often reduced to sentiment scores or tags, can dehumanise customer experiences. (Ramaswamy & Ozcan, 2019). Empathy, when over-quantified, risks becoming hollow. It is not programmable; it must be practised with presence and attentiveness (Shumanov & Johnson, 2021). In today's screen-mediated service landscape, empathy is essential, not optional. It helps organisations "see" their customers, acknowledge their emotions, and respond with authenticity (Weißhaar & Huber, 2016).

2.3 Empathy in the Age of Digital Interfaces

Let's dive into this idea of empathy. We're looking at how digital spaces, especially those automated chatbots, shake up how we show and feel empathy these days. With customer service increasingly digitised, there's this real worry about losing that emotional connection. Those standard replies from chatbots? Sure, they're quick, but they often miss the mark regarding the subtleties and context that today's customers want. People are still hunting for speedy answers and real understanding and acknowledgement of their feelings when they're reaching out for help. It's like, can we get both the efficiency and the warmth (Blümel et al., 2023).

Customer-service dialogue is a sequential set of interdependent messages between customers and service agents. Social contacts and interactions can be examined at an atomic level, which examines the emotion of each message, and at a cumulative level, which looks at the overall degree and nature of emotion across the whole interaction. The closest proxy that the industry standard currently has for the quality of a service is the cumulated emotion score. However, that approach does not have the granularity needed to catch real-time emotional shifts and risks missing emotionally important moments (Yom-Tov et al., 2018).

In contrast, analysis of message-level emotions can facilitate real-time identification of the customer's emotional state at any point of an exchange, providing much more profound insights on critical touchpoints such as service failure and recovery. It could result in emotionally tone-deaf responses even though the outcome of a service is satisfactory. Genuine digital empathy will only be possible if systems are designed to discern emotional cues as they emerge rather than after the fact, ensuring the acknowledgement of users as they travel through the service touchpoints, even with automated agents. This is consistent with the growing emphasis on creating emotionally coherent responses within human-like dialogue systems (Wei et al., 2019). Digital empathy, therefore, is not about mimicking emotion artificially; rather, it is about enabling systems to recognise emotional contexts and respond in a way that preserves human dignity and trust.

The study emphasises that digital empathy should be built into the architecture of customer service technologies. Automated engagements should, therefore, occur not based on demographic context but situational elements of the customer's emotional state, as failure to do so can turn dismissive or refuse to accommodate escalation parameters that ultimately lower customer satisfaction (Wang et al., 2023). Ultimately, emotional intelligence in digital interfaces must be more than sentiment analysis: it must enable emotionally attuned, context-sensitive engagement at all interaction points.

2.4 Barriers to Sustaining Empathy in Digital CRM

Despite its recognised importance, empathy in digital CRM is often hindered by systemic challenges (Ciarli et al., 2021). Over-standardisation through scripted responses limits agents' ability to adapt emotionally in real time. Volume pressure, driven by speed and efficiency targets, compresses the opportunity for meaningful, empathetic engagement. Additionally, limited emotional feedback mechanisms reduce the system's ability to learn and recalibrate responses. At the same time, misinterpretation risks remain high due to a lack of contextual cues, causing even well-intentioned replies to come across as impersonal or tone-deaf.

Though central to effective CRM and user-centred design, empathy is constrained by operational and systemic barriers. In traditional and digital contexts, time constraints, protocol-driven communication, and the need to balance professional boundaries with emotional attentiveness have been identified as key obstacles to empathy (Derksen et al., 2016). Within digital CRM systems specifically, communication gaps and resource limitations further restrict the ability of designers and service agents to engage empathically with users (Grech et al., 2023). Overcoming these challenges requires both structural and technological strategies. Emerging technologies such as Virtual Reality (VR) provide innovative tools for enhancing empathy among designers by immersing them in simulated customer environments and emotional states (X. Hu et al., 2021). This immersive capability helps designers understand users' perspectives better, narrowing the empathy gap between creators and users. For example, Grech et al. (2023) developed a fully immersive VR environment that allows designers to experience scenarios from multiple user viewpoints, fostering cognitive and affective empathy. Similarly, Hu et al. (2021) explored VR-based empathic design approaches, showing how designers can be virtually placed in users' environments to gain deeper insight into their challenges, emotions, and needs. These studies affirm VR's potential to cultivate empathic engagement in the design process and improve user-centred outcomes in CRM.

2.5 Empathy as a Differentiator in Digital Commerce

Digital platforms are working hard to weave empathy into how they operate. It's all about connecting with customers better and keeping them happy. Netflix, for example, personalises what customers watch by picking up on emotional cues from how they use the service. Likewise, big e-commerce players like Zalando and Amazon are exploring new tech that can sense emotions, especially when customers are going through the return process. Which is often super frustrating or regretful (Dogadkina, 2023). A global survey by leading research company and consultancy firm PwC in 2022 showed that a whopping 59% of people think that companies have lost that "human touch." So, adding a bit of digital empathy could set businesses apart in today's marketplace and can act as a critical differentiator (Puthiyamadam & Reyes, 2022).

Recent research further underscores the importance of emotional and empathic elements in e-commerce strategy. It emphasises the necessity of balancing AI automation and human engagement to maintain emotional connection and customer trust (Jakkula, 2023). Emotional content, particularly in user-generated reviews, has also influenced consumer perceptions and behaviours. When online reviews include emotional expression and are perceived as empathetic, they significantly enhance perceptions of product quality and positively impact purchase intentions by reducing information asymmetry (Ye & Motoki, 2024).

When shopping online, many choices are influenced by emotions rather than purely rational considerations. Researchers have explored various tech tools that can help support deeper personalisation and emotional resonance in online retail experiences. This way, they can create a more personalised shopping experience that really connects on an emotional level. These findings show that when businesses put empathy at the heart of their digital commerce, it's not just about improving things for the customer. It also leads to higher satisfaction and loyalty, a massive win in today's digital marketplace. So, embedding that kind of understanding can give companies an edge over leg up over their competitors (Bielozorov et al., 2019).

2.6 Integrating Empathy into CRM Strategy

It is vital to design empathy at every touchpoint of the CRM systems (Sinansari et al., 2023), from the design of interfaces, training of the service personnel, model calibration for the AI-based systems, feedback mechanisms, etc. Empathic design addresses needs and not just grievances (De Keyser et al., 2020). This is where AI-powered tools enter the frame because they now assist businesses in identifying signs of frustration, such as several requests for assistance or frequent clicks on return policies (Suri et al., 2020). These triggers can kick off some empathetic actions, such as proactive customer service outreach or FAQs specially tailored to their needs. It's all about making customers feel genuinely "being heard." (Shin et al., 2017). Starbucks, for example, has done a remarkable job using AI in their mobile app. The app picks up on signals of frustration or irritation, such as a customer having trouble with payments. Starbucks has made a name for itself in the cutthroat world of global coffee shops by using big data and AI. They're not just throwing out casual offers; they're digging deep to understand their customers better and personalise those experiences by harnessing data from its expansive customer base, as it's all about keeping people engaged and coming back for more.

2.6.1 Data-Driven Personalization

Starbucks gets a lot out of its mobile app and loyalty program, which are the primary channels for customer data collection. They collect information, from what customers buy to what they like and even their locations, to customise their marketing strategies. They send out tailored promotions and suggest products that fit their customers' unique tastes and preferences. This personalisation doesn't just make the experience better for customers, but it also builds loyalty and helps boost sales.

2.6.2 AI-Powered Decision Making

For Starbucks, it's not just about marketing; it uses AI algorithms for different operational decisions, such as where to put new stores and how to manage their inventory. By looking at who lives in an area, how much traffic there is, and what customers are up to, AI really helps determine the best spots for new locations. Plus, it ensures each store has the right products to match people's expectations. This innovative use of AI helps things run smoothly and ensures they're offering the products preferred by its customers in the area.

2.6.3 Impact on Customer Engagement

Starbucks has impressively embraced big data and AI. They've shifted gears from reacting to what customers want to be ahead of the game, as they can now even figure out what the customer might need before ordering it. For Starbucks, it's not just about serving coffee; it's about creating tailored experiences that feel personal, even when they're happening on a large scale. This strategy makes Starbucks stand out in a crowded market. It helps build an emotional bond between customers and brands. It is a perfect example of how smartly using big data and AI can completely change how businesses manage their CRM.

By prioritising personalisation and leveraging technology to anticipate and meet customer needs, Starbucks has set a benchmark in the industry for enhancing customer relationships through innovative technological integration. My Starbucks Barista, through the Starbucks mobile app, allows you to place an order through voice command or messaging to a virtual barista using artificial intelligence algorithms behind the scenes. Since there are so many nuances to an individual order, it's quite an accomplishment for an artificial intelligence engine to provide a seamless customer experience. When identified, the app triggers a helpful message or flags the interaction to a live barista who can offer support at the next visit, bridging digital and physical empathy (Bernard, 2018).

2.7 Measuring the ROI of Empathy

Empathy is now a measurable KPI. Emotional metrics such as trust index, churn signals, and sentiment variation complement traditional CRM analytics. Far from being a "soft" or intangible metric, empathy has measurable outcomes. According to a study by Deloitte (2023), brands rated high on empathy outperformed competitors in customer loyalty by 30%, customer lifetime value by 27%, and Net Promoter Scores (NPS) by 40% (Delloitte, 2023). This evidence shows that weaving empathy into CRM strategies can increase emotional satisfaction and outstanding business results. Today, businesses are not just looking at the usual KPIs but also keeping an eye on emotional metrics, like customer sentiment, trust levels, and how often complaints are escalated.

3. THE ROLE OF EMOTION AI IN BUILDING EMPATHY

Emotional AI technology (also called affective computing) is a subdomain of artificial intelligence that can detect distress signals through vocal tone, language patterns, facial expressions, and even physiological signals like heart rate. These technologies are becoming more popular as tools to bring empathy into the workplace. Organisations can become more efficient and boost productivity by using them (N. Thomas & Docherty, 2025). Emotion AI opens new avenues for crafting emotionally driven customer experiences. It gives machines the ability to pick up on how people are feeling. In the domain of CRM, where building trust is super important, this technology allows companies to extend empathy in ways that go beyond just what one human can offer to another. It has enabled machines to bridge that gap, making interactions feel more personal and understanding, even if they're not human.

3.1 Core Technologies Behind Emotion AI

Artificial Emotional Intelligence (AEI) is a specialised subset of AI designed to recognise, interpret, and respond to human emotions. It enables machines to simulate emotional awareness in interaction scenarios (Erol et al., 2020). These systems rely on a convergence of multiple technologies. Together, these tools construct an emotional profile that allows organisations to adjust real-time interactions.

Figure 1. Emotion Detection Engines[1]

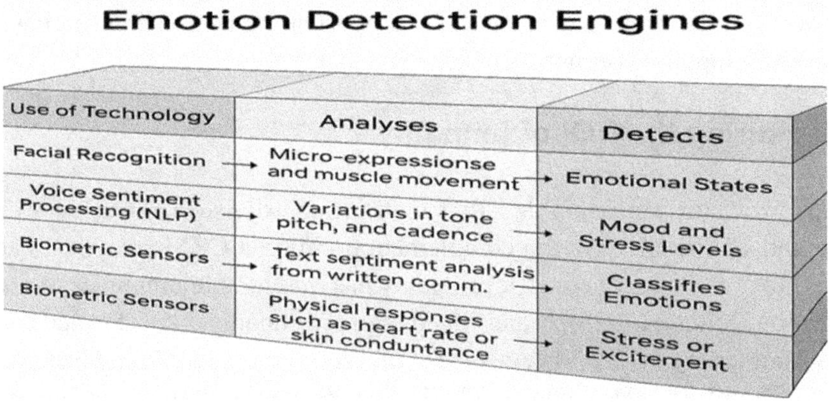

3.2 Real-World Applications in Customer Service and Support

Emotion AI is rapidly being integrated into contact centres and virtual support environments. Empirical research suggests that customer experience in chatbot-assisted service environments is influenced by a triad of factor categories: chatbot-related, user-related, and context-specific dimensions (Følstad & Brandtzæg, 2017; Gnewuch et al., 2017) . Among these, chatbot-related elements are often subdivided into functional capabilities (e.g., the ability to execute tasks accurately), system performance (e.g., interface responsiveness and reliability), and anthropomorphic features that enhance perceived human likeness and relational cues (Chaves & Gerosa, 2021). These diverse factors interact to shape customer perceptions and emotional responses, which may range from satisfaction and trust to frustration or scepticism, depending on the quality of the interaction (Sidaoui et al., 2020).

The outcomes of these experiences are typically reflected in user behaviour, either through increased engagement with the chatbot such as continued usage and receptiveness to automated recommendations or through behavioural intentions toward the brand, including purchase decisions and brand advocacy (Gkinko &

Elbanna, 2023). Studies consistently identify response relevance and effective problem resolution as the most influential factors contributing to positive customer evaluations (Chung et al., 2020).

These factors are directly associated with increased customer satisfaction, which subsequently enhances the likelihood of repeat usage, product purchases, and recommendations to others(Araujo, 2021). According to a 2023 report by Gartner, 40% of global customer service operations have already incorporated some form of emotion recognition to enhance interaction quality (O'Brien, 2023).

Case Study: Cogito in Healthcare Call Centres

Cogito, a U.S.-based firm, uses real-time voice analysis to provide emotional intelligence support in call centres. Their software tracks vocal patterns for signals of agitation or disengagement. It then offers live prompts to agents, such as "slow down" or "show empathy," resulting in increased customer satisfaction and reduced call escalations (Leonard, 2023). Aetna Health, which integrated Cogito, reported a 14% improvement in agent empathy ratings within six months.

Case Study: BMW's Emotion-Aware Interface

BMW has implemented Emotion AI in its in-car assistant to detect driver moods using facial scanning and voice tone. Based on detected stress or fatigue, the assistant can adjust lighting, suggest breaks, or offer music. This emotionally aware experience fosters brand loyalty and enhances customer interaction even outside the showroom (Schmelzer, 2019).

3.3 Emotion AI in Digital Marketing and Personalization

Emotion AI is also transforming digital marketing by enabling real-time emotional targeting. For instance, Unilever used facial analysis to evaluate consumer reactions to different advertisements before releasing them. By measuring micro-emotional responses, the company could tailor campaigns to evoke the desired emotional impact and improve engagement (Barnett, 2018).

Amazon's Alexa and Google Assistant have begun integrating voice sentiment detection to adapt interactions based on user tone (D'Cruze, 2025). A frustrated tone may trigger a calming response or alternative suggestions, reflecting AI's potential to deliver personalised, empathetic responses (Guo et al., 2025; Tuhin, 2025) .

3.4 Enhancing Employee-Customer Empathy Through Emotion AI

Emotion AI not only assists in customer interaction but also augments employee training. Walmart, for example, has adopted virtual reality (VR) simulations combined with emotion detection to train employees in empathy. The VR environment places employees in various customer conflict scenarios, while emotion sensors provide feedback on their emotional responses and decision-making patterns (Heikkilä & Douglas, 2022). This approach has resulted in enhanced employee confidence and improved customer satisfaction scores.

Table 1. Industry-Specific Use Cases of Emotion AI in CRM

Industry	Application	Outcome
Healthcare	Voice analysis in patient calls	Improved empathy and trust (Leonard, 2023)
Automotive	Driver emotion detection	Enhanced safety and user experience (Schmelzer, 2019)
Retail	In-store mood tracking via facial recognition	Personalised shopping suggestions (D'Cruze, 2025)
Hospitality	Emotion-aware concierge services	Higher guest satisfaction ratings (Tuhin, 2025)

Source: Compiled from Research Studies

3.5 Limitations and Calibration Challenges

Despite its promise, Emotion AI faces significant challenges. Emotions are inherently subjective and can be influenced by culture, context, and individual variation. Misinterpretation risks are amplified when systems are not trained on diverse datasets. For example, facial emotion recognition tools have been shown to underperform on non-white faces due to biased training sets (Buolamwini & Gebru, 2018).

Furthermore, detecting emotion does not guarantee understanding. Affective responses must be followed by meaningful action to avoid the risk of "performative empathy", interactions that seem caring but are emotionally shallow (Shen et al., 2024a).

3.6 Ethical Considerations in Emotion AI Deployment

The deployment of Emotion AI raises ethical questions around consent, privacy, and transparency. Do customers know their emotions are being analysed? Have they consented to this data collection? Is there a clear opt-out option?

The European Union's AI Act has proposed strict regulations on biometric emotion recognition, categorising it as a high-risk AI system (EU AI Act, 2023). Companies operating in these regions must ensure compliance through clear disclosures, data anonymisation, and fair algorithmic practices. As part of its broader digital strategy, the European Union has prioritized the regulation of artificial intelligence (AI) to foster conditions that support both innovation and responsible use. Recognizing

AI's transformative potential across multiple domains, including healthcare, transportation, energy, and manufacturing. The EU aims to harness these benefits while safeguarding fundamental rights and public trust.

In April 2021, the European Commission introduced a landmark proposal for the Artificial Intelligence Act, marking the world's first attempt to develop a comprehensive legal framework for AI governance. The proposal adopts a risk-based classification model, in which AI systems are categorized based on the level of risk they present to users and society. These categories range from minimal risk to unacceptable risk, with corresponding regulatory obligations. High-risk AI systems, for instance, are subject to stricter compliance requirements, including transparency, accountability, and human oversight measures. This tiered approach ensures that the intensity of regulation aligns with the potential impact of the AI application (*EU AI Act*, 2023).

4. ROBOTICS AI AND HUMAN-CENTRIC AUTOMATION

Robotics AI, a branch of artificial intelligence focused on autonomous service capabilities, has become a transformative force in customer relationship management (Selesi-Aina et al., 2024). While often associated with physical robots, in the CRM context, Robotics AI includes both tangible service robots and intelligent software agents; designed to interact, assist, and support customers while maintaining a human-centred approach (Xiao & Kumar, 2021). These systems enable businesses to improve efficiency, reduce human workload, and sustain emotionally resonant interactions even in high-volume environments (Wirtz et al., 2023).

4.1. From Chatbots to Conversational AI Agents

Early automation in CRM began with basic rule-based chatbots. These systems could answer frequently asked questions but lacked contextual understanding. Modern Robotics AI, however, leverages advancements in machine learning, NLP, and contextual memory to evolve into **conversational agents** capable of holding nuanced, multi-turn dialogues(Adamopoulou & Moussiades, 2020).

Case Example: IKEA's AI Assistant

IKEA uses a conversational AI named Billie to handle customer inquiries about returns, deliveries, and product information. Unlike scripted bots, Billie can understand intent, redirect complex queries to human agents, and even escalate emotional responses to supervisor attention. This blend of automation and empathy demonstrates how Robotics AI supports Human-AI synergy (Marr, 2024).

4.2. Service Robots in Physical Retail and Hospitality

In addition to software-based agents, physical robots are being deployed in customer-facing environments. Service robots enhance engagement by offering consistent, multilingual, and nonjudgmental support (Mahr et al., 2024).

Case Example: Pepper by SoftBank Robotics

Pepper, a humanoid robot developed by SoftBank, has been deployed in banks, airports, and retail stores across Asia and Europe (Guizzo, 2014). Equipped with emotion detection and natural conversation capabilities, Pepper helps customers check in, locate products, and access services. In Japan's Mizuho Bank, Pepper assisted customers during the COVID-19 pandemic to reduce human contact while maintaining service quality (Mizuho Bank, 2024).

Case Example: Hilton's "Connie" Concierge Robot

Hilton collaborated with IBM Watson to develop "Connie," a robot concierge who provides guests with hotel information, dining suggestions, and tourist tips (Volpicelli, 2016). Connie learns from interactions and improves over time, acting as an engaging supplement, not a replacement for human staff.

4.3. Human-AI Collaboration Models in CRM

Robotics AI should not be viewed as a substitute for human agents but as a co-worker that extends human capabilities. The most effective CRM systems follow a collaborative AI model, where tasks are divided based on complexity and emotional need. Recent studies underscore the transformative potential of human-AI collaboration across diverse sectors. There is enough research evidence which suggests that hybrid teams combine human intuition with algorithmic precision and they can outperform both human-only and AI-only systems in special contexts, one of them being customer service (Molina et al., 2024). In the healthcare sector, AI-enabled tools are increasingly being adopted to support clinical documentation and integrity specialists. Apart from functioning as valuable although imperfect, still as assistants (Reddy et al., 2019).

Mutual understanding and collaboration can be achieved between human operators and AI agents, with the help of a key enabler, which is by establishment of Shared Mental Models (SMMs) (Andrews et al., 2023). As in the aviation sector, AI systems are capable of enhancing human decision-making by processing large volumes of data, delivering real-time insights (Lopes et al., 2024), and managing routine or repetitive tasks (Khattak et al., 2023). Although, the effectiveness of such collaboration depends on key design factors, including flexible system integration, support for human sensemaking, and the preservation of human oversight and control (Abduljabbar et al., 2019).

Despite its many advantages, integrating AI into collaborative workflows also introduces challenges. Ethical concerns, system limitations, and unforeseen risks must be proactively addressed to ensure safe, efficient, and responsible operation (Molina et al., 2024). Moving forward, research should prioritise advancements in transparency, explainability, and real-world testing to fully realise the promise of human-AI collaboration (Mennella et al., 2024).

Table 2. Human-AI Role Division in Customer Relationship Management

Task Type	Ideal Agent	Rationale
Routine FAQs and Order Inquiries	AI (Conversational Agent)	Efficient handling of repetitive tasks with quick, scalable responses (Adamopoulou & Moussiades, 2020)
Product Returns and Delivery Questions	AI with Escalation Capability	AI handles common queries while complex or emotional issues are escalated to human agents (Marr, 2024).
Emotional Customer Complaints	Human-Agent	Requires empathy, active listening, and nuanced judgment not yet fully replicable by AI (Wirtz et al., 2023).
Hotel Check-ins and Concierge Assistance	Service Robot (AI)	Provides 24/7 support in multiple languages, enhancing convenience and consistency (Volpicelli, 2016).

Source: Adapted from Research Studies

4.4. Benefits of Human-Centric Automation

A human-centric automation strategy maximizes the strengths of both AI and people. Human-centric automation leverages the strengths of both AI systems and human judgment to optimize organizational performance. Digital transformation (DX), when aligned with user-centred design and employee engagement, enhances workflows, decision-making, and overall operational efficiency. Tools designed with usability in mind reduce cognitive overload and promote better data-driven insights. The Human-Oriented Process Enhancement (H.O.P.E.) model exemplifies this approach by integrating digital maturity with human factors. Applied within a litigation management project at an automotive manufacturing plant, the model demonstrated improved compliance, risk mitigation, and streamlined legal processes. As a result, the Digital Maturity and Human Factors (DMHF) Index was developed to align digital adoption strategies with organisational goals while maintaining a strong focus on human needs (Chrusciak et al., 2025). The study by Chrusciak et al. (2025) suggests the benefits such as scalability (AI handles volume; humans handle complexity), 24/7 Support (Robotic-AI ensures uninterrupted customer engagement), Data-Driven Empathy (AI identifies stress patterns), humans respond personally, Learning Loops (AI enhances process quality). Further, collaboration improves decision-making and elevates organisational performance. improve an organisation's competitiveness and social sustainability, which is earned from human decisions to improve responses over time (Williamson & Prybutok, 2024).

Case Example: Bank of America's "Erica"

Bank of America's virtual financial assistant "Erica" uses voice and text to help customers manage transactions, monitor accounts, and learn about financial health. With over 1 billion interactions logged by 2023, Erica represents a scalable form of empathic service. However, customers can seamlessly switch to human support when nuance is needed, ensuring the emotional depth of service is preserved (Cocheo, 2025).

4.5. Design Principles for Empathic Robotics AI

Robotics AI must be designed with principles of ethical, cultural, and emotional intelligence (Barajas et al., 2025). The significant design considerations include:

Table 3. Principle for Empathic Robotics AI (Source: Research Papers)

Sl. No	Principle	Target
1.	**Transparency**	Disclose when customers are interacting with AI.
2.	**Fallback to Humans**	Always allow escalation to a human agent.
3.	**Tone Sensitivity**	Adjust language and responses based on emotional cues.
4.	**Cultural Localization**:	Use localized language models and etiquette rules.

Researchers argue that a successful robotic CRM experience requires not only **functional design** but also **emotional acceptability**, a place where customers feel acknowledged, respected, and supported (Cohen & Aulbach, 2024).

5. SENTIMENT ANALYSIS FOR EMOTIONAL INSIGHTS

In the age of digital-first communication, organisations are inundated with customer feedback through social media, reviews, emails, and chat interactions. **Sentiment analysis**, a key application of natural language processing (NLP), allows businesses to extract emotional tone, intentions, and satisfaction levels from this unstructured data. By classifying text as positive, negative, or neutral, sentiment analysis enables organizations to uncover actionable insights, personalize experiences, and refine marketing strategies with empathy. This capability plays a crucial role in Emotion-AI systems, where accurately interpreting affective states enhances customer engagement and supports emotionally intelligent automation (Abdi et al., 2019)

5.1. What Is Sentiment Analysis?

Sentiment analysis (SA), also known as opinion mining/analysis, is a computational study of people's opinions, attitudes, and emotions which are expressed in text, and are considered an important sub-domain of Natural Language Processing (NLP) (Liu & Zhang, 2012). During the last ten years, massive growth has been observed in internet usage and has led to a surge in user-generated content such as reviews, comments, and social media posts. The understanding of public sentiment expressed in this digital feedback system has gained significance in sustaining social awareness, informed decisions, and even social stability (Birjali et al., 2021).

SA enables systems to process and respond to human emotions at a large scale and thus plays a significant role in advancing artificial intelligence (Chaturvedi et al., 2018). It categorizes sentiment into fundamental divisions such as positive, neutral, and negative, but more nuanced emotional categories like joy, trust, fear, sadness, disgust, surprise, anticipation, and anger are also increasingly being incorporated (Bose et al., 2020). This allows for a deeper emotional understanding of textual data.

SA is not limited to a single language. It has been successfully applied across multiple semantic contexts, including English (Rodríguez-Ibánez et al., 2023), Chinese (Peng et al., 2017), Arabic (Oueslati et al., 2020), German (Remus et al., 2010), Portuguese (Pereira, 2021), Russian (Smetanin, 2020), and Spanish (Angel et al., 2021), highlighting its global relevance and adaptability (Mao et al., 2024).

5.2 Tools and Techniques

Modern sentiment analysis employs:

- **Lexicon-based methods**: Compare words to pre-defined sentiment dictionaries (e.g., SentiWordNet) (Ojeda-Hernández et al., 2023).
- **Machine learning models**: Supervised algorithms like SVM, logistic regression, or Naïve Bayes (Shedthi B & Shetty, 2024).
- **Deep learning models**: Use recurrent neural networks (RNNs), convolutional neural networks (CNNs), and transformer models in sentiment analysis tasks incorporating long short-term memory (LSTM) and transformers like BERT to detect contextual sentiment (Sahoo et al., 2023).

The strengths and limitations of the techniques are given in the Table No. 4

Table 4. Comparing Sentiment Analysis Techniques in CRM

Method	Strengths	Limitations	Example Use Case
Lexicon-based	Simple, fast, easy to implement	Limited context, rigid	A quick scan of the FAQ tone
ML (e.g., SVM)	Learns from labelled data	Needs retraining, lacks nuance	Tagging emails and chats
Deep Learning (BERT)	Captures sarcasm, context	Requires large data & compute	Social media listening, ad testing

Source: The Cases (mentioned below in section 5.2 & 5.3) and referred Research Studies (given in section 5.1)

Real-World Platform Example: Sprinklr

Sprinklr's sentiment engine processes social and customer feedback from 30+ channels. It detects spikes in negativity and flags service teams for proactive intervention used by brands like Samsung, Microsoft, and McDonald's (Amazon, 2021).

5.3 Use of Sentiment Analysis in CRM and Customer Support

Sentiment analysis is now embedded in leading CRM platforms such as Salesforce, Zoho, and HubSpot. These systems tag messages based on emotional tone and urgency. For instance:

- **Urgent + Negative** = escalated directly to human support.
- **Positive sentiment** = logged for loyalty rewards or testimonial requests.

Case Example: Zappos

Zappos integrates sentiment detection in their email responses. Emails with a high emotional tone (positive or negative) are prioritized. This approach allows service agents to react quickly, resolve conflicts early, and send appreciation where it's due (Amazon Q., 2020).

5.4 Marketing, Product Insights & Innovation

Sentiment analysis also plays a crucial role in brand management and product development. Companies monitor real-time reactions to product launches, ad campaigns, or service failures.

Case Study: Netflix

Netflix uses sentiment mining on Twitter and Reddit to analyze user reactions to show releases (J. Thomas et al., 2024). For example, after the release of *Stranger Things Season 4*, spikes in positive sentiment around certain characters were used to inform merchandise campaigns (Pahalyants, 2021).

5.5 From Insights to Action

Merely detecting sentiment is insufficient. Effective CRM systems turn insights into actions:

- Flagging frustration before it becomes churn.
- Rewarding promoters to encourage brand advocacy.
- Feeding insights into product design, UX, and pricing strategies.

6. ETHICAL IMPLICATIONS AND CHALLENGES IN EMPATHIC AI FOR CRM

As organizations increasingly deploy AI technologies that simulate empathy, such as Emotion AI, Robotics AI, and sentiment analysis, they face complex ethical considerations. While these systems promise personalized and emotionally aware customer engagement, they also raise questions of **consent, transparency, bias, privacy**, and **manipulation**. Without careful governance, empathic AI risks crossing the line from connection to coercion (Maiti et al., 2025).

6.1. Privacy and Consent

Emotion AI technologies often rely on biometric data, such as facial expressions, tone of voice, or even heart rate raising significant privacy concerns. Customers may not be fully aware that their emotional states are being recorded or analysed (Ashok et al., 2022). Hence, customers should receive clear notice and give informed consent before their emotional data is captured.

Case Example: Walmart's Facial Emotion Detection Pilot

In 2022, Walmart tested cameras equipped with facial analysis to detect frustration at self-checkout counters. Despite claims of anonymized data use, a backlash followed over the lack of clear customer consent. The program was quietly paused amid criticism about transparency (Anderson, 2017).

Ethical Guideline: Algorithmic Bias and Discrimination

Emotion AI systems trained on non-diverse datasets may misread or misclassify emotions, particularly across gender, race, and culture (de Fine Licht & de Fine Licht, 2020).

Research Study (in the above context):

Buolamwini and Gebru (2018) found that facial recognition systems misclassified Black women's expressions at rates over 34%, while white male expressions were misclassified less than 1% of the time. Such disparities can lead to **biased service quality**, with marginalized groups potentially experiencing more rejections or escalations in customer service.

Solution: Organizations must audit AI models regularly and train them on **culturally diverse, representative datasets**.

6.2. Manipulation vs. Empathy

When AI detects emotions, it can be used to support and **influence** customer behaviour (Purdy et al., 2019). When companies optimize interactions to drive sales through emotional vulnerability, this becomes ethically murky (Shen et al., 2024b).

Case Example: Cambridge Analytica Scandal

Though not CRM-specific, this case demonstrates how emotion-mined data (via Facebook likes) was used to tailor political messages based on user psychological profiles raising concerns over emotional manipulation (M. Hu, 2020).

6.3. Regulatory Landscape

Governments are beginning to define legal boundaries for emotional and biometric AI.

- **EU AI Act (2023 Draft):** This act classifies emotion recognition systems as high-risk and requires risk assessment, transparency, and human oversight (*EU AI Act*, 2023).
- **California Privacy Rights Act (CPRA):** This act expands consumer rights over **sensitive personal data**, including inferred emotions ("California Privacy Rights Act, 2020 (CPRA)," 2020) .
- **OECD AI Principles:** Advocate for fairness, accountability, and human-centric AI development (OECD, 2019).

Table 5. Emerging Regulations Impacting Empathic AI in CRM

Region	Regulation	Implication for CRM AI
European Union	EU AI Act (2023 draft)	Mandatory disclosures, risk classification
United States	CPRA	Consent for inferred emotional/biometric data
Global	OECD AI Principles	Emphasis on transparency and human oversight

6.4. Organizational Ethics and Accountability

Beyond legal compliance, businesses must define internal policies for **responsible AI governance**. This includes:

- Creating AI ethics committees
- Establishing red lines (e.g., "never manipulate emotion for upselling")
- Training staff on ethical tech usage

Example: Microsoft's Responsible AI Standard

Microsoft implemented an internal framework that requires product teams to assess ethical risks, document mitigation steps, and apply fairness guidelines. This framework serves as a model for operationalizing AI ethics (Microsoft, 2023).

7. CASE STUDIES AND BEST PRACTICES IN EMPATHIC AI-DRIVEN CRM

While theory and models are essential, the true impact of Emotion AI, Robotics AI, and Sentiment Analysis in CRM is best understood through real-world application. This section presents five case studies across industries, illustrating how leading organizations use empathic AI to strengthen customer relationships, improve retention, and scale emotional intelligence.

7.1. Zappos: Prioritizing Emotional Signals in Service Workflows

Zappos, renowned for its customer service culture, integrates sentiment detection into its email and chat workflows. The system identifies messages with heightened emotional tones whether joyful or frustrated and automatically prioritizes them in the customer support queue.

- **Technology Used**: NLP-based sentiment scoring and emotional tagging
- **Results**: 20% reduction in complaint escalation time, 15% increase in customer satisfaction (CSAT) for emotional cases, and enhanced brand perception for being "attentive" and "understanding"

This illustrates a key best practice: emotion-aware routing, where digital empathy supports operational efficiency (Forbes, 2023).

7.2. Starbucks: Behaviour-Based Empathy in the Mobile App

Starbucks implemented behaviour-sensitive features in its mobile app using AI. If a customer experiences repeated failed transactions or app crashes, the system detects frustration through behavioural cues (e.g., rapid retry attempts) and:

- Triggers an apology message or offers a coupon
- Alerts the barista during the next store visit to offer support
- Suggests app assistance via chatbot with a softer tone
- **Technology Used**: Predictive analytics, behavioural signal recognition, the mobile feedback loop
- **Result**: App churn rate dropped by 12%,, Loyalty program engagement increased by 18% among frustrated users

This shows how emotion prediction can bridge digital and in-store empathy (Wiley Business Review, 2022).

7.3. BMW: In-Car Emotion Recognition for Safer Driving

BMW's in-car assistant detects driver emotions using facial recognition and vocal tone analysis. If stress, fatigue, or frustration is detected:

- The car suggests playing relaxing music
- Adjusts lighting and cabin temperature
- Recommends a break if on a long drive
- **Technology Used**: Real-time facial analytics, acoustic emotion detection
- **Result**: Driver satisfaction was boosted by 22%, and brand association with safety and care strengthened (Forbes Auto, 2023). BMW's use of AI highlights how empathy can be embedded in non-traditional CRM touchpoints.

7.4. Bank of America: Erica as an Empathic Financial Assistant

"Erica," Bank of America's voice- and text-enabled virtual assistant, provides emotional support in financial contexts where stress is common.

- Erica uses tone and text sentiment to escalate concerns (e.g., overdraft anxiety)
- Proactively recommends credit education tips if distress signals are detected
- Escalates complex emotional queries to human bankers
- **Tech Stack**: Sentiment analysis, intent detection, emotional context mining
- **Results**: 92% of users felt Erica "understood their financial needs", 1 billion+ customer interactions handled since 2023

This case demonstrates how **empathetic automation can enhance trust** in sensitive domains.

7.5. Cogito: Real-Time Empathy Coaching in Call Centres

Cogito's software, adopted by Aetna and Humana, uses real-time voice analytics to coach agents during live calls. It flags:

- Impatience or disengagement in tone
- Recommends behaviours like slowing speech, showing more empathy
- Tracks improvement over time
- **Benefits**: Empathy rating improved by 14%, reduced call durations by 12%, and an 18% increase in first-call resolution (Cogito, 2023).

In this way, Cogito exemplifies how AI can augment not replace human empathy.

Table 6. Summary of Best Practices from Case Studies

Company	Empathic Feature	Outcome
Zappos	Emotion-prioritized ticket routing	Faster resolution, improved CSAT
Starbucks	Behaviour-based frustration detection	Higher app retention, loyalty boost
BMW	In-car emotion adaptation	Safer driving, stronger brand attachment
Bank of America	Sentiment-aware financial guidance	High trust in digital assistant
Cogito	Real-time empathy coaching	Enhanced service quality and efficiency

Sources[2]

8. FUTURE DIRECTIONS: THE EVOLUTION OF EMPATHIC AI IN CRM

As firms respond to intensifying margin pressures, accelerated strategy cycles, and rising customer expectations, AI technologies, amplified by data analytics are reshaping how brands interact with their audiences. These advancements not only optimize operational efficiency but also unlock new dimensions of emotional resonance and personalized engagement, laying the groundwork for deeper, more durable customer-brand relationships (Evans & Ghafourifar, 2019). As artificial intelligence continues to evolve, so too does its potential to reshape the landscape of customer relationship management. Empathic AI technologies such as Emotion AI, Robotics AI, and Sentiment Analysis are still in their early stages. Still, the direction is clear: the future belongs to emotionally intelligent machines that can build trust, personalize experiences, and ethically connect with customers on a human level.

8.1. From Reaction to Anticipation: Predictive Empathy

Future CRM platforms will not only detect how customers feel but also predict how they **will feel**. With advances in **predictive analytics** and behavioural modelling, systems will be able to anticipate frustration, confusion, or delight before it manifests in explicit feedback.

- **Example:** A customer who browses refund policies and FAQs in quick succession may trigger proactive live chat support preventing churn before it happens.

This evolution from reactive to anticipatory empathy will become a competitive advantage.

8.2. Multimodal Emotion Recognition

Next-generation Emotion AI will combine multiple data channels voice, text, facial expressions, and physiological data to build a richer, more accurate emotional profile.

- **Emerging Tools:** Affectiva's in-vehicle sentiment platform uses in-cabin cameras and voice analytics to assess stress and fatigue.
- **Application:** In omnichannel CRM, multimodal emotion tracking will ensure seamless emotional continuity across phone, email, and in-person touchpoints (Affectiva, 2023).

8.3. Rise of Affective Robotics

The integration of tactile sensors, expressive avatars, and socially aware robotics will make AI systems more emotionally expressive and socially intelligent.

- **Forecast:** By 2030, retail and healthcare sectors are expected to see significant deployment of **effective robots** for customer interaction and support (McKinsey, 2024).
- **Potential Use:** Robots that read customer body language in physical stores to offer relevant help or direct human staff to intervene.

8.4. Embedded Ethics and Algorithmic Audits

As regulation tightens and customer awareness grows, ethical transparency will become a **core feature**, not an afterthought.

- **Future systems will include:**
    - o Real-time **AI explainability dashboards** for agents and customers
    - o Built-in **bias detection mechanisms**
    - o AI behaviour logs for compliance audits

These features will make AI more trustworthy, especially in sectors like banking, insurance, and public services.

8.5. Strategic Recommendations

To prepare for the future of empathic AI in CRM, organizations should:

- Invest in **cross-functional AI teams** (tech + psychology + ethics)
- Design with **"human fallback" options** in all AI systems
- Conduct regular **bias audits and emotional impact assessments**
- Promote **AI literacy** among staff and customers to demystify emotional tech
- Embrace **global inclusivity** in emotion models training AI on data from diverse geographies and cultures

CONCLUSION

In a world where digital interactions increasingly define customer experiences, empathy is no longer a human-only trait it is a strategic capability powered by artificial intelligence. This chapter has explored how Emotion AI, Robotics AI, and Sentiment Analysis are transforming CRM by enabling emotionally intelligent engagement at scale. These technologies offer various benefits to organisations, from understanding customer emotions to proactively responding to distress signals. These technologies offer organizations a new frontier in building trust, loyalty, and personalized value. Nonetheless, every opportunity comes with some responsibilities, and so does E-AI. Empathic AI must be designed and utilised with ethical prevision, cultural understanding, and a clear dedication to human dignity. The future of CRM does not lie in replacing human empathy but in scaling it up and using technology as a bridge, not a barrier (Singh, 2024).

REFERENCES

Abdi, A., Shamsuddin, S. M., Hasan, S., & Piran, J. (2019). Deep learning-based sentiment classification of evaluative text based on Multi-feature fusion. *Information Processing & Management, 56*(4), 1245–1259. DOI: 10.1016/j.ipm.2019.02.018

Abduljabbar, R., Dia, H., Liyanage, S., & Bagloee, S. A. (2019). Applications of Artificial Intelligence in Transport: An Overview. *Sustainability (Basel), 11*(1), 1. Advance online publication. DOI: 10.3390/su11010189

Adamopoulou, E., & Moussiades, L. (2020). Chatbots: History, technology, and applications. *Machine Learning with Applications, 2*, 100006. DOI: 10.1016/j.mlwa.2020.100006

Amazon, Q. (2020). *Zappos Case Study* [Research Reports]. Amazon Web Services, Inc. https://aws.amazon.com/solutions/case-studies/zappos-case-study/

Amazon. (2021). *Sprinklr Case Study | AWS Inferentia | AWS* [Research & Case Study]. Amazon Web Services, Inc. https://aws.amazon.com/solutions/case-studies/sprinklr-case-study-inf1/

Anderson, G. (2017, July 27). *Walmart's Facial Recognition Tech Would Overstep Boundaries*. Forbes. https://www.forbes.com/sites/retailwire/2017/07/27/walmarts-facial-recognition-tech-would-overstep-boundaries/

Andrews, R. W., Lilly, J., Srivastava, D., & Feigh, K. M. (2023). The role of shared mental models in human-AI teams: A theoretical review. *Theoretical Issues in Ergonomics Science, 24*(2), 129–175. DOI: 10.1080/1463922X.2022.2061080

Angel, S. O., Negrón, A. P. P., & Espinoza-Valdez, A. (2021). Systematic literature review of sentiment analysis in the Spanish language. *Data Technologies and Applications, 55*(4), 461–479. DOI: 10.1108/DTA-09-2020-0200

Araujo, A. L. (2021). *Museums and Atlantic Slavery* (p. 122). *Scopus*. Advance online publication. DOI: 10.4324/9781003080121

Ashok, M., Madan, R., Joha, A., & Sivarajah, U. (2022). Ethical framework for Artificial Intelligence and Digital technologies. *International Journal of Information Management, 62*, 102433. DOI: 10.1016/j.ijinfomgt.2021.102433

Bahadur, W., Aziz, S., & Zulfiqar, S. (2018). Effect of employee empathy on customer satisfaction and loyalty during employee–customer interactions: The mediating role of customer affective commitment and perceived service quality. *Cogent Business & Management, 5*(1), 1491780. DOI: 10.1080/23311975.2018.1491780

Barajas, N. C., Rouchitsas, A., & Broo, D. G. (2025). Examining Human–Robot Interactions: Design Guidelines for Trust and Acceptance. In Malakhatka, E., & Wiberg, M. (Eds.), *Human-Technology Interaction: Interdisciplinary Approaches and Perspectives* (pp. 117–137). Springer Nature Switzerland., DOI: 10.1007/978-3-031-78357-9_4

Barnett, M. (2018, March 7). *Unilever trials in-store facial recognition technology*. Marketing Week. https://www.marketingweek.com/unilever-in-store-facial-recognition/

Bernard, M. (2018, May 28). Starbucks: Using Big Data, Analytics And Artificial Intelligence To Boost Performance. *Forbes.Com*. https://www.forbes.com/sites/bernardmarr/2018/05/28/starbucks-using-big-data-analytics-and-artificial-intelligence-to-boost-performance/

Bielozorov, A., Bezbradica, M., & Helfert, M. (2019). The Role of User Emotions for Content Personalization in e-Commerce: Literature Review. In Nah, F. F.-H., & Siau, K. (Eds.), *HCI in Business, Government and Organizations. eCommerce and Consumer Behavior* (pp. 177–193). Springer International Publishing., DOI: 10.1007/978-3-030-22335-9_12

Birjali, M., Kasri, M., & Beni-Hssane, A. (2021). A comprehensive survey on sentiment analysis: Approaches, challenges and trends. *Knowledge-Based Systems, 226*, 107134. DOI: 10.1016/j.knosys.2021.107134

Blackader, B., Buesing, E., Amar, J., Raabe, J., Mehndiratta, M., & Gupta, V. (2025, March 19). The right mix of humans and AI in contact centers | McKinsey. *McKinsey & Company*. https://www.mckinsey.com/capabilities/operations/our-insights/the-contact-center-crossroads-finding-the-right-mix-of-humans-and-ai

Blümel, J. H., Zaki, M., & Bohné, T. (2023). Personal touch in digital customer service: A conceptual framework of relational personalization for conversational AI. *Journal of Service Theory and Practice, 34*(1), 33–65. DOI: 10.1108/JSTP-03-2023-0098

Bose, R., Dey, R. K., Roy, S., & Sarddar, D. (2020). Sentiment Analysis on Online Product Reviews. In Tuba, M., Akashe, S., & Joshi, A. (Eds.), *Information and Communication Technology for Sustainable Development* (Vol. 933, pp. 559–569). Springer Singapore., DOI: 10.1007/978-981-13-7166-0_56

Buolamwini, J., & Gebru, T. (2018). Proceedings of Machine Learning Research. *Conference on Fairness, Accountability and Transparency, 81*, 77–91.

California, P. R. A. 2020 (CPRA). (2020). *Consumer Privacy Act*. https://www.consumerprivacyact.com/california-privacy-act-2020-cpra/

Chakrabarti, B., & Baron-Cohen, S. (2006). Empathizing: Neurocognitive developmental mechanisms and individual differences. *Progress in Brain Research*, *156*, 403–417. DOI: 10.1016/S0079-6123(06)56022-4 PMID: 17015093

Chaturvedi, I., Cambria, E., Welsch, R. E., & Herrera, F. (2018). Distinguishing between facts and opinions for sentiment analysis: Survey and challenges. *Information Fusion*, *44*, 65–77. DOI: 10.1016/j.inffus.2017.12.006

Chaves, A. P., & Gerosa, M. A. (2021). How Should My Chatbot Interact? A Survey on Social Characteristics in Human–Chatbot Interaction Design. *International Journal of Human–Computer Interaction*. https://www.tandfonline.com/doi/abs/10.1080/10447318.2020.1841438

Chrusciak, C. B., Szejka, A. L., & Canciglieri, O.Junior. (2025). Integrating digital transformation with human-centric factors strategies to enhance organisational process performance: The H.O.P.E. model. *Journal of Industrial Information Integration*, *44*, 100785. DOI: 10.1016/j.jii.2025.100785

Chung, M., Ko, E., Joung, H., & Kim, S. J. (2020). Chatbot e-service and customer satisfaction regarding luxury brands. *Journal of Business Research*, *117*, 587–595. DOI: 10.1016/j.jbusres.2018.10.004

Ciarli, T., Kenney, M., Massini, S., & Piscitello, L. (2021). Digital technologies, innovation, and skills: Emerging trajectories and challenges. *Research Policy*, *50*(7), 104289. DOI: 10.1016/j.respol.2021.104289

Cocheo, S. (2025, March 12). You Won't Find GenAI Under The Hood of BofA's Erica... Yet. *The Financial Brand*. https://thefinancialbrand.com/news/customer-experience-banking/hy-bank-of-americas-erica-virtual-assistant-has-a-human-touch-187354

Cohen, H., & Aulbach, L. (2024). Beyond artificial intelligence ethics: Exploring empathetic ethical outcomes for artificial intelligence. In Caballé, S., Casas-Roma, J., & Conesa, J. (Eds.), *Ethics in Online AI-based Systems* (pp. 279–295). Academic Press., DOI: 10.1016/B978-0-443-18851-0.00017-2

Crolic, C., Thomaz, F., Hadi, R., & Stephen, A. T. (2022). Blame the Bot: Anthropomorphism and Anger in Customer–Chatbot Interactions. *Journal of Marketing*, *86*(1), 132–148. DOI: 10.1177/00222429211045687

D'Cruze, D. (2025, February 27). Amazon unveils Alexa+: Voice assistant gets AI boost to take on Google Assistant, Apple's Siri. *Business Today*. https://www.businesstoday.in/technology/news/story/amazon-unveils-alexa-voice-assistant-gets-big-ai-boost-to-take-on-google-assistant-apples-siri-466043-2025-02-27

de Fine Licht, K., & de Fine Licht, J. (2020). Artificial intelligence, transparency, and public decision-making. *AI & Society*, *35*(4), 917–926. DOI: 10.1007/s00146-020-00960-w

De Keyser, A., Verleye, K., Lemon, K. N., Keiningham, T. L., & Klaus, P. (2020). Moving the Customer Experience Field Forward: Introducing the Touchpoints, Context, Qualities (TCQ) Nomenclature. *Journal of Service Research*, *23*(4), 433–455. DOI: 10.1177/1094670520928390

Delloitte. (2023). *Humanizing the Customer Experience: Empathy in the Age of AI* (p. 20). Deloitte Insights. https://www.deloittedigital.com/content/dam/digital/global/documents/insights-20240624-cx-roundtable.pdf

Derksen, F. A., Olde Hartman, T. C., Bensing, J. M., & Lagro-Janssen, A. L. (2016). Managing barriers to empathy in the clinical encounter: A qualitative interview study with GPs. *The British Journal of General Practice*, *66*(653), e887–e895. DOI: 10.3399/bjgp16X687565 PMID: 27884917

Dogadkina, O. (2023, October 16). How AI-Powered Personalization Is Reshaping Online Shopping And Beyond. *Forbes*. https://www.forbes.com/councils/forbestechcouncil/2023/10/16/how-ai-powered-personalization-is-reshaping-online-shopping-and-beyond/

Erol, B. A., Majumdar, A., Benavidez, P., Rad, P., Choo, K.-K. R., & Jamshidi, M. (2020). Toward Artificial Emotional Intelligence for Cooperative Social Human–Machine Interaction. *IEEE Transactions on Computational Social Systems*, *7*(1), 234–246. DOI: 10.1109/TCSS.2019.2922593

EU AI Act: First regulation on artificial intelligence. (2023, June 8). Topics | European Parliament. https://www.europarl.europa.eu/topics/en/article/20230601STO93804/eu-ai-act-first-regulation-on-artificial-intelligence

Evans, M., & Ghafourifar, A. (2019). *Build A 5-star customer experience with artificial intelligence*. https://www.forbes.com/sites/allbusiness/2019/02/17/customer-experience-artificial-%20intelligence/#1a30ebd415bd

Fischer, A. H. (2018). Comment: The Emotional Basis of Toxic Affect. *Emotion Review*, *10*(1), 57–58. DOI: 10.1177/1754073917719327 PMID: 30034516

Følstad, A., & Brandtzæg, P. B. (2017). Chatbots and the new world of HCI. *Interactions (New York, N.Y.)*, *24*(4), 38–42. DOI: 10.1145/3085558

Gkinko, L., & Elbanna, A. (2023). Designing trust: The formation of employees' trust in conversational AI in the digital workplace. *Journal of Business Research*, *158*, 113707. DOI: 10.1016/j.jbusres.2023.113707

Gnewuch, U., Morana, S., & Maedche, A. (2017). Towards Designing Cooperative and Social Conversational Agents for Customer Service. *International Conference on Information Systems, 2017*. https://www.researchgate.net/publication/320015931_Towards_Designing_Cooperative_and_Social_Conversational_Agents_for_Customer_Service

Goleman, D. (1995). *Emotional intelligence: Why it can matter more than IQ* (8. print). Bantam Books.

Grech, A., Wodehouse, A., & Brisco, R. (2023). Designer Empathy in Virtual Reality: Transforming the Designer Experience closer to the User. *IASDR Conference Series*. https://dl.designresearchsociety.org/iasdr/iasdr2023/fullpapers/119

Guizzo, E. (2014, December 26). *How Aldebaran Robotics Built Its Friendly Humanoid Robot, Pepper—IEEE Spectrum*. https://spectrum.ieee.org/how-aldebaran-robotics-built-its-friendly-humanoid-robot-pepper

Guo, Y., Xu, L., & Wang, C. (2025). Exploring the effect of empathic response and its boundaries in artificial intelligence service recovery. *Journal of Retailing and Consumer Services*, *82*, 104065. DOI: 10.1016/j.jretconser.2024.104065

Heikkilä, M., & Douglas, W. H. (2022, December 23). What's next for AI. *MIT Technology Review*. https://www.technologyreview.com/2022/12/23/1065852/whats-next-for-ai/

Hu, M. (2020). Cambridge Analytica's black box. *Big Data & Society*, *7*(2), 2053951720938091. DOI: 10.1177/2053951720938091

Hu, X., Nanjappan, V., & Georgiev, G. V. (2021). SEEING FROM THE USERS' EYES: AN OUTLOOK TO VIRTUAL-REALITY BASED EMPATHIC DESIGN RESEARCH. *Proceedings of the Design Society*, *1*, 2601–2610. DOI: 10.1017/pds.2021.521

Huang, M.-H., & Rust, R. T. (2024). The Caring Machine: Feeling AI for Customer Care. *Journal of Marketing*, *88*(5), 1–23. DOI: 10.1177/00222429231224748

Jakkula, A. R. (2023). The Human Element in AI: Balancing Automation with Empathy in E-Commerce. *Journal of Artificial Intelligence & Cloud Computing*, *2*(4), 1–2. DOI: 10.47363/JAICC/2023(2)351

Khattak, A., Chan, P.-W., Chen, F., & Peng, H. (2023). Time-Series Prediction of Intense Wind Shear Using Machine Learning Algorithms: A Case Study of Hong Kong International Airport. *Atmosphere (Basel)*, *14*(2), 2. Advance online publication. DOI: 10.3390/atmos14020268

Lee, S., Comer, L. B., Dubinsky, A. J., & Schafer, K. (2011). The Role of Emotion in the Relationship Between Customers and Automobile Salespeople. *Journal of Managerial Issues*, *23*(2), 206–226. https://www.jstor.org/stable/23209226

Lehnert, L., & Kuehnl, C. (2025). Empathy at the heart of customer experience: A holistic framework for understanding and enhancing consumer empathy through the lens of customer experience. *Psychology and Marketing*, *42*(2), 332–358. DOI: 10.1002/mar.22130

Leonard, M. (2023, August 9). *Customer Advisory Board Recap: Discussing the Future of AI for CX & EX*. Cogito. https://cogitocorp.com/blog/the-future-of-ai-for-ex-and-cx/

Liu, B., & Zhang, L. (2012). A Survey of Opinion Mining and Sentiment Analysis. In Aggarwal, C. C., & Zhai, C. (Eds.), *Mining Text Data* (pp. 415–463). Springer US., DOI: 10.1007/978-1-4614-3223-4_13

Lopes, N. M., Aparicio, M., & Neves, F. T. (2024). *Challenges and Prospects of Artificial Intelligence in Aviation: Bibliometric Study*. Data Science and Management., DOI: 10.1016/j.dsm.2024.11.001

Lynch, S. (2024, December 23). *Predictions for AI in 2025: Collaborative Agents, AI Skepticism, and New Risks | Stanford HAI*. https://hai.stanford.edu/news/predictions-for-ai-in-2025-collaborative-agents-ai-skepticism-and-new-risks

Mahr, D., Odekerken-Schröder, G., & Steins, M. (2024). Service robots and innovation: An ecosystem approach. *Journal of Product Innovation Management*, jpim.12756. Advance online publication. DOI: 10.1111/jpim.12756

Maiti, M., Kayal, P., & Vujko, A. (2025). A study on ethical implications of artificial intelligence adoption in business: Challenges and best practices. *Future Business Journal*, *11*(1), 34. DOI: 10.1186/s43093-025-00462-5

Manthiou, A., Hickman, E., & Klaus, P. (2020). Beyond good and bad: Challenging the suggested role of emotions in customer experience (CX) research. *Journal of Retailing and Consumer Services*, *57*, 102218. DOI: 10.1016/j.jretconser.2020.102218

Mao, Y., Liu, Q., & Zhang, Y. (2024). Sentiment analysis methods, applications, and challenges: A systematic literature review. *Journal of King Saud University. Computer and Information Sciences, 36*(4), 102048. DOI: 10.1016/j.jksuci.2024.102048

Markovic, S., Iglesias, O., Singh, J. J., & Sierra, V. (2018). How does the Perceived Ethicality of Corporate Services Brands Influence Loyalty and Positive Word-of-Mouth? Analyzing the Roles of Empathy, Affective Commitment, and Perceived Quality. *Journal of Business Ethics, 148*(4), 721–740. DOI: 10.1007/s10551-015-2985-6

Marr, B. (2024, April 4). *The Amazing Ways IKEA Is Using Generative AI.* Forbes. https://www.forbes.com/sites/bernardmarr/2024/04/05/the-amazing-ways-ikea-is-using-generative-ai/

Mennella, C., Maniscalco, U., De Pietro, G., & Esposito, M. (2024). Ethical and regulatory challenges of AI technologies in healthcare: A narrative review. *Heliyon, 10*(4), e26297. DOI: 10.1016/j.heliyon.2024.e26297 PMID: 38384518

Microsoft. (2023). *Responsible AI Principles and Approach | Microsoft AI.* https://www.microsoft.com/en-us/ai/principles-and-approach

Mizuho Bank. (2024). *Behind the Scenes during the COVID–19 Pande... | Mizuho Bank.* https://www.mizuhogroup.com/bank/what-we-do/lessons

Molina, D. A., Kharlov, V., & Chen, J.-S. (2024). Towards Effective Human-AI Collaboration in Decision-Making: A Comprehensive Review and Conceptual Framework. *2024 Portland International Conference on Management of Engineering and Technology (PICMET)*, 1–6. DOI: 10.23919/PICMET64035.2024.10653303

Murray, J., Elms, J., & Curran, M. (2019). Examining empathy and responsiveness in a high-service context. *International Journal of Retail &amp. International Journal of Retail & Distribution Management, 47*(12), 1364–1378. DOI: 10.1108/IJRDM-01-2019-0016

Narimisaei, J., Naeim, M., Imannezhad, S., Samian, P., & Sobhani, M. (2024). Exploring emotional intelligence in artificial intelligence systems: A comprehensive analysis of emotion recognition and response mechanisms. *Annals of Medicine and Surgery (London), 86*(8), 4657–4663. DOI: 10.1097/MS9.0000000000002315 PMID: 39118764

O'Brien, K. (2023, December 7). *The Future of AI in Customer Service | IBM* [Article]. https://www.ibm.com/think/insights/customer-service-future

OECD. (2019). *AI principles.* OECD. https://www.oecd.org/en/topics/ai-principles.html

Ojeda-Hernández, M., López-Rodríguez, D., & Mora, Á. (2023). Lexicon-based sentiment analysis in texts using Formal Concept Analysis. *International Journal of Approximate Reasoning*, *155*, 104–112. DOI: 10.1016/j.ijar.2023.02.001

Oueslati, O., Cambria, E., HajHmida, M. B., & Ounelli, H. (2020). A review of sentiment analysis research in Arabic language. *Future Generation Computer Systems*, *112*, 408–430. DOI: 10.1016/j.future.2020.05.034

Pahalyants, V. (2021, March 24). Netflix: A streaming giant's big data approach to entertainment. *Harvard Business School's Digital Innovation and Transformation*. https://d3.harvard.edu/platform-digit/submission/netflix-a-streaming-giants-big-data-approach-to-entertainment/

Peng, H., Cambria, E., & Hussain, A. (2017). A Review of Sentiment Analysis Research in Chinese Language. *Cognitive Computation*, *9*(4), 423–435. DOI: 10.1007/s12559-017-9470-8

Pereira, D. A. (2021). A survey of sentiment analysis in the Portuguese language. *Artificial Intelligence Review*, *54*(2), 1087–1115. DOI: 10.1007/s10462-020-09870-1

Powell, P. A., & Roberts, J. (2017). Situational determinants of cognitive, affective, and compassionate empathy in naturalistic digital interactions. *Computers in Human Behavior*, *68*, 137–148. DOI: 10.1016/j.chb.2016.11.024

Purdy, M., Zealley, J., & Maseli, O. (2019, November 18). The Risks of Using AI to Interpret Human Emotions. *Harvard Business Review*. https://hbr.org/2019/11/the-risks-of-using-ai-to-interpret-human-emotions

Puthiyamadam, T., & Reyes, J. (2022). *Experience is everything: Here's how to get it right. PricewaterhouseCoopers Global Consumer Insights Survey*. PriceWaterhouseCooper. https://www.pwc.com/us/en/services/consulting/library/consumer-intelligence-series/future-of-customer-experience.html

Ramaswamy, V., & Ozcan, K. (2019). Digitalized Interactive Platforms: Turning Goods and Services into Retail Co-Creation Experiences. *NIM Marketing Intelligence Review*, *11*(1), 18–23. DOI: 10.2478/nimmir-2019-0003

Reddy, S., Fox, J., & Purohit, M. P. (2019). Artificial intelligence-enabled healthcare delivery. *Journal of the Royal Society of Medicine*, *112*(1), 22–28. DOI: 10.1177/0141076818815510 PMID: 30507284

Remus, R., Quasthoff, U., & Heyer, G. (2010). SentiWS-A Publicly Available German-language Resource for Sentiment Analysis. *LREC*. https://www.researchgate.net/profile/Gerhard-Heyer/publication/220746590_SentiWS_-_A_Publicly_Available_German-language_Resource_for_Sentiment_Analysis/links/5f7afa5fa6fdcc0086573ff5/SentiWS-A-Publicly-Available-German-language-Resource-for-Sentiment-Analysis.pdf

Rodríguez-Ibánez, M., Casánez-Ventura, A., Castejón-Mateos, F., & Cuenca-Jiménez, P.-M. (2023). A review on sentiment analysis from social media platforms. *Expert Systems with Applications*, *223*, 119862. DOI: 10.1016/j.eswa.2023.119862

Sahoo, C., Wankhade, M., & Singh, B. K. (2023). Sentiment analysis using deep learning techniques: A comprehensive review. *International Journal of Multimedia Information Retrieval*, *12*(2), 41. DOI: 10.1007/s13735-023-00308-2

Schmelzer, R. (2019, August 18). BMW's Increasing Investment in AI - Insights from an Interview with Sam Huang of BMW iVentures. *Forbes*, *2023*, •••. https://www.forbes.com/sites/cognitiveworld/2019/08/18/bmws-increasing-investment-in-ai-insights-from-an-interview-with-sam-huang-of-bmw-iventures/

Selesi-Aina, O., Obot, N. E., Olisa, A. O., Gbadebo, M. O., Olateju, O. O., & Olaniyi, O. O. (2024). The Future of Work: A Human-centric Approach to AI, Robotics, and Cloud Computing. *Journal of Engineering Research and Reports*, *26*(11), 62–87. DOI: 10.9734/jerr/2024/v26i111315

Seyitoğlu, F., & Ivanov, S. (2024). Robots and emotional intelligence: A thematic analysis. *Technology in Society*, *77*, 102512. DOI: 10.1016/j.techsoc.2024.102512

Shedthi, B. S., & Shetty, V. (2024). 1 - Role of machine learning in sentiment analysis: Trends, challenges, and future directions. In Hemanth, D. J. (Ed.), *Computational Intelligence Methods for Sentiment Analysis in Natural Language Processing Applications* (pp. 1–21). Morgan Kaufmann., DOI: 10.1016/B978-0-443-22009-8.00011-2

Shen, J., DiPaola, D., Ali, S., Sap, M., Park, H. W., & Breazeal, C. (2024a). Empathy Toward Artificial Intelligence Versus Human Experiences and the Role of Transparency in Mental Health and Social Support Chatbot Design: Comparative Study. *JMIR Mental Health*, *11*, e62679. DOI: 10.2196/62679 PMID: 39321450

Shen, J., DiPaola, D., Ali, S., Sap, M., Park, H. W., & Breazeal, C. (2024b). Empathy Toward Artificial Intelligence Versus Human Experiences and the Role of Transparency in Mental Health and Social Support Chatbot Design: Comparative Study. *JMIR Mental Health*, *11*(1), e62679. DOI: 10.2196/62679 PMID: 39321450

Shin, H., Ellinger, A. E., Mothersbaugh, D. L., & Reynolds, K. E. (2017). Employing proactive interaction for service failure prevention to improve customer service experiences. *Journal of Service Theory and Practice*, *27*(1), 164–186. DOI: 10.1108/JSTP-07-2015-0161

Shumanov, M., & Johnson, L. (2021). Making conversations with chatbots more personalized. *Computers in Human Behavior*, *117*, 106627. DOI: 10.1016/j.chb.2020.106627

Sidaoui, K., Jaakkola, M., & Burton, J. (2020). AI feel you: Customer experience assessment via chatbot interviews. *Journal of Service Management*, *31*(4), 745–766. DOI: 10.1108/JOSM-11-2019-0341

Sinansari, P., Salsabila, S. H., Hanoum, S., Lopatka, A., & Wlodarski, W. (2023). Identify Customer Element Through Empathy Map and User Persona. *Procedia Computer Science*, *225*, 4148–4156. DOI: 10.1016/j.procs.2023.10.411

Singh, V. (2024, January). *Tech-Enabled Collaboration: Building And Leading Teams Effectively*. Forbes. https://www.forbes.com/councils/forbestechcouncil/2024/01/19/tech-enabled-collaboration-building-and-leading-teams-effectively/

Smetanin, S. (2020). The Applications of Sentiment Analysis for Russian Language Texts: Current Challenges and Future Perspectives. *IEEE Access : Practical Innovations, Open Solutions*, *8*, 110693–110719. DOI: 10.1109/ACCESS.2020.3002215

Suri, S., Sharma, K., & Papneja, S. (2020). Frustration Detection On Reviews Using. *Machine Learning*, *1–5*, 1–5. Advance online publication. DOI: 10.1109/INCET49848.2020.9153975

Thomas, J., Sateesh Kumar, T. K., Menon, V. A., & Thomas, L. P. (2024). Analyzing Sentiment in Netflix User Opinions: A Statistical Examination. In Choudrie, J., Mahalle, P. N., Perumal, T., & Joshi, A. (Eds.), *ICT for Intelligent Systems* (pp. 587–601). Springer Nature., DOI: 10.1007/978-981-97-6678-9_51

Thomas, N., & Docherty, N. (2025). "Empathy Code": The Dangers of Automating Empathy in Business. *Business & Society*, *00076503241310142*, 00076503241310142. Advance online publication. DOI: 10.1177/00076503241310142

Tuhin, M. (2025, January 23). *Study Finds AI Outperforms Humans in Generating Empathetic Responses*. Science News Today. https://www.sciencenewstoday.org/study-finds-ai-outperforms-humans-in-generating-empathetic-responses

Volpicelli, G. (2016, March 24). *IBM Watson now powers a Hilton hotel robot concierge—Ars Technica*. https://arstechnica.com/gadgets/2016/03/ibm-watson-hilton-robot-connie/

Wang, X., Hwang, Y. H., & Guchait, P. (2023). When Robot (Vs. Human) Employees Say "Sorry" Following Service Failure. *International Journal of Hospitality & Tourism Administration*, *24*(4), 540–562. DOI: 10.1080/15256480.2021.2017812

Wei, W., Liu, J., Mao, X., Guo, G., Zhu, F., Zhou, P., & Hu, Y. (2019). Emotion-aware Chat Machine: Automatic Emotional Response Generation for Human-like Emotional Interaction. *Proceedings of the 28th ACM International Conference on Information and Knowledge Management*, 1401–1410. DOI: 10.1145/3357384.3357937

Weißhaar, I., & Huber, F. (2016). Empathic Relationships in Professional Services and the Moderating Role of Relationship Age. *Psychology and Marketing, 33*(7), 525–541. *Scopus.* Advance online publication. DOI: 10.1002/mar.20895

Williamson, S. M., & Prybutok, V. (2024). Integrating human-centric automation and sustainability through the NAToRM framework: A neuromorphic computing approach for resilient industry 5.0 supply chains. *International Journal of Information Management Data Insights*, *4*(2), 100278. DOI: 10.1016/j.jjimei.2024.100278

Wirtz, J., Hofmeister, J., Chew, P. Y. P., & Ding, X. D. (2023). Digital service technologies, service robots, AI, and the strategic pathways to cost-effective service excellence. *Service Industries Journal*, *43*(15–16), 1173–1196. DOI: 10.1080/02642069.2023.2226596

Xiao, L., & Kumar, V. (2021). Robotics for Customer Service: A Useful Complement or an Ultimate Substitute? *Journal of Service Research*, *24*(1), 9–29. DOI: 10.1177/1094670519878881

Ye, Y., & Motoki, K. (2024). Effects of emotions on the helpfulness of online reviews of healthy and unhealthy food in Japan and the United States. *Journal of Consumer Behaviour*, *23*(5), 2611–2623. DOI: 10.1002/cb.2364

Yoo, J. W., Park, J., & Park, H. (2024). The impact of AI-enabled CRM systems on organizational competitive advantage: A mixed-method approach using BERTopic and PLS-SEM. *Heliyon*, *10*(16), e36392. DOI: 10.1016/j.heliyon.2024.e36392 PMID: 39253149

Yun, J., & Park, J. (2022). The Effects of Chatbot Service Recovery With Emotion Words on Customer Satisfaction, Repurchase Intention, and Positive Word-Of-Mouth. *Frontiers in Psychology*, *13*, 922503. DOI: 10.3389/fpsyg.2022.922503 PMID: 35712132

ENDNOTES

1 Self-generated

2 Mentioned along with the case studies

Chapter 11
A.I. for Emotional Storytelling in Advertisements

Mitrajit Biswas
https://orcid.org/0000-0002-0409-867X
O.P. Jindal Global University, India

ABSTRACT

In the now intensely competitive commercial world, where the attention of the consumer is heavily sought out but a fleeting treasure, advertising has transformed from simply promoting a product. It has shifted into an art form where persuasive storytelling intertwines with marketing one's product. Consumers these days are savvier and bolder. They demand more than information; they demand emotions that reflect their aspirations. The advertising campaigns that are the most memorable and impactful are those that skillfully appeal to the emotions of their target audience. Such campaigns ensure lasting brand loyalty and provoke sustainable behavioural change. But, the shifting of the advertising world has largely changed during this period, owing to Artificial Intelligence (AI). The complicated and complex understanding of humans' underlying motivations, empathizing, and inspiration were all parts of human craftsmanship, but now these resources are altered, supplemented or challenged on many levels by algorithms.

INTRODUCTION

In the now intensely competitive commercial world, where the attention of the consumer is heavily sought out but a fleeting treasure, advertising has transformed from simply promoting a product. It has shifted into an art form where persuasive

DOI: 10.4018/979-8-3373-1867-7.ch011

storytelling intertwines with marketing one's product. Consumers these days are savvier and bolder. They demand more than information; they demand emotions that reflect their aspirations. The advertising campaigns that are the most memorable and impactful are those that skilfully appeal to the emotions of their target audience. Such campaigns ensure lasting brand loyalty and provoke sustainable behavioural change. But, the shifting of the advertising world has largely changed during this period, owing to Artificial Intelligence (AI). The complicated and complex understanding of humans' underlying motivations, empathizing, and inspiration were all parts of human craftsmanship, but now these resources are altered, supplemented or challenged on many levels by algorithms. The algorithms can study enormous volumes of data, identifying and recognizing complex behavioural patterns, and even producing emotionally compelling content. AI is much more than a tool for automating repetitive tasks. It is swiftly evolving into a highly engaged partner in crafting and distributing emotionally powerful brand stories and, therefore, transforming storytelling in advertising in its entirety.

This section focuses on the complex mix between AI and ad storytelling regarding how new technologies are changing the craft and science of brand communication. We'll examine specific examples of AI innovations that are changing the face of advertising such as **Enhanced Emotion Tracking and Hyper Audience Segmentation**. AI goes beyond standard demographics by allowing brands to scan the target audience's emotional landscape in real-time. Brands can analyse social media activity, online reviews and other digital footprints to detect timely shifts in emotions, emerging triggers, and favourable sentiment. In turn, these highly delicate audience profiles make it possible for specific emotional messaging to be sent out rather than generalized.

AI-Focused Narrative Creation and Contextual Storytelling: AI algorithms now go beyond analysing existing narratives to create their own narratives that modify themselves depending on how the audience responds to them. These algorithms make use of NLP and ML to construct a specific set of characters and events that can alter when feedback from the audience is received. In that way, every narration becomes an evolving story to its audience.

The personalization of dynamic content and contextual relevance: AI empowers every brand to streamline content delivery across multiple touchpoints throughout the brand's journey. Each consumer gets to experience a brand story tailored specifically to their preferences, emotional triggers, and how they consume media. AI can also use contextual data like location, time of the day, and even previous locational browsing history to deliver more relevant messages that resonate with them on a deeper spectrum. AI powered visual and auditive storytelling as well as sensory immersion: With the use of AI, the development of visual and auditive content has changed drastically. Brands can now create advertisements AI powered trend forecasting: AI systems can now analyse massive amounts of multilayered data and determine the most important emotional and cultural trends that would influence consumer behavior in the upcoming future. Brands are tapping into the emerging culture and emotions, enabling them to construct campaigns that are ahead of the curve.

The swift incorporation of AI within advertising storytelling does present some problems which need to be evaluated further. These issues include the following:

The Authenticity of AI-Generated Emotion: AI relies on data and algorithms. Does that mean it can replicate human emotions? Can it portray empathy, nostalgia, joy or any other useful emotion when creating an impactful story? Will AI require a human touch to help it produce something that is not formulaic and emotionally manipulative in nature?

The Ethical Implications of Emotional Targeting: AI can analyse emotions of humans proficiently. However, how far can this capability be exploited? What are the dangers of using AI to target consumers through emotional manipulation in advertising? Where is the line IIAI's use boundary lies when it comes to performing such tasks while protecting the consumer's privacy and autonomy

The Balance Between Automation and Human Creativity: Considering the level to which AI can automate the storytelling process, what is the Human role? To what extent can one AI's use in advertising change the human element? How can the AI and humans be made to work together for exceptional advertising narrative?

The concern with AI potentially constructing emotion-centric filter bubbles: Will AI satiate one's desire to the extent that customers are only allowed to view emotions or beliefs that they already possess?

This chapter's purpose is to analyse these critical issues in a comprehensively balanced way, paying due attention to the great possibilities as well as the accompanying challenges posed using AI in advertising storytelling. Through various case studies, industry analyses, and established theories, we will attempt to illustrate the possibility of the ethical and effective use of AI by advertisers in the future to help brands connect with their consumers at a deeper level in a more complex and fragmented world of advertising.

LITERATURE REVIEW: HOW AI IS UTILIZED FOR ENGAGEMENT AND NARRATIVES IN ADVERTISING

Advertising is among the most heavily influenced sectors by the emerging artificial intelligence (AI) which affects its storytelling and emotional engagement aspects. The intersection of new advanced AI technologies with the art of storytelling allows brands to create sophisticated and data-driven narratives that deeply connect with the consumers. This literature review attempts to focus on primary literature on the foundations of advertising storytelling, the critical importance of emotional engagement to consumer actions, and the newer uses of AI in emotionally charged brand stories. Additionally, it analyses the ethical dilemmas and self-constraints of AI storytelling which gives the review the right balance of depth and breadth on the research and practice issues.

The Art of Connection: Storytelling's Role in Advertisements

Telling stories enables businesses to advertise effectively, helping them shape their audience perception as well as foster brand loyalty. In addition to building relationship with consumers, Escalas (2004) claims that narratives act as mental structures that facilitate specific brand recall that increases emotional intimacy and engagement. This connection is necessary, as previous studies show that captivating brand stories improve the retention of the message, foster positive brand image, and purchasing intention (Pulizzi, 2012).

In advertising, storytelling is the most effective form of communication as it can touch the emotions of the audience, which is one of the key reasons for acting. "Emotional connections to brands greatly aid brand recall, adding the ad to the list of things that people remember." This claim has also been made by Thompson et al. (2011). Stated differently, Pine & Gilmore (1999) proclaim that these brand commercials that are attached to stories foster personal immersion into the brand, which translates to continued usage of them. According to the narrative transportation theory postulated by Green & Brock (2000), consumers are more likely to hold favourable attitudes and behavioural intentions towards brands when they become cognitively and emotionally engrossed in the brand's story. In the past, advertising methods depended solely on creativity, intuition, and empathy to build stories that resonate with the audience. The introduction of AI technology has created a new form of advertising strategies in which brands leverage data to tell stories, allowing them to optimize and personalize ads like never before.

Emotional Engagement and Consumer Behavior: The Heart of the Matter

Emotional engagement is a critical factor in making any advertisement compelling enough to persuade customers, with emotion being a determinant of their self-perception, personal preferences, and decisions related to purchases. It is widely acknowledged that advertisements driven by emotion with no rationale are far more persuasive than those based on rational appeals (Bagozzi et al., 1999). The AIDA model (Attention, Interest, Desire, Action) is a classic model for marketing proposed by Strong (1925) and highlights the importance of the initial encounter via rational stimulus. Further modern studies have placed attention to the role of emotional engagement in marketing using emotions as a centre of branding. Brakus et al. (2009), for example, propose four experiential dimensions - sensory experience, affective experience, intellectual experience, and behavioural experience - that drive brand loyalty and satisfaction. Schmitt (2010) agrees, further claiming that modern consumers do not merely consume marketing messages, but actively participate in the brand's marketing through thinking, feeling, and sensing. The increased focus on emotional engagement has accelerated the uptake of AI technologies that allow AI to mine huge datasets on consumer preferences, emotions, and behaviours. This makes it possible for brands to enhance their advertising stories to reach all audience segments more effectively and efficiently.

A New Era in Advertising with the Introduction of AI Technology

With the increase in technology, advertising will never be the same. The way brands study consumers, create content, and strategize their storytelling is completely different now. AI systems can study and analyse vast data sets and recognize different levels of engagement and attitudes and how this impact purchasing behavior from the consumers (Kumar et al. 2016). Some of the newly adopted technologies that have had the greatest impact on the world of advertising include:

Sentiment Analysis: AI systems can analyse the emotions of consumers through text, voice, and even facial expressions with great accuracy. This allows brands to measure the audience's reception of the message and modify it in real-time (Pang & Lee 2008).

Predictive Consumer Preferences: Segmentation of AI audiences can now give the prediction of emotional triggers and buying decisions. This allows brands to create and present highly specific content that aligns with the expectations and desires of the consumer (Chaffey, 2019).

Adaptive Content Personalization: AI systems can modify stories in real-time depending on how an individual user interacts with the product, making sure the storytelling is as relevant and impactful as possible by modifying it based on the specific user interaction and preference (Davenport et al., 2020).

AI Powered Trend Analysis: AI can scan and sift through social media and other data sources to track changes in consumer's emotional state and cultural trends.

With the accelerating integration of AI in branding, the prospect of crafting emotionally engaging narratives that are backed by data is transforming advertising.

AI Created Content and Engagement: The New Horizon

Perhaps the most innovative aspect of AI in advertising is to be able to develop content and narratives which elicit an emotional response. Machine learning models such as NLG (natural language generation) and deep learning based creative tools enable forms of automated storytelling containing narratives, dialogues, videos and images tailored to the consumer's data (Garnier et al., 2020).

Here are some instances of AI storytelling:

AI Video Ad Generation: AI can produce personalized video ads to users by altering style, tone, images and even content to capture the user's attention and maximize their emotional response (Smith, 2019).

Emotionally Intelligent Chatbots and Conversational AI: Using emotionally intelligent chatbots that can simulate human interactions allows brands to improve customer service by actively engaging with consumers and building a stronger relationship through empathetic communication (Huang & Rust, 2021).

Neural Storytelling Models: Analysing AI generated fictitious narratives, brands, and promotional ads relies on successful storytelling patterns to create content that aligns with emotion provoking storytelling technique (McCormick et al., 2021).

Nonetheless, this increased focus on AI's contribution to content generation raises a critical question around the authenticity and emotional depth underlying AI's narrative. AI can replicate emotions, but how far does its understanding of human cultures and emotions go is still open for dialogue.

Ethical Considerations and Limitations: The Nuance of Navigating

As brands use AI based narratives, they will need to steer through the ethical implications of customer data privacy, security issues, and biased algorithms. AI in consumer insights is a double-edged sword that Zuboff (2019) suggests will cut deeply by invading the user's privacy and contributing to data exploitation. Can

Sundar & Marathe's (2010) said AI truly understand a person's emotion or is it algorithmic pattern recognition mimicry? This brings forward a question of trust.

Further challenges include:

Algorithmic Bias: Content generation through exclusionary AI model datasets will bring the birth of division culture fuelled by destruction of wider inclusion.

Loss of Human Creativity: Storytelling begins and ends with AI, which not only raises issues of originality, but dwindles the creativity of the human race.

Emotional Manipulation: Concerns about AIs exploiting a person's emotion and using it for monetary purposes, where advertising strategies become manipulative are problematic.

To further understand this issue: An AI that only supplies information which it predicts will elicit a specific positive emotional outcome from a user, may only feed an individual it's 'positive' perspective.

To tackle these issues, it is imperative to have responsible AI governance, frameworks to regulate advertising, and policies aimed at protecting consumers so as to mitigate the absence of AI in the storytelling process.

Responsible Narrative Engagement, Integrity Preservation Innovation

The rest of the literature explains how AI can be a transformative marketing tool in storytelling and emotional engagement in advertising. With the introduction of AI, brands can craft tailored narratives, assess emotions, and create customized content changing the whole landscape of advertising. But all these have the issues of authenticity, ethics, and algorithmic bias that remain unsolved and problematic.

As the advertising era moves into the usage of AI, the need for greater human input alongside technological innovation is paramount. The future of storytelling using AI tools will need to be used in such a way that they are ethical, emotionally relevant, and engaging, all while building true trust and relevance with consumers.

METHODS: A MULTI-METHOD APPROACH ON THE ROLE OF AI IN EMOTIONAL ADVERTISING STORYTELLING

This study applies multi-method strategies, which include a literature review, an industry use case, focused case studies, and a discussion examination that deals critically with the issue at hand. Such a comprehensive framework guarantees a robust analysis of AI's ability to foster emotional engagement through advertising storytelling and its theoretical and practical ramifications.

Literature Review and Theoretical Synthesis: Mapping the Intellectual Landscape

Scope and Rationale:

The researcher performed an integrative and systematic literature review to ensure a robust theoretical base. The goal of the research was to analyse and argue advertising storytelling, emotional engagement, and AI in marketing, particularly how those concepts associate together.

The goal was to identify and interpret important theoretical frameworks, empirical evidence, and emerging issues that provide an understanding of the current body of knowledge and decide what literature is available for future research.

Sources and Search Strategy:

The search covered a peer-reviewed academic journal, proceedings of conferences, industry reports, market research white papers, and books and thus was expansive.

Databases derived from Scopus, Web of Science, Google Scholar, as well as IEEE Xplore for AI-related marketing and communication were utilized.

Although, Marketing journals and periodicals such as Journal of Consumer Research, Journal of Advertising Research and International Journal of Market Research have also been used.

Some of the important keywords from the topic's scope were included in the search phrases, such as:

"Artificial Intelligence in advertising"

"Emotional engagement in consumer marketing"

"AI powered storytelling"

"Consumer behavior in the context of AI technologies"

"Sentiment analysis for advertising emotional engagement"

"Marketing texts and natural language processing"

"Bias in algorithms used for advertisements"

"Emotional targeting by AI, ethical concerns"

Data Extraction and Synthesis

A systematic framework for reviewing and classifying pertinent studies was developed. This framework primarily focused on:

The conceptual frameworks dealing with Emotional engagement in Advertising.

Applied artificial intelligence in sentiment analysis, audience segmentation, and content customization.
AI advertising: techniques and effectiveness.
AI limitations in advertising.

Information was extracted and put into tables so that findings can easily be compared or contrasted. This included citations, publication dates, methods, and conclusions. The analysis was integrated using the thematic approach, where patterns, similarities, and differences were distilled to offer a coherent critique of the pre-existing knowledge.

2. How Ai is Applied to Advertising: Applications in the Industry Practice Ai in Advertising Industry Advertising Strategies Implementation. A use study search was performed on real-world applications of machine learning advertising algorithms. The search was restricted to advertising campaigns of selected brands that used AI to improve emotional storytelling, innovative content creation, audience engagement, advertisement performance, and AI analytics-based sentiments and emotion forecasting.

3. In-Depth Case Analysis: Every case study was studied comprehensively regarding goals and objectives of advertising. The purposes and aims of the campaign and the results of AI-enabled storytelling for the campaign. AI technologies used: The AI technologies utilized in content development, personalization, and engagement tracking. Creative execution: The embedding of AI-generated stories into advertisement formats and the creativity employed in advertisement to garner emotional appeal. Consumer impact: The story-driven AI's impact on audience sentiment, attitudes, and behavior as measured through consumer's response data and feedback. Each case study encompassed the review of the campaigns strategy, campaign execution and the results, as well as the scrutiny of the strengths and weaknesses.

4. Critical Discussion and Ethical Considerations: The Ethical Frontier in Focus Synthesizing key insights. The amalgamation of the literature survey, industry review and case study analyses provided comprehensive insights on the role of AI in advertising emotion. Comparative remarks were made over the AI produced brand storytelling versus conventional narration by a human and the relative benefits and limitations of using AI in brand storytelling. Identification of research gaps: The major questions that need an answer are: The boundaries of the algorithm's ability to reproduce real human feelings. The potential harm or risks of emotional manipulation using AI.

ANALYSIS AND FRAMEWORK: STORYTELLING WITH AI IN ADVERTISING AND ENGAGEMENT

This part looks at advertising's AI-powered storytelling from an analytical perspective, advanced by solid theoretical models and other examples. The discussion is execution through four main views – the theoretical bases of storytelling and emotional engagement, the construction AI technology application analytical framework, evaluation of AI powered case studies, and evaluation of success factors and limitations as well as future attention areas.

Storytelling, AI, and Emotional Cognition: Deep Dive

Adverts and the story they tell have their basis in consumer behavioural psychology and persuasive communication strategies. Numerous research models that build on how and why advertising narratives work have shown that stories in advertising cause emotional and cognitive decisions to be made. This section integrates the following frameworks, expanding on their relevance and application:

Elaboration Likelihood Model (ELM) (Petty & Cacioppo, 1986)

This model argues that people get persuaded through two routes. These include the central route (Receptive, critical thinking towards message content) and the peripheral route (Emotional content and a more simplistic heuristic decision making). For example: An Artificial Intelligence (AI) powered augmented reality allows the consumer to virtually project a product in their abode and narrates a tailored tale around the product depending on the customer's previous activity online. With the help of AI, brands are now able to develop multi-sensory offerings to fulfil deeper emotional expectations from them.

Analytical Framework: The "4Es" of Experiential AI Advertising

In regard to advertising in particular, the unique brand storytelling enabled by AI is investigated using the "4Es" Experiential Marketing model delineated by Pine & Gilmore, 1998, which denotes brand experience with four parameters.

Education:

> AI provides knowledge-sharing and informational storytelling by:
> Tracking the consumer's vehicle of choice and tailoring the narratives in advertisements during the educational phase.

Offering AI-powered voice-assisted storytelling and product recommendation systems through chatbots in interactive ads.
AI can provide personalized tutorials, product demos, and other forms of educational resources tailored to specific consumer demands.

Entertainment:

AI enhances brand narratives by creating engaging and emotionally immersive content through:
AI-Generated music, personalized video storytelling, and advertisement gamification.
Automatic video editing powered by deep neural networks alongside video narration with synthetic voices are examples of modern branded content tools powered by AI.

For instance, AI technology allows one to compose an original score for a commercial that matches the desired tone of the video advertisement ordesign interactive gaming applications that feature branded characters.

ANALYSIS

Planning the Integration of Artificial Intelligence in the Storytelling Framework of Emotion-Centric Advertising Campaigns

This chapter attempts to describe the systematic and important research the study undertook to address the issue of how Artificial Intelligence (AI) can impact emotional involvement in advertising storytelling. It incorporates concepts from theory and practice, employing a critical literature review, analysis of AI use cases, and case study research of the selected examples.

1. Literature Review and the Theoretical Framework: Building the Base

Establish the theoretical framework storytelling within advertising, emotional involvements, and the application of AI in digital marketing through systematic reviews of scholarly literature and industry reports This review sought to answer how the intersection between AI and emotional advertising works, alongside the concepts, theories, and key empirical findings that support it. The method involved scrutiny of peer-reviewed journal articles, conference papers, marketing whitepapers, industry reports, and case studies published in reputable journals on advertising and AI. Boolean operators were utilized to eliminate irrelevant search results alongside forward and backward chaining to locate more pertinent works.

For the analysis, two primary theoretical models are considered in this chapter which serve to give a strong analytical foundation:

The Elaboration Likelihood Model (ELM), Petty and Cacioppo (1986)

This model offers a very insightful categorization of how audience processes and reacts to persuasive information through the distinction of a central route that is a very deeply cognitive process and a peripheral route which is an emotional trigger or the engagement on the surface. AI's role in storytelling is evaluated since rational as well as emotional involvement, evaluating how AI can tailor the message to both cognitive and affective processing. In addition, the case of ELM is analysed to find out, under what circumstances and context the central or the peripheral route is most likely to be utilized.

The Technology Acceptance Model (TAM), Davis (1989)

This framework focuses on how consumers understand and accept AI based storytelling advertising narratives in relation to factors like perceived usefulness, ease of engagement with the content, and AI's credibility. Of particular interest will be the level of trust towards an AI because it is one of the most crucial issues with modern technology. These theories help analyse the impacts of AI storytelling in a more focused manner, specifically with respect to the effectiveness, consumer engagement, and brand loyalty.

Construction of An Analytical Framework

The 4E's of Experiential Marketing (Education, Entertainment, Esthetic, and Escape) by Schmitt (1999) serves as the theoretical basis upon which new frameworks and models are built towards the needs of today and the future. These were the criteria used to evaluate and classify advertising narratives for AI fostered marketing:

Educate: AI tutorials or product demonstrations enable for more interaction and deeper understanding which fosters increased consumer knowledge.

Entertain: AI generated stories that foster engagement on a more intense emotional level through music, interactive games, videos and many other mediums.

Esthetic: AI generated sensory rich images, videos and other interactive media that are visually pleasing and are of superior quality.

Escape: AI immerses consumers in augmented reality (AR) and virtual reality (VR) stories rendering a feeling of immersion and escapism.

Reflecting on the Implications Synthesis of Findings: The review of the case studies, literature, and AI framework studies was synthesized to understand the role of AI in emotion-based storytelling in greater depth. This construction brings out the theoretical and practical aspects of: Primary success factors for emotional advertising through AI. Novel developments in AI-based storytelling. Existing problems in using AI in advertising stories. Evaluation of Success Factors and Limitations: The analysis points out a few success factors of AI storytelling like: Ability of AI to interpret emotions and engagement with content. The greater need for personalized and customizable stories in digital advertising. The ability of AI to customize and produce different types of content for little money and low scale. Nonetheless, these factors led to several limitations and ethical issues as well: Absence of true emotional intelligence: AI can predict and simulate emotions but lacks genuine human empathy and the ability to interpret cultural and emotional subtlety. Data privacy problems: Personalized AI solutions are fundamentally based on consumer data, which brings into question privacy, consent, and AI discrimination issues. Authenticity difficulties: Narratives produced by AI are perceived to be inauthentic or overly mechanistic, which may disconnect consumers that seek authenticity from the brand.

CONCLUSION AND FUTURE RESEARCH DIRECTIONS

This chapter presents a detailed analysis of an AI innovation in the context of digital advertising storytelling including a review of its Pros and cons. AI can augment emotional engagement, upgrade storytelling techniques, and promote personalization to the highest imaginable degree. Nonetheless, ethical issues, the question of authenticity, and the AI emotionality bottleneck must be dealt with. Human-AI Collaboration: Investigating the possibility of AI in the enhancement of human creativity in a hybrid approach. AI and Cultural Sensitivity: Investigation of AI-adaptation to different cultures storytelling and traditions. Measuring Long-Term AI Impact: The consequences of AI-assisted storytelling on brand loyalty and consumers' emotional relationship with the brands over a long period of time.

AI-DRIVEN TRANSFORMATION OF STORYTELLING IN ADVERTISING: OPPORTUNITIES AND CHALLENGES

In the fiercely competitive commercial landscape, where the attention of the consumer is heavily sought out but a fleeting treasure, advertising has transformed from simply promoting a product. It has shifted into an art form where persuasive storytelling intertwines with marketing one's product. Consumers these days are savvier and bolder. They demand more than information; they demand emotions that reflect their aspirations. The advertising campaigns that are the most memorable and impactful are those that skilfully appeal to the emotions of their target audience. Such campaigns ensure lasting brand loyalty and provoke sustainable behavioural change. But, the shifting of the advertising world has largely changed during this period, owing to Artificial Intelligence (AI). The complicated and complex understanding of humans' underlying motivations, empathizing, and inspiration were all parts of human craftsmanship, but now these resources are altered, supplemented or challenged on many levels by algorithms. The algorithms can study enormous volumes of data, identifying and recognizing complex behavioural patterns, and even producing emotionally compelling content. AI is much more than a tool for automating repetitive tasks. It is swiftly evolving into a highly engaged partner in crafting and distributing emotionally powerful brand stories and, therefore, transforming storytelling in advertising in its entirety.

This section focuses on the complex mix between AI and ad storytelling regarding how new technologies are changing the craft and science of brand communication. We'll examine specific examples of AI innovations that are changing the face of advertising such as Enhanced Emotion Tracking and Hyper Audience Segmentation. AI goes beyond standard demographics by allowing brands to scan the target audience's emotional landscape in real-time. Brands can analyse social media activity, online reviews and other digital footprints to detect timely shifts in emotions, emerging triggers, and favourable sentiment. In turn, these highly delicate audience profiles make it possible for specific emotional messaging to be sent out rather than generalized (Chaffey, 2019).

AI-Focused Narrative Creation and Contextual Storytelling: AI algorithms now go beyond analysing existing narratives to create their own narratives that modify themselves depending on how the audience responds to them. These algorithms make use of NLP and ML to construct a specific set of characters and events that can alter when feedback from the audience is received. In that way, every narration becomes an evolving story to its audience (Davenport et al., 2020). The personalization of dynamic content and contextual relevance: AI empowers every brand to streamline content delivery across multiple touchpoints throughout the brand's journey. Each consumer gets to experience a brand story tailored specifically to their preferences, emotional triggers, and how they consume media. AI can also use contextual data like location, time of the day, and even previous locational browsing history to deliver more relevant messages that resonate with them on a deeper spectrum. AI powered visual and auditive storytelling as well as sensory immersion: With the use of AI, the development of visual and auditive content has changed drastically. Brands can now create advertisements AI powered trend forecasting: AI systems can now analyse massive amounts of multilayered data and determine the most important emotional and cultural trends that would influence consumer behavior in the upcoming future (Kumar et al. 2016). Brands are tapping into the emerging culture and emotions, enabling them to construct campaigns that are ahead of the curve.

ETHICAL IMPLICATIONS, HUMAN CREATIVITY, AND THE FUTURE OF AI IN ADVERTISING STORYTELLING

Despite the benefits, the integration of AI in advertising storytelling presents several challenges that require careful evaluation. These include concerns about the authenticity of AI-generated emotion and whether AI can truly replicate human empathy, nostalgia, or joy. There are also ethical implications related to emotional targeting and the potential for manipulation, raising questions about the boundaries of AI's use in protecting consumer privacy and autonomy (Zuboff, 2019). The role of human creativity in an increasingly automated storytelling process must also be considered (Sundar & Marathe, 2010). This paper seeks to analyse these critical issues, balancing the opportunities with the accompanying challenges. Through case studies, industry analyses, and established theories, it explores the possibility of ethical and effective AI use by advertisers to connect with consumers at a deeper level. Furthermore, the ethical aspects of how AI systems in advertising can create emotion-centric filter bubbles that reinforce beliefs that they already possess also pose issues for discussion. This study will use multi-method strategies, including literature review, industry use case, and focused case studies (Green & Brock, 2000).

- **Emphasis on Emotional Connection:** The article stresses the importance of emotionally engaging narratives in advertising and examines how AI can be used to create more resonant and personalized content that connects with consumers on a deeper level (Escalas, 2004; Pine & Gilmore, 1999; Thompson et al., 2011).
- **Ethical Concerns Regarding Emotional Manipulation:** The ethical implications of using AI to target consumers through emotional manipulation are a central theme, raising questions about consumer privacy, autonomy, and the potential for harm (Zuboff, 2019; Sundar & Marathe, 2010).

- **The Role of Human Creativity:** The article explores the balance between automation and human creativity, questioning how AI's use in advertising might change the human element and how AI and humans can work together for exceptional advertising narratives.
- **Creation of Emotion-Centric Filter Bubbles:** The creation of filter bubbles is a core issue. An AI that is only providing specific content that is predicted to give a positive emotional outcome may negatively impact the user (Zuboff, 2019).

In summary, the article presents AI as a transformative technology with the potential to revolutionize advertising storytelling and emotional engagement. However, it emphasizes the need for careful consideration of ethical implications, the preservation of human creativity, and the protection of consumers' well-being in this new era of AI-driven advertising (Bagozzi et al., 1999; Brakus et al., 2009; Schmitt, 2010).

REFERENCES

Bagozzi, R. P., Gopinath, M., & Nyer, P. U. (1999). The Role of Emotions in Marketing. *Journal of the Academy of Marketing Science, 27*(2), 184–206. DOI: 10.1177/0092070399272005

Brakus, J. J., Schmitt, B. H., & Zarantonello, L. (2009). Brand Experience: What Is It? How Is It Measured? Does It Affect Loyalty? *Journal of Marketing, 73*(3), 52–68. DOI: 10.1509/jmkg.73.3.052

Chaffey, D. (2019). *Digital Marketing: Strategy*. Implementation and Practice.

Davenport, T. H., Guha, A., Grewal, D., & Bressgott, T. (2020). How artificial intelligence is changing marketing. *Journal of the Academy of Marketing Science, 48*(1), 24–42. DOI: 10.1007/s11747-019-00696-0

Escalas, J. E. (2004). Narrative Processing: Building Consumer Connections to Brands. *Journal of Consumer Psychology, 14*(1-2), 168–179. DOI: 10.1207/s15327663jcp1401&2_19

Green, M. C., & Brock, T. C. (2000). The role of transportation in the persuasiveness of public narratives. *Journal of Personality and Social Psychology, 79*(5), 701–721. DOI: 10.1037/0022-3514.79.5.701 PMID: 11079236

Kumar, V., Petersen, J. A., & Leone, R. P. (2016). Driving Profitability by Encouraging Customer Referrals: Who, When, and How. *Journal of Marketing, 80*(5), 1–17. DOI: 10.1509/jmkg.74.5.001

Pine, B. J., & Gilmore, J. H. (1999). *The Experience Economy: Work Is Theatre & Every Business a Stage*. Harvard Business Press.

Schmitt, B. (2010). *Customer Experience Management: A Revolutionary Approach to Connecting with Your Customers*. John Wiley & Sons.

Smith, G. (2019). *Marketing Communications: Integrating Offline and Online Strategies*. Kogan Page Publishers.

Strong, E. K. (1925). *The psychology of selling and advertising*. McGraw-Hill Book Company.

Sundar, S. S., & Marathe, S. S. (2010). Personalization and its discontents: A meta-analysis of the effects of customization on perceived source credibility. *New Media & Society, 12*(6), 875–892.

Thompson, D. V., Hamilton, R. W., & Rust, R. T. (2011). Feature Fatigue: More May Be Too Much. *JMR, Journal of Marketing Research*, *42*(4), 431–442. DOI: 10.1509/jmkr.2005.42.4.431

Zuboff, S. (2019). *The Age of Surveillance Capitalism: The Fight for a Human Future at the New Frontier of Power*. PublicAffairs.

Chapter 12
A Critical Analysis of the Emotive Content in Customer Speech for a Robust CRM

Sippee Bharadwaj
https://orcid.org/0000-0002-7808-7156
International School of Business and Research, Bangalore, India

Nila Chotai
https://orcid.org/0009-0003-6563-4071
International School of Business and Research, Bangalore, India

Manish Kothari
International School of Business and Research, Bangalore, India

Chinmoy Bharadwaj
https://orcid.org/0000-0002-3027-6812
University of Science and Technology, Meghalaya, India

ABSTRACT

Emotion analysis plays a vital role in Customer Relationship Management (CRM), impacting customer interactions and brand perception. While previous CRM systems have relied mostly on text-based sentiment analysis, the expanding usage of audio-based feedback underlines the necessity for speech-based emotion recognition. This research analyses vowel prosodic qualities in customer speech to better emotion prediction, boost customer retention, and expedite corporate procedures. By concentrating on vowel-level emotion extraction, this study intends to boost the accuracy and reliability of emotion classifiers via an analysis of consumer audio

DOI: 10.4018/979-8-3373-1867-7.ch012

feedback from online purchase platforms. Advanced speech analytics approaches are applied to decipher the emotional content of utterances by studying prosodic variables such as F0, MFCC, ZCR, STE, and formants. The results contribute to translating sentiment research into practical CRM solutions, developing customer-centric business strategies, and boosting the usefulness of business analytics.

INTRODUCTION

It is well known that emotion plays a crucial part in human communication and can provide information about customer satisfaction. Customers frequently express their opinions about freshly bought products more efficiently via voice communications in their native language than in written form. Increasingly, customers offer their comments on social media platforms and online purchasing websites in both text and audio formats. Some customers prefer expressing their feedback and reviews in voice and video messages, which typically represent genuine opinions about the products. Processing these unstructured customer utterances for emotion recognition delivers a more accurate depiction of thoughts compared to text-based evaluations. Customer service executives may better comprehend complaints, resolve customer difficulties, and make brand strategies by identifying emotions in customers' speech. Emotions have a key influence in determining consumer social relationships. To accurately analyze and interpret the emotive content is difficult when the variances are manually understood in the customer's voice which takes time and needs sophisticated language abilities. This difficulty has prompted the development of automated speech-emotion recognition (ASER) systems, which are designed to interpret and extract emotions effectively. Applications such as online games, contact centers, customer care services, online classrooms, and online commerce increasingly rely on voice input to understand customer behavior and their pattern of purchase for Customer Relationship Management (CRM). These systems analyze speech signals to detect emotions like happiness, sadness, anger, and surprise or mixed emotions, enabling businesses to respond effectively to customer needs. The current study focuses on an in-depth analysis of the affective content inherent in customer speech, with a special emphasis on the significance of vowels. Prior research has underlined the strategic value of emotion-evocative communication, enabled the prediction of emotional reactions, and produced emotive effects. Voice-only communication has been proven to boost empathic accuracy since speech signals provide precise information on the emotional states of speakers (Kraus, 2017). Within the context of customer contacts, assessing the emotional content in speech offers the potential to boost customer-centric activities and establish meaningful connections. Emotional tones carried through speech often reflect the speaker's sentiments about a product or service, making it an invaluable tool for customer engagement. ASER systems can harness this speech data to deliver actionable insights for companies, driving improvements in service quality and customer satisfaction. The proposed work identified vowels as essential factors in the perception of emotion through speech. Vowels are the most sonorous elements of speech, allowing for greater acoustic variation that conveys emotional subtleties. For instance, the high-pitched and prolonged /i:/ vowel sound is often associated with happiness and excitement, whereas the lower-pitched and shorter /o:/ sound may indicate negative emotions such as sadness or disappointment. By analyzing patterns and variations in vowel articulation, the study explores the relationship between vowel use and the transmission of emotional states. Such insights are likely to increase the knowledge of consumer experiences by highlighting how vowels work as indications of emotional content in speech. This vowel-specific analysis, along with other acoustic data such as pitch, intensity, and speech rate, provides for a more detailed assessment of client attitudes. Recognizing the importance of emotional content in speech signals, particularly through prosody properties of vowel phonemes, can improve emotion recognition systems dramatically. Prosody comprises the rhythm, stress, and intonation patterns of speech, which are crucial in communicating emotions. For instance, rapid speech rates combined with higher pitch levels may signal excitement or urgency, while slower rates with lower pitch can indicate calmness or sadness. These prosodic features, when combined with spectral characteristics such as formants and Mel Frequency Cepstral Coefficients (MFCCs), provide a comprehensive framework for emotion recognition. MFCCs, in particular, capture the fine-grained details of the speech spectrum, making them invaluable for identifying subtle emotional cues in customer interactions. This improvement is especially essential for customer relationship management (CRM) systems, because assessing consumer emotional states may enhance service quality and satisfaction. By integrating Speech Emotion recognition (SER) capabilities, CRM platforms can personalize interactions, predict customer needs, and proactively address concerns. For example, real-time emotion detection in contact centers can alert agents to distressed customers, enabling them to adapt their responses accordingly. A more engaging and sympathetic experience may be produced by emotion-aware virtual assistants that adjust their vocabulary and tone to the user's emotional state.

The major purpose of this chapter is to study the relationship between emotions and vowels, leveraging current improvements in emotion recognition for CRM. The sections within the chapter are outlined as follows. The study's primary premise is articulated, emphasizing the enhancement of product quality through the evaluation of customers' emotional states as a strategic goal. A summary of the literature is presented, outlining current research on customer emotions in CRM through speech analytics. Studies such as those by Rummer and Schweppe (2019) and Fahad et al.

(2021) provide foundational insights into vowel-specific emotion recognition, demonstrating its potential to improve CRM outcomes. These findings are supplemented by research on prosody and spectral features, which collectively form the basis for advanced SER systems. The significance of this research is subsequently examined, highlighting its implications for both academic research and practical applications. The subsequent sections delineate the data-gathering approach, methodology, and experimental results. Data collection involves acquiring speech samples from diverse customer interactions, ensuring a representative dataset that captures a wide range of emotional states. Methodological frameworks incorporating feature extraction approaches have been also applied. A comparative study has been done depending on the location of vowels, based on gender-related emotions concerning customers. This emphasizes the benefits of adding vowel-specific and prosodic characteristics into SER systems. The chapter concludes with implications, limits, and scopes for future study. Key implications include the potential for SER systems to change CRM by allowing more empathic and successful customer interactions. Limitations, such as the necessity for big, labeled datasets and difficulty in cross-cultural emotion recognition, indicate opportunities for further study. Additionally, investigating the nucleus level in speech signals gives the potential for additional breakthroughs in emotion recognition. To improve the accuracy of emotion recognition, it is very necessary to integrate multimodal data, such as combining text, speech, and facial expressions to find accurate emotive content.

The conclusion synthesizes the results, underlining the vital relevance of emotional content in customer interactions and the transformational potential of AI-driven SER systems in CRM. Understanding and utilizing emotional content in speech is crucial for strengthening CRM systems. By concentrating on vowel-specific analysis and prosody traits, researchers may work on present limits in emotion identification, opening the door for more compassionate and successful CRM solutions. This holistic strategy not only enhances customer pleasure but also develops brand loyalty, offering a competitive advantage for businesses in an increasingly customer-centric industry. The integration of modern SER systems into CRM platforms represents a huge step forward, integrating technology progress with the underlying human desire for meaningful, emotion-driven interactions.

BACKGROUND

The fast growth of Artificial Intelligence (AI) and big data analytics is drastically transforming Customer Relationship Management (CRM). AI-driven solutions allow firms to examine massive consumer information fast, giving nuanced insights into customer behavior and preferences. By integrating predictive analytics and machine learning (ML) approaches, organizations may reach new levels of customization, boost customer engagement, and improve resource allocation (Pendyala & Lakkamraju, 2024). The key technical breakthroughs, like as Natural Language Processing (NLP), have further altered CRM via technologies like chatbots and sentiment analysis systems, which dramatically boost responsiveness and empathy during client interactions. Additionally, the intersection of AI, blockchain, and big data analytics allows enterprises to build more personalized products and services, promoting better levels of consumer happiness, loyalty, and retention. While these technology improvements bring enormous advantages, they also raise concerns, notably around data privacy and regulatory compliance. Robust data protection mechanisms are required to minimize these risks (Khan et al., 2024). The incorporation of AI into CRM systems has led to enhanced automation and efficiency. AI-powered solutions that automate decision-making with little human involvement have revolutionized personalization. Technologies such as machine learning and predictive analytics have changed conventional marketing and sales techniques by providing focused outreach and exact client segmentation. Additionally, virtual assistants and chatbots have greatly increased query response times, thus enhancing customer satisfaction (Pramod Gavade, 2024). Large volumes of consumer data may be tapped by businesses using AI-based CRM systems, yielding actionable insights that support long-term customer engagement strategies. Advances in AI-based CRM have been divided into four primary areas by recent studies such as data-driven CRM strategies, strategic deployment, AI applications in CRM operations, and improving the customer experience (Ozay et al., 2024). Strategic deployment entails employing AI to match CRM objectives with larger company goals, ensuring that technical deployments contribute to long-term value generation. The goal of data-driven CRM methods is to use extensive data analytics to comprehend consumer behavior and forecast future requirements. AI applications in CRM functions entail the integration of techniques like NLP, computer vision, and voice recognition to expedite operations and enhance customer interactions. Lastly, customer experience improvement stresses the role of AI in offering tailored, smooth, and sympathetic encounters that promote loyalty and retention. Despite these advantages, substantial difficulties persist. Some of the issues including data security, privacy issues, and ethical difficulties are recurring impediments in the adoption of AI-driven CRM systems (Sheshadri Chatterjee & Ranjan Chaudhuri, 2022). Regulatory compliance, particularly with frameworks like GDPR, adds a layer of complexity, forcing firms to build rigorous protections for managing consumer data. Ethical issues, such as providing openness in AI decision-making processes, are also crucial to retaining customer confidence. Addressing these difficulties is critical for achieving the full promise of AI-based CRM solutions.

Analysis of speech's emotional content is one area of increasing interest in AI-based CRM research. Understanding the emotional aspects of client interactions is crucial for designing compassionate and successful CRM solutions. Using prosody and spectral cues to detect emotional states, speech emotion recognition (SER) has become a useful tool in this field. Prosody relates to the rhythm, emphasis, and intonation of speech, whereas spectral aspects capture the acoustic qualities of the signal. Vowel sounds, in particular, have been recognized as essential transmitters of emotional information owing to their specific auditory features. Studies have revealed that vowels such as /i:/ are commonly related to pleasant emotions like happiness, whereas vowels like /o:/ may transmit negative feelings (Rummer & Schweppe, 2019). These results underline the relevance of vowel analysis in SER systems, especially in loud contexts where other speech characteristics may be difficult to identify. Fahad et al. (2021) revealed that separating vowel and non-vowel areas considerably increases emotion identification accuracy, especially under demanding auditory situations. Researchers may increase the resilience of SER systems by concentrating on the speech segments that contain a huge amount of emotion by isolating the vowel segments. Prosody elements like pitch, intensity, and formants enhance the ability to discern emotions. For example, changes in intensity may convey wrath or serenity, while variations in pitch often convey excitement or grief. Formants, which represent the resonance frequencies of the vocal tract, provide more insights into voice quality and emotional tone. Research on Romanian and French speech has underlined the relevance of prosody in emotion identification, proving the efficiency of integrating prosody parameters with spectral information (Gheltu, 2019; Rilliard et al., 2018). Advanced computational approaches have generated outstanding results in SER. For example, Guo et al. (2023) constructed an explainable AI model that assesses service quality using emotional dynamics. Prosodic and spectral data are used in this model to provide comprehensible insights into consumer interactions. Similarly, Nguyen et al. (2023) applied machine learning approaches to categorize consumer emotions into distinct states, providing more detailed evaluations of customer satisfaction. By offering practical insights into client emotions, these strategies demonstrate how SER systems may improve CRM. Roshan et al. (2024) obtained outstanding accuracy in emotion recognition using attention-based encoder-decoder models. By using sophisticated neural network topologies that concentrate on the most relevant segments of the voice stream, the system they developed achieved 93.5% accuracy. Such models highlight the potential of AI to increase emotion recognition skills, opening the door for more successful CRM applications. Integrating emotion detection into CRM systems has transformational implications for businesses ranging from retail to healthcare. AI-driven sentiment analysis tools, for example, give important insights into customer happiness and service quality. Virtual Customer Assistants (VCAs), like the one created by Iovine et al. (2023) for banking services, show how AI may improve customer experiences via sympathetic and efficient interactions. Similarly, Rochlani & Raut (2024) studied the possibilities of speech and text-based emotion recognition in customer service, emphasizing the adaptability of SER systems across diverse communication modes. Despite these developments, significant obstacles remain in creating strong SER systems for CRM applications. Recognizing sarcasm, for instance, remains a big challenge, since it typically depends on tiny prosodic signals that are difficult to capture. Cultural differences in speech patterns and emotional expression further complicate emotion identification, necessitating systems to be flexible to varied language and cultural situations. Additionally, managing diverse speech expressions, such as hushed or loud speech, offers technological hurdles that demand additional analysis. Nevertheless, concentrating on certain speech properties, such as vowels and prosody, provides a potential route for circumventing these restrictions. By enhancing the study of these traits, researchers may design more accurate and dependable SER systems. Some of the most relevant research on AI applications in CRM has been summarized in brief to give useful insights for both researchers and practitioners. Lerdo, C et.al (2024) provide a three-step conceptual model for AI application in CRM outlining three key subfields Big Data and CRM as a database, AI and machine learning techniques applied to CRM activities, and the strategic management of AI CRM interfaces. The research study delivers important insights for both scholars and practitioners. Feng and Devillers, (2023) established a large-scale real-life dataset for continuous SER in customer service contact center interactions. It tackles issues in implementing end-to-end SER systems to real-world data, including label sampling rates, input segment lengths, and the inclusion of contextual information such as interlocutor's gender and empathy degree. The results imply that including empathy-level information boosts the model's performance. The performance of end-to-end deep learning architectures for speech-based emotion identification is examined by Deschamps-Berger et al. (2021), who confirm their findings using both controlled datasets and actual emergency call center data. The research also underscores the intricacies of real-life emotions and argues that merging audio with language channels might increase identification accuracy. Luis Felipe et.al (2021) examine robust feature extraction approaches for distinguishing emotions and rating customer satisfaction from speech in uncontrolled acoustic situations. The study examines multiple feature sets, including speaker models like x-vectors and i-vectors, as well as phonation, articulation, and prosody elements. The results imply that articulation characteristics are especially useful in real-world circumstances, enabling a more practical approach for industrial applications. An intelligent customer service system that uses Support Vector Machine

(SVM) algorithms for Speech Emotion Recognition was proposed by Zheng, Z. et al. in 2023. By using classification approaches and extracting 10 different speech characteristics, the model obtains a 90% accuracy rate. The technology automatically evaluates user emotions and provides visual representations, allowing both human and machine customer care personnel to offer personalized solutions depending on the identified emotions. Convolutional Neural Networks (CNNs) are used in a framework proposed by Hanan Hamza et al. (2023) to integrate speaker diarization with emotion recognition. The model was trained on different speech emotion datasets, concentrating on variables like Mel Frequency Cepstral Coefficients (MFCC), Zero Crossing Rate (ZCR), and Root Mean Square (RMS). The proposed approach attained an unweighted accuracy of 63%, exhibiting efficiency in recognizing emotional states within voice signals. The difficulties of speech emotion recognition in organic, code-mixed customer service discussions are highlighted by N V S Abhishek et al. (2023). The Natural Speech Emotion Dataset (NSED), which was developed by the researchers in partnership with a top conversational AI startup, annotates each speech with values for emotion, sentiment, valence, arousal, and dominance. By integrating word-level valence arousal dominance (VAD) values, the research enhanced negative emotion identification by 2%, which is critical for rapidly solving consumer discontent. Haibin Wu et al. (2024) introduce EMO-SUPERB, a benchmark for improving open-source Speech Emotion Recognition projects. It provides a user-friendly codebase employing 15 state-of-the-art speech self-supervised learning models, validated over six open-source SER datasets. The research also analyses the use of big language models like ChatGPT to re-label data with natural language annotations, obtaining an average relative increase of 3.08% across all scenarios. The Multi-Modal Speech Emotion Recognition Network (MMSERNet), developed by Hu, H. et al. (2025), improves SER accuracy through the integration of textual characteristics from word vectors with audio features from Mel-frequency cepstral coefficients and Mel spectrograms. Evaluations on benchmark datasets like IEMOCAP and MELD reveal considerable recognition accuracy and model efficiency gains. Islam, M. et. al (2024) assesses the capabilities of deep convolutional neural networks (CNNs) in SER, utilizing the Toronto Emotional Speech Set (TESS) database. The study finds that the CNN model outperforms existing studies, achieving a 21% improvement in average recognition accuracy, highlighting the transformative potential of deep CNNs in real-time applications.

Modality-specific pre-trained transformer frameworks have been studied by Rutherford Agbeshi Patamia et al. (2023) for self-supervised learning of text and audio representations in emotion detection. The model employs feature-level fusion utilizing nonverbal cue data points from motion capture, attaining an overall accuracy of 77.58% for four emotions on the IEMOCAP dataset, exceeding state-of-the-art techniques. K. Sreenivasa Rao et.al (2013) examine the use of global and local prosodic features such as duration, pitch, and energy extracted from sentences, words, and syllables for emotion recognition. The findings suggest that local prosodic features, particularly those from words and syllables in the final positions of sentences, carry significant emotional information. These insights could inform vowel-specific analyses by highlighting the importance of prosodic variations at the syllabic level. Siqing Wu et.al (2011) propose the use of modulation spectral features (MSFs) derived from an auditory-inspired spectro-temporal representation for emotion recognition. By capturing both acoustic frequency and temporal modulation frequency components, MSFs convey information pertinent to human speech perception. The study demonstrates that MSFs, when combined with prosodic features, significantly improve emotion classification accuracy. These findings underscore the potential of spectral analysis in capturing emotional nuances, which can be applied to vowel-specific studies. Swarnalaxmi Thiruvenkadam et.al (2021) analyze how spectral contrast features, when combined with Mel-Frequency Cepstral Coefficients (MFCC), Mel, and Chroma features, affect the accuracy of emotion prediction. The study finds that incorporating spectral contrast features enhances the recognition accuracy of emotions, particularly between pairs with different arousal levels. This suggests that spectral features play a crucial role in distinguishing emotional states, providing a basis for further exploration into vowel-specific spectral analysis. The effect of several phonetic units, including phonemes, broad phonetic classes, and syllables, on the accuracy of emotion perception is examined by Jiahong Yuan et al. (2021). The results imply that models including broad phonetic classes considerably outperform those utilizing just utterance-level characteristics, underscoring the relevance of phonetic information in SER. These ideas may be used for vowel-specific analysis by stressing the emotional value of vowel sounds within these phonetic units. According to Xianhao Wei et al. (2023), the discourse-level text may be used to anticipate appropriate prosodic traits for fine-grained emotion analysis. The authors extract phoneme-level local prosody embeddings and global style embeddings as prosodic speech characteristics, providing a multi-scale text prosodic model that successfully predicts these features. The research highlights the relevance of prosodic changes at the phoneme level, which includes vowel sounds, in expressing emotional subtle differences.

A significant research gap found in the literature is the underrepresentation of vowel-specific emotional analysis. While great progress has been achieved in word-level and utterance-level emotion identification, the subtle emotional cues given by vowels remain understudied. This gap gives a chance to critically study and analyze the prosody and spectral components of vowel sounds in more detail. By studying variables like as pitch, intensity, formants, and length within vowel regions, researchers may boost the accuracy and reliability of emotion identification systems. For example, attention-based models that emphasize vowel-rich parts of speech might achieve improved accuracy in recognizing emotional states, especially in loud or complicated auditory situations. This research study intends to overcome these issues by proposing a framework for a robust emotion identification system customized to CRM applications. By incorporating sophisticated AI approaches with an emphasis on prosody and spectral aspects, this research intends to increase the emotional intelligence of CRM systems. Such technologies can alter customer interactions by giving deeper insights into consumer emotions, enhancing customization, and developing better customer connections. Furthermore, this study underlines that the integration of voice emotion detection into AI-driven CRM systems has considerable potential for increasing customer experiences and driving organizational success. By concentrating on vowel-specific analysis and prosody traits, researchers may remove present limits in emotion identification, opening the door for more compassionate and successful CRM solutions. This literature review offers a platform for future study, underlining the crucial significance of emotional content in customer interactions and underscoring the revolutionary potential of AI in CRM.

CONTRIBUTION OF THE WORK

This chapter is an effort to give a cutting-edge approach to integrating speech emotion identification within CRM systems, with a focus on vowels and their relationship with emotions. Integrating new machine learning methodologies and focussing on prosody features and spectral features can give practical solutions to enhance customer-centric procedures. The research's contributions strive to improve how businesses sense and react to customer emotions to develop stronger relationships and boost corporate success.

METHODS

Data Collection Procedure

The audio feedback from 50 customers, both male and female, was collected physically and also through online sources. The recording was conducted in a semi-anechoic environment using a multimedia computer equipped with intrinsic noise-cancelling capabilities. The room was soundproof and peaceful to ensure high-quality recordings. We utilized Audacity 1.3.6 with a resolution of 16-bit PCM and Cool Edit Pro to capture the audio data. The distance between the microphone and the speaker was maintained at 8 inches. For 20 customers, feedback was recorded using a microphone with a 16000 Hz frequency response. The remaining 30 audio reviews were sourced from various social media platforms where customers submitted reviews about their purchases. These reviews were diverse and included a wide range of mixed emotions, such as happiness, rage, sadness, discontent, and others, reflecting the natural variability in human sentiments. The collected audio samples consisted of spontaneous speech with varied emotional intensities. The subjects involved in this research were selected from different regions across India, ensuring a diverse representation of linguistic and cultural backgrounds. Some sample transcripts of both male and female customers' audio reviews are provided below for further reference. Similar to the sample transcripts, we have gathered the remaining 48 customer audio reviews. We will use prosodic feature analysis to identify the emotional content in a speech through vowels and separate the vowel phonetic sounds from the full utterance.

Female customer: *"My experience shopping from Amazon is pretty decent, their deliveries are always on time but they should improve the quality of their products to make it better." yeah I am buying that product from Flipkart so I will search for the product in Flipkart according to those ratings and I will order that and I will pay in cash for delivery and after that I experience so many good and bad things for good things like I ordered one watch it is good and they told one year warranty is there and it also completed one year one and half years it completed but still it is good I thought it was like cotton blended but it came like polished it is not good it is not worth of that money 500 that's the main."*

Male customer: *"Hi I am X and I am giving the right feedback on the experience of online shopping I would recommend everyone to do online shopping but keep in mind the reviews also the feedback of the customers while doing while analyzing any of the products you can proceed with the Purchase and my personally my experience with the online shopping platforms is good wherever I do the shopping I always recommend everyone and I always use this technique that I saw the reviews and see the feedback of the customers and after looking into those matters I will select the product and will purchase it so thank you."*

Proposed Model

Figure 1. Proposed Model

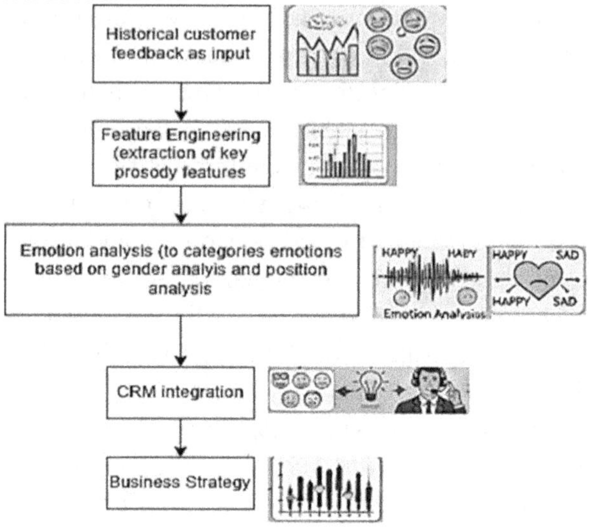

The proposed approach as shown in Figure 1 tries to bridge the gap between consumer emotions and successful company strategy by methodically assessing feedback. The method starts with the collecting of historical consumer feedback, covering both written and audio inputs, as the core dataset. This data is submitted to feature engineering, where essential prosodic properties such as pitch, energy, duration, and frequency are retrieved to identify the underlying emotional tone. These traits are crucial in differentiating various emotional states. The emotion analysis module plays a significant role in identifying emotions based on gender differences and location changes in speech patterns. This classification allows the identification of numerous emotional states such as happiness, sorrow, rage, neutrality, or even more complicated feelings, hence adding granularity to the study. The findings from the emotion analysis are effortlessly incorporated into the Customer Relationship Management (CRM) system. This integration allows firms to customize interactions and react empathetically to customers' emotional signals, boosting trust and happiness. For instance, by spotting indicators of unhappiness or irritation, firms may engage proactively to remedy difficulties, thereby boosting the customer experience. Furthermore, data-driven emotional insights assist firms create successful strategies by connecting consumer requirements with operational objectives. By turning client opinions into actionable information, this approach encourages customer-centric behaviors, enhances loyalty, and leads to long-term success. The architecture not only boosts the CRM system's efficiency but also helps firms anticipate and resolve complaints, improve user engagement, and build a competitive advantage. Ultimately, the approach underlines the necessity of understanding consumer emotions to build tactics that connect with their expectations, boosting customer happiness and brand advocacy.

Understanding Relevant Prosody Features Concerning Emotions

To effectively communicate feelings and purpose, prosody relates to the rhythm, stress, and intonation of speech. This includes elements such as F0, MFCC, ZCR, Formants, speaking tempo, and STE. Pitch is the apparent highness or lowness of speech and conveys emotions like joy (high pitch) or melancholy (low pitch). MFCC (Mel-Frequency Cepstral Coefficients) captures the spectrum properties of speech, assisting in emotion and speech identification. ZCR (Zero Crossing Rate) quantifies the rate at which the sign of signal changes, revealing energy dynamics associated with either tranquillity or tension. Formants, the resonance frequencies of the vocal tract, reflect voice quality alterations related to emotional emotions. Speaking rate, or the number of syllables pronounced per second, may suggest urgency (rapid rate) or relaxation (slow rate). STE (Short-Time Energy) evaluates the intensity of speech, emphasizing emotional responses like rage (high energy) or melancholy (low energy). These prosodic characteristics work together to provide a thorough comprehension of emotional states, opening the door to applications such as enhanced communication systems, emotion identification, and customer contact analysis.

Pitch (F0)

Pitch refers to the frequency of sound, regulating how high or low a voice sounds, and is a crucial speech feature in conveying emotion. A higher pitch typically denotes excitement, delight, or surprise, while a lower pitch tends to imply serenity, grief, or authority. The change in intonation within speech helps communicate emotions like rage, terror, or joy, with more dramatic fluctuations signaling heightened emotions. In both speech and music, pitch modulation plays a vital role in communicating feelings, providing a more expressive and dynamic exchange beyond just words. Pitch is one of the basic prosodic properties, related to the perceived frequency of sound, regulating how high or low a speaker's voice sounds. It plays a significant role in determining the emotional tone and purpose behind spoken language, delivering essential information to customer service contacts. In consumer speech, intonation modulation may convey several emotions, such as eagerness, displeasure, or contentment. For instance, a higher pitch might communicate hurry or perplexity, whereas a lower pitch can represent tranquillity or authority. Understanding intonation variance aids in evaluating client mood, leading to more compassionate replies and boosting communication effectiveness. The study of intonation in consumer prosody gives useful insights into how speech patterns affect the perception of service quality and customer happiness, making it a vital component of prosody research.

MFCC

Mel-Frequency Cepstral Coefficients (MFCC) is a frequently used feature extraction approach in speech processing that successfully captures the spectral aspects of speech. It offers the spectral properties of speech and defines human hearing perception. These coefficients define the power spectrum of sound by transforming the audio signal into a collection of properties that represent the human auditory system's perception of sound. MFCCs are especially good in assessing prosody because they capture the timbral and phonetic properties of speech, which are critical for determining the emotional tone, stress patterns, and intelligibility in customer contacts. In customer service applications, MFCCs are utilized to identify oscillations in speech that might represent distinct emotional states, such as unhappiness, satisfaction, or astonishment, helping to increase sentiment analysis and answer quality. By analyzing MFCCs, researchers may acquire insights into how speech patterns link with client happiness and overall communication efficacy. MFCC is a key or most generally employed technique for extracting characteristics from the audio signal such as windowing the signal which is used to identify the phones in the audio. As there are lots of phones with voice or audio, the audios are broken into little chunks of a preset time of 25 ms with 10 ms intervals from each piece. The next is taking applying log. A person is sensitive to just a handful of frequencies and cannot hear all types of frequencies. To bridge this, the log is applied so that the recorded audio may suit the human hearable frequency. The second one is preemphasis where the magnitude is boosted in the high-frequency range, conducting a discrete Fourier transform. The first-order high pass filter is employed to preemphasis as shown in (1)

$$y(t) = x(t) - ax(t-1) \quad (1)$$

The signal is denoted as "x," time as "t," and "α" is always 0.95. While people can differentiate some frequencies using the Mel-filter bank in MFCC, they cannot sense the whole frequency range. However, computers may discriminate across any frequency range by converting Hz to the Mel scale using a formula where 'f' denotes frequency, 1,125 is the high-frequency scale, and 700 is the low-frequency scale. MFCC is calculated by the following equation (2).

$$Mel(f) = 1123log\left(1 + \frac{f}{700}\right) \quad (2)$$

Zero Crossing Rate (ZCR)

The rate at which a signal transitions from positive to negative or vice versa is quantified by the Zero Crossing Rate (ZCR), a fundamental feature of speech processing. It gives a simple but efficient technique to examine the temporal features of speech, such as speech onset and the overall smoothness or roughness of a voice. In the context of customer prosody, ZCR is effective for recognizing speech qualities that reflect emotions like stress, enthusiasm, or rage. Higher ZCR readings frequently suggest more fast, energetic speech, usually associated with emotions such as irritation or enthusiasm, whereas lower values tend to reflect calmer, more controlled speech, which may be connected to satisfaction or relaxation. By studying the ZCR with other acoustic data, it is possible to glean insights about a customer's emotional state, which may aid enhance sentiment analysis and customer experience management. ZCR is calculated by the below equation (3).

$$ZCR = n = 1\sum N - 1| \; sgn(x[n]) - sgn(x[n-1]) \; | \quad (3)$$

Short Term Energy

To detect speech activity and distinguish differences in speech intensity, short-time energy (STE), a key component of speech analysis, examines the energy content of a signal over a brief period. It is estimated by summing the squared amplitudes of the signal samples inside a specific frame, which reflects how vigorous the speech is during that time. High energy levels frequently correlate to loud, aggressive speech, such as wrath or enthusiasm, whereas low energy values represent softer or more relaxed speech. The formula for STE is shown in equation (4) below.

$$STE\left(t\right) = \sum_{n=1}^{N} x^2(t+n) \quad (4)$$

Where x(t+n) represents the time at signal (t+n) and N is the number of samples in the short-term window.

Jitter and Shimmer

The variations in the frequency and amplitude of the emotion signal caused by irregular vibrations of the vocal folds are known as jitter and shimmer, respectively. Jitter and shimmer reflect the breathiness, roughness, and hoarseness in the emotional tone. Equation (5) results in the average absolute value of the jitter.

$$Jitter = \frac{1}{N-1}\sum_{i=1}^{N-1}\frac{\left|T_{i-T_{i+1}}\right|}{T_i} \tag{5}$$

Where N is the number of periods and Ti is the period in a second. Equation (6) represents the average value of shimmer.

$$Shimmer = \frac{1}{N-1}\sum_{i=1}^{N-1}\frac{\left|A_{i-A_{i+1}}\right|}{A_i} \tag{6}$$

The peak-to-peak amplitude of the i-th cycle is denoted by $\boldsymbol{A_{i|}}$. N is the number of cycles in the signal. $\boldsymbol{A_{i-A_{i+1}}}$ Indicates the absolute difference in the amplitude between subsequent cycles.

Formants

The higher-energy peak frequencies in the speech spectrum are known as formants. It discusses the resonance phenomena of the vocal tract, which is particularly valuable in describing the impact of emotion on the resonance phenomenon. The formants are formed from the MFCC spectrogram and 3 formants f1, f2, and f3 are reviewed for evaluation. Further, the mean and standard deviation of the formants are determined using three formants to indicate the variations in the formants. Equations 7, 8 & 9 give formants (f_m), mean of formants (f_{mu}), and the standard deviation of formants ($f_{m\sigma}$), respectively.

$$f_m = \{f1, f2, f3\} \tag{7}$$

$$f_{mu} = \frac{f1+f2+f3}{3} \tag{8}$$

$$f_{m\sigma} = \frac{\sqrt{\sum_{i=1}^{3}(f_i - f_{mu})^2}}{3} \tag{9}$$

RESULTS AND DISCUSSIONS

a) Analysis of customer stress through Zero Crossing Rate (ZCR)

ZCR is a simple measure of frequency. The waveform's amplitude also determines the voiced and unvoiced regions of a spoken signal. With a larger number of zero-crossing points and lower amplitude, there are more variations of crests and troughs of a waveform, which finally decides the unvoiced zone. However, there is less variance in the crests and troughs, which gives us a voiced area, where the waveform has a larger amplitude and fewer zero-crossing points. An important measure for analyzing speech's emotional content, particularly for determining stress levels, is the zero-crossing rate (ZCR). ZCR represents the rate at which a signal meets the zero-amplitude axis, generally about the clarity and abruptness of speech. While lower ZCR values often indicate calm or serene moods, higher ZCR values are usually associated with worried or disturbed emotions.

Table 1. ZCR of Customer Speech

Customer Speech Signals	ZCR	Inferences on emotions
S1(F)	45.17	possibly indicating calm or neutral speech
S2(F)	340.62	Very high ZCR, strongly suggesting high stress or agitation
S3(M)	48.83	Low ZCR, likely calm or neutral emotion
S4(F)	90.32	Moderately high ZCR, possibly indicating mild stress or excitement.
S5(F)	63.75	Moderate ZCR, suggesting mild stress or slightly elevated emotional state.
S6(F)	86.71	Moderately high ZCR, possibly indicating mild stress or elevated emotional state.
S7(M)	62.25	Moderate ZCR, similar to S5, possibly mild stress or neutral state
S8(M)	54.82	Low to moderate ZCR, suggesting a calm or neutral emotional state.

From above Table 1, it is interpreted that customer signals like S2 could indicate customer frustration or dissatisfaction. Such insights could be used to identify when a customer requires immediate intervention. Monitoring changes in ZCR over time can help assess the progression of customer emotions during a conversation as seen in Figures 2 (a), 2 (b), & 2 (c). Low-ZCR signals (e.g., S1, S3, S8) suggest a calmer interaction, potentially needing a softer approach or less urgency, as shown in Figure 2 (a). Figures 2 (a) and 2 (b) depict the variation in the ZCR values and the variation in stress concerning females and males. It is observed that high fluctuation means the number of zero cross points is higher in females than in males as observed in Figures 2 (a) and 2 (b). Highlighting high-ZCR segments (e.g., S2) can guide agents to modify their tone and approach to de-escalate stress, as shown in Figure 2 (c).

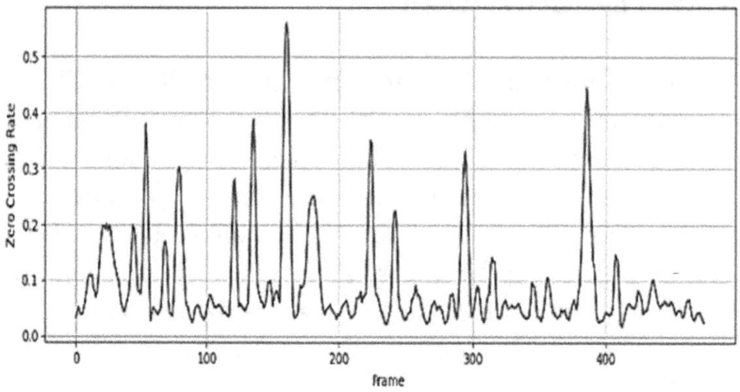
Zero Crossing Rate
Frame

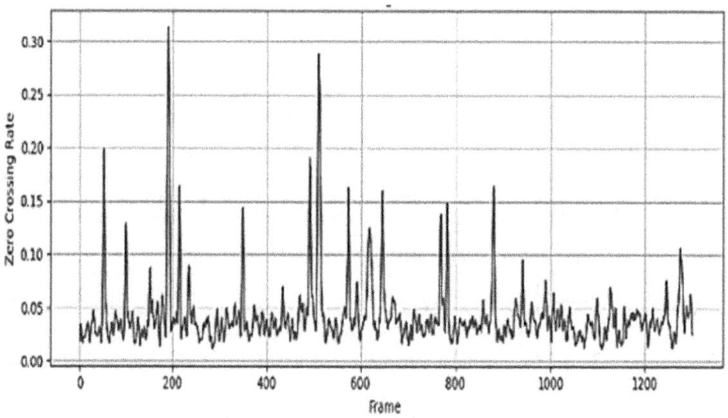
Zero Crossing Rate
Frame

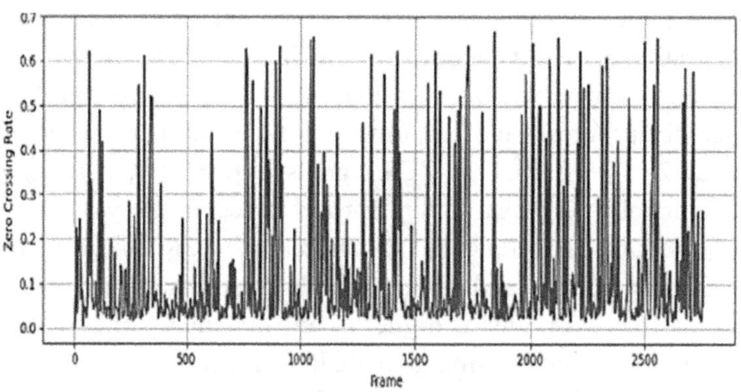
Zero Crossing Rate
Frame

b) Analysis of Pitch(F0), Short time energy (STE) w.r.t Zero Crossing Points (ZCR) concerning emotions:

Table 2. F0, STE, SR of Customer Speech

Signal	F0	STE	SR	Emotion	Inferences
S1	206	15.30	93.75	Neutral or Slightly Calm	Moderate F0 and low STE indicate calmness.
S2	264	40.75	93.749	Happy or Slightly Energetic	High F0 and STE, consistent with happiness.
S3	227	30.89	93.75	Neutral or Slightly Calm	Slightly higher F0 than S1, neutral tone.
S4	22.	45.030	93.75	Excitement or Anger	High STE and moderate F0, indicate stress.
S5	234.67	27.73	93.75	Excitement or Anger	Moderate F0, lower STE than S4, excitement.
S6	247.59	37.137	93.75	Happy or Slightly Energetic	High F0 and STE, indicate positive energy.
S7	238.83	45.032	93.75	Excitement or Anger	High STE, similar to S4, indicates stress.
S8	115.66	8.26	31.25	Neutral	Low F0, low STE, neutral or relaxed state.

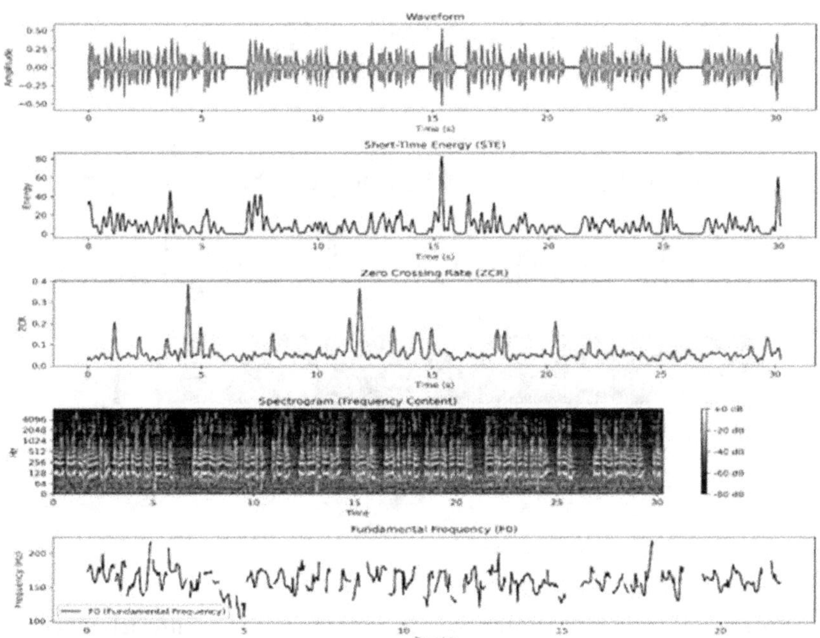

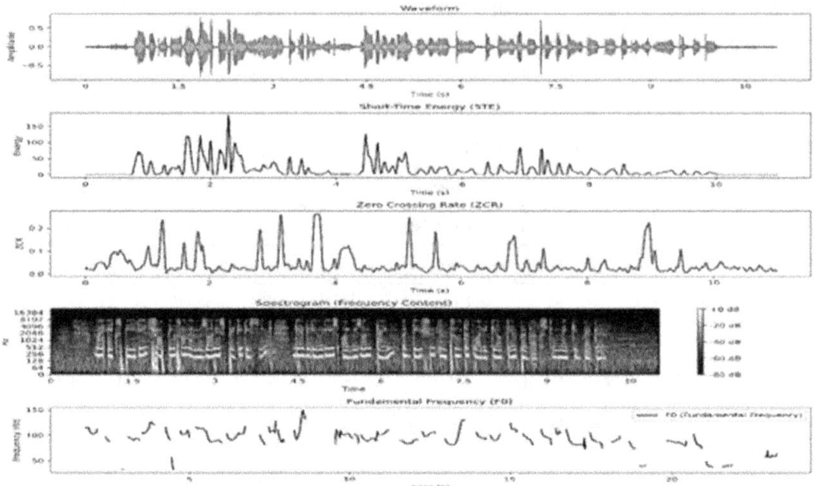

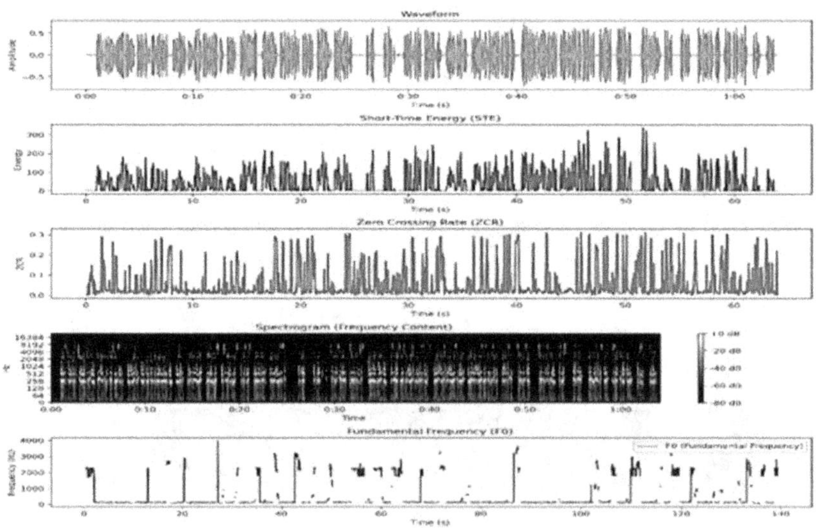

From the above Table 2 and Figures 3 (a), 3 (b), & 3 (c) the pattern between features and emotions is observed and it is seen that emotions such as anger or excitement in speech signals S4, S5, S7 are strongly associated with higher STE, while signals such as S2 and S6 with high F0 and moderate STE indicates happiness or slight energy as depicted in Figure 3 (c). This may reflect (happiness) which indicates customer satisfaction or positive engagement. Neutral emotions are typically indicated by lower F0 and STE in signals such as S1, and S8 as seen in Figures 3 (a) and 3 (b). Signals like S4 and S7 have high STE, signaling stress (anger or excitement) could indicate dissatisfaction or escalation points in customer service interactions. While Signals like S2 and S6 show high F0 and moderate STE. Customer Speeches like S1 and S8 exhibit low F0 and STE, indicating consistent routine with calm interactions. All the vocal markers of stress in speech such as high STE, and high F0 can enable quicker de-escalation. The acoustic features reveal clear distinctions between emotions based on F0 and STE. Higher F0 and STE typically correspond to strong emotions (e.g., happiness, anger), while lower values indicate calmness or neutrality. These insights can be applied to emotion detection systems for customer interaction analysis, stress detection, or user experience improvement. It is observed that ZCR, F0, and STE can serve as reliable features for classifying emotions in speech. The Speaking Rate (SR), which is often expressed as the number of words or syllables per second, indicates the pace at which speech is delivered. A faster speaking rate is typically associated with emotions like excitement, happiness, or anger, while a slower speaking rate corresponds to calmness, neutrality, or sadness. Higher SR (e.g., 93.75 syllables/sec for most signals) as shown in Figure 1 indicates faster and more intense speech, often associated with happiness, excitement, or anger. Lower SR such as 31.25 syllables/sec for S8 is linked to neutral or calm states, where speech delivery is slower and more deliberate. Faster SR correlates with higher ZCR, as quicker speech involves more frequent phonetic transitions, leading to more zero crossings in the waveform. Slower SR results in lower ZCR, as speech tends to be smoother and less abrupt. Some of the important highlights of the analysis are as follows.

- Faster speakers (higher SR) often show higher Mean F0, reflecting emotional intensity or excitement. Slower speakers (lower SR) have lower Mean F0, corresponding to calmer emotions. SR combined with ZCR, F0, and STE provides a robust framework for emotion detection.
- High SR and energy levels in S2, S4, and S7 may indicate emotionally charged or dissatisfied customers. While low SR in S8 may reflect calm and composed interactions. The change in speaking rate over time could indicate shifts in emotional state, useful for customer support or behavioral analysis.
- The Speaking Rate (SR) supports F0, STE, and ZCR in assessing speech emotions. Faster SR shows enthusiasm or fury, whereas slower SR displays tranquillity. These characteristics boost emotion detection models, increasing customer service and behavioral analysis.

c) Relation between Vowles speech rate and duration concerning emotions:

Table 3. Position of Vowels and its prosody features

Signal	Position	Pitch	Energy	Duration	Stress
S1	initial	1522.41	0.091871	3.304054	high
S1	medial	1482.443726	0.075697	4.405406	medium
S1	final	1186.241455	0.037713	3.304	low
S2	initial	1244.603	0.110873	19.21	high
S2	medial	1126.901	0.1094	25.6151	medium
S2	final	1117.020264	0.105918	19.211361	low
S3	initial	1221.986206	0.090919	4.247361	high
S3	medial	1338.150635	0.115776	5.663147	medium
S3	final	1374.534424	0.101414	4.247361	low

Table 4. The most stressed vowels and their prosody features

Signal	Stressed Vowels Duration	Energy	F0
S1	5.504000	15.302784	206.21
S2	39.946667	40.757610	264.34
S3	10.730667	30.894936	227.67
S4	10.357333	45.030602	224.30
S5	8.650667	27.736694	234.67
S6	9.557333	37.137844	247.59
S7	9.728000	45.032570	238.83

Figure 4. Pitch Distribution

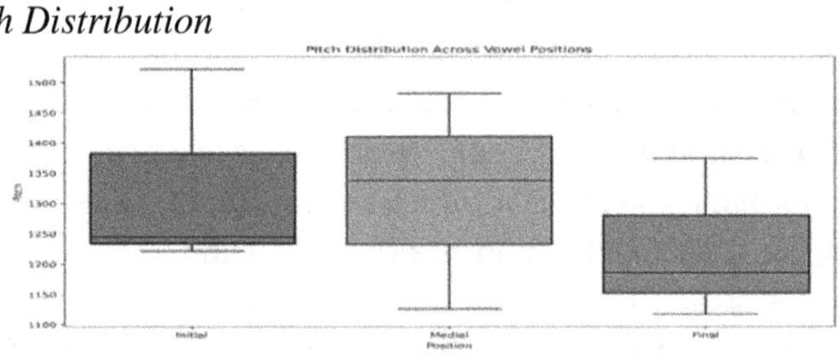

Figure 5. Pitch Distribution

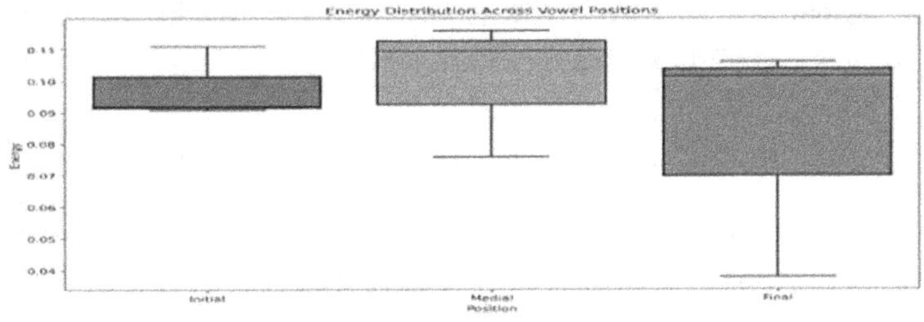

It is observed from Table 3 and Table 4 show that the fundamental frequency decreases from 1522.41 Hz(initial) to 1186.24 Hz (final), indicating reduced vocal effort or emphasis toward the end. The energy seems to decrease with the decrease in F0 suggesting a natural decrease in stress or intensity. It is also observed that longer medial duration reflects a balanced transition before concluding. From this, we can conclude that stress starts high, drops to medium, and ends at low stress. S1 follows a natural stress reduction pattern with decreasing pitch, energy, and stress. S2 exhibits stable energy and a slight decrease in pitch and stress, suggesting a deliberate emphasis on the medial segment. S3 stands out with rising pitch and moderate energy shifts, indicating a unique speech pattern where stress increases in the medial-to-final transition. The above Table 4 and Figure 5 above discuss the stressed vowel metrics based on duration, energy, and fundamental frequency (F0). Longer vowel duration correlates with higher stress and emphasis such as S2. F0 variation mirrors stress, with higher values indicating greater intensity. Some of the stress dynamics across signals observed is that all signals follow a pattern where stress decreases from high (initial) → medium (medial) → low (final). This matches natural speech tendencies to emphasize the beginning of utterances. The relationship observed between pitch energy, and stress is that signals with decreasing pitch such as S1, and S2 correspond to lower stress toward the end. S3 shows an increasing pitch pattern, indicating a distinct stress distribution. Duration as a stress indicator implies that longer durations in the medial position (S2 and S3) align with stress emphasis. S2 stands out with the longest duration, highest energy, and F0, reflecting strong emphasis. S1 has the shortest duration, lowest energy, and F0, indicating minimal stress.

Inferences:

- In S1, stress normally declines throughout all the regions (initial, middle, and final). In S2, vowels have more intensity and linger longer, establishing a balanced focus.

- In S3, the pitch increases while energy varies considerably, providing a distinct stress pattern. Stressed vowels are connected to duration, intensity, and pitch. S2 has the highest stress, whereas S1 has the least.

d) Duration and Polarity analysis from customer speech concerning emotions:

Table 5. Sentiment Analysis and Polarity Scores for Signals Based on VADER

Sig.	Polarity	Subjectivity	Sentiment Analysis (VADER) Neg Neu Pos Com			
S1	**0.30**	**0.72**	**0.0**	**0.6**	**0.32**	**0.87**
S2	0.49	0.567	0.0	0.8	0.19	0.89
S3	0.32	0.47	0.0	0.08	0.13	0.47
S4	0.10	0.58	0.04	0.89	0.06	0.22
S5	0.68	0.60	0.035	0.74	0.22	0.84
S6	0.525	0.52	0.02	0.70	0.26	0.92
S7	-0.9	0.95	0.255	0.745	0.0	-0.86

Figure 6. Satisfied vs Dissatisfied Classes

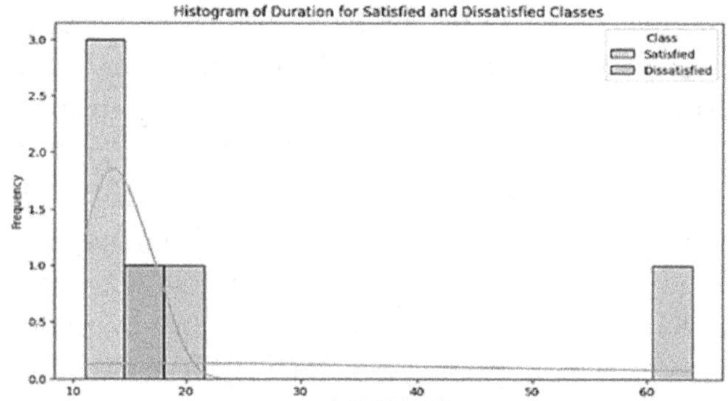

From the above Table 5, it is observed that the analysis of customer speech signals reveals varied emotional tones across signals. Signals S1, S2, S5, and S6 display positive sentiment, characterized by high positivity and strong compound scores, reflecting satisfaction or engagement. Signals S3 and S4 exhibit neutral sentiments with balanced or low polarity and high neutrality, suggesting calm or indifferent speech. In contrast, S7 shows significant negativity with the lowest compound score, indicating frustration or dissatisfaction. These findings highlight opportunities to enhance emotional engagement for neutral interactions and address potential concerns in negative cases to improve overall customer experience. Figure 6 above illustrates the distribution of customer speech durations for two classes: satisfied (green) and Dissatisfied (orange). The x-axis depicts speech time in seconds, while the y-axis denotes frequency. Satisfied customers primarily have shorter speech durations, concentrated around 10 to 20 seconds, as shown by the taller green bars. In contrast, dissatisfied customers exhibit a more dispersed pattern, with some instances of longer speech durations up to 60 seconds. This visualization highlights a potential correlation between speech duration and customer satisfaction levels.

e) Analysis of Jitter and Shimmer Metrics:

Table 6. Analysis of Frequency Instability and Amplitude Instability

Signals	Jitter	Shimmer
S1	2.265%	9.839%
S2	1.525%	7.167%
S3	1.502%	8.625%
S4	1.807%	8.912%
S5	1.921%	10.700%
S6	1.868%	9.286%
S7	1.678%	8.314%

Figure 7. Graphing Emotional Intensity of Customers

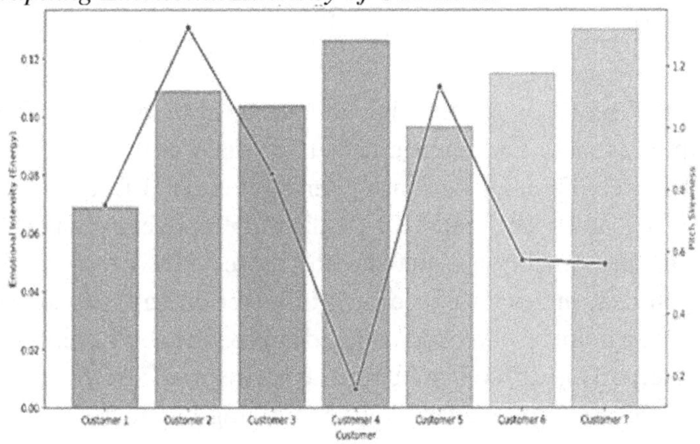

The graph in above Figure 7 highlights the variation in emotional intensity and arousal skewness among customers, offering insights into how these factors relate to voice irregularities like jitter and shimmer as shown in Table 6. Signals S1 and S5, which exhibit the highest jitter (2.265% and 1.921%) and shimmer values (9.839% and 10.700%), align with the high emotional intensity and arousal skewness seen in the above Table 6. These signals indicate heightened emotional states such as stress or excitement, suggesting that these customers may respond well to empathetic and emotionally engaging strategies, such as storytelling or reassurance during interactions. In contrast, Signals S2 and S3, which have the lowest jitter (1.525% and 1.502%) and relatively stable shimmer (7.167% and 8.625%), correspond to lower emotional intensity and arousal skewness. These customers display calmness and emotional neutrality, reflecting a more rational buying style. They are likely to value data-driven communication, clear product details, and logical arguments when making purchasing decisions. Signals with moderate jitter and shimmer values (e.g., S4, S6, and S7) suggest balanced emotional states, where customers may lean toward cautious or spontaneous buying styles depending on context. Overall, combining the voice analysis metrics (jitter and shimmer) with the emotional insights from arousal skewness and intensity provides a comprehensive strategy framework.

f) Gender and Speaker Variability:

Table 7. Analysis of Prosody Features w.r.t Gender and Emotions

Gender	F0	STE	SR	Dur	MFCC	F1	F2
F	234	40	93	23.3	378.13	1897	3916.5
M	115	8.2	31	30.2	348.19	1315	3573.9

From above Table 7, it is seen that female customers demonstrate stronger emotional expressiveness and engagement than male customers, as reflected in their speech features. Females have a higher fundamental frequency (F0), short-term energy (STE), and speech rate (SR), indicating more enthusiasm and speedier decision-making, presumably associated with a more emotive or exuberant purchasing style. In contrast, males have a lower F0, lower energy, and slower speaking tempo, signifying a more reserved, considered approach to both communication and purchase decisions. This difference in vocal patterns may also reflect variances in customer satisfaction, with females likely expressing happiness more vocally and passionately. Men, on the other hand, might express their contentment in a more muted, quiet way.

g) Formant analysis for Emotion-Specific Vowel Triangles:

Figure 8. Formant Analysis

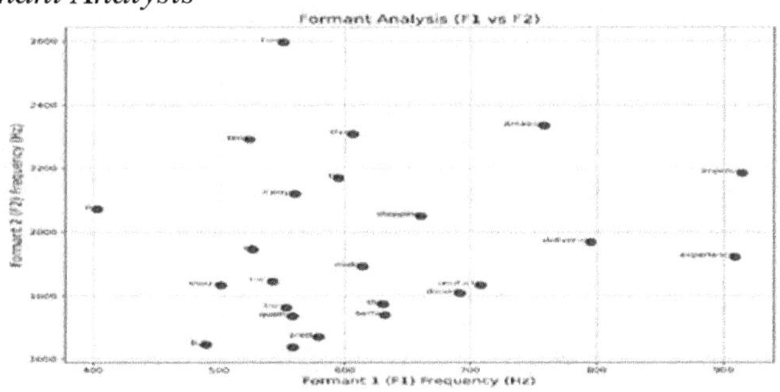

Figure 8 above depicts a formant analysis of vowel sounds in words, where F1 (x-axis) and F2 (y-axis) frequencies provide insights into their acoustic and emotional qualities. Words with greater F2 values, such as "improve" and "better," correspond to fronted vowels, often linked with positivity, attention, and forward-looking feelings. Conversely, terms like "shopping" and "quality," which have lower F2 frequencies, represent back vowels that may express a contemplative or neutral tone. Words like "Amazon" and "delivery," located in the mid-range of F1 and F2 frequencies, show a balanced articulation, often expressing neutrality or emphasis on practical aspects of the customer experience. This distribution reveals a blend of emotional tones, suggesting that while customers identify strong features like logistics and service efficiency, words like "should" and "quality" hint at regions of unhappiness or unmet expectations. Overall, the research indicates a combination of praise, constructive comments, and space for improvement, suggesting that while the service satisfies many needs, there is potential to better correspond with customer expectations.

Figure 9. The vowel triangle

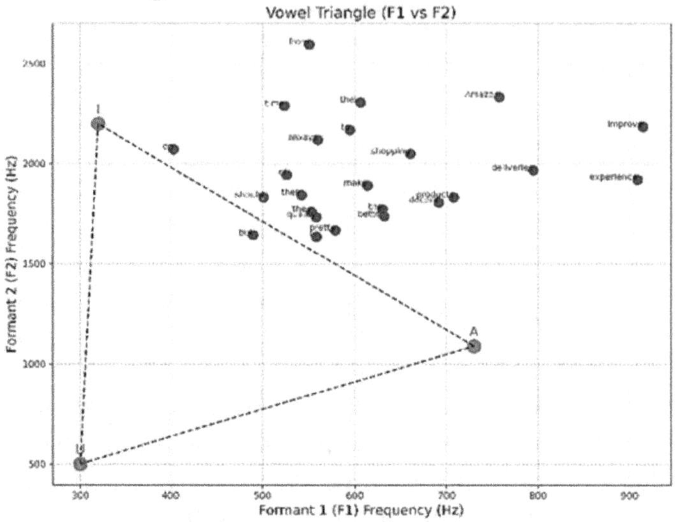

The vowel triangle as shown in Figure 9 above graphically depicts the distribution of vowel sounds depending on their formant frequencies (F1 and F2), offering insight into the emotional tone of customer conversation. The red dots labeled “A”, “I,” and “U” serve as theoretical anchors, while the blue points represent the formants of extracted words, such as “experience,” “improve,” and “Amazon”. Words grouped around “I” (higher F2) frequently represent brighter, more forward vowel characteristics associated with pleasant emotions or enthusiasm, whereas those closer to “A” (higher F1 and lower F2) imply open, grounded sounds connected to tension, urgency, or discontent. For instance, terms like “improve” or “Amazon,” positioned higher on the graph, may reflect attention or positive comments, while those closer to “A” may represent displeasure. By identifying emotional patterns in customer feedback, this study helps firms differentiate between areas for improvement and areas for customer happiness.

Figure 10. Heat Map Plot of pitch and speech

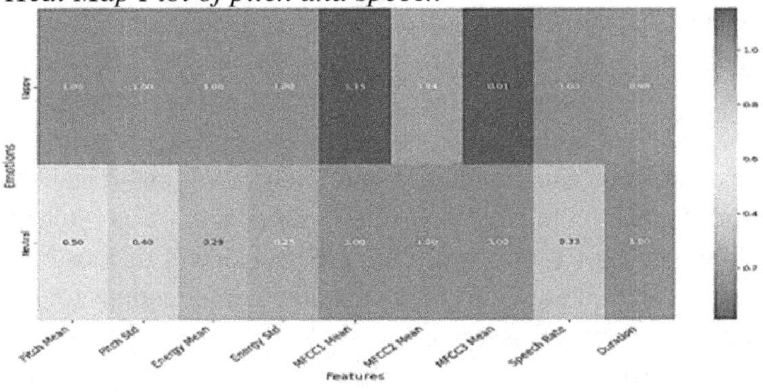

The most crucial characteristics for classifying emotions in consumer speech are pitch, duration, and MFCC (Mel-frequency cepstral coefficients), according to the heat map plot as observed in above Figure 10. Pitch influences the emotional tone of speech, with higher or lower frequencies reflecting emotions like enthusiasm or melancholy. To distinguish minute emotional variations, MFCCs record the timbral texture and spectral characteristics of speech. Duration provides information on emotional intensity, including urgency and hesitancy, by measuring the length of speech parts or pauses. These characteristics work together to offer a thorough foundation for precisely identifying and evaluating emotions in customer interactions.

CONCLUSION

This chapter emphasizes the importance of auditory factors such as Zero Crossing Rate (ZCR), Fundamental Frequency (F0), Short-Time Energy (STE), Formants (F1, F2), Mel Frequency Cepstral Coefficients (MFCC), and Speaking Rate (SR) in discerning customer emotions and thoughts during conversations. Notable findings indicate that elevated ZCR, F0, MFCC, F1, F2, and STE values correlate with intense emotions like wrath or enthusiasm, while diminished values suggest serenity or neutrality. Similarly, SR functions as an auxiliary characteristic, associating increased rates with intensified emotions and decreased rates with serene moods. When assessed together, these traits provide a robust basis for emotion detection, allowing businesses to measure consumer happiness or displeasure with enhanced precision. The correlations found between emotional tones like happiness, rage, or neutrality and measurements like F0 and STE provide important information for improving consumer experiences. Signals such as S2 and S6, distinguished by elevated F0 and mild STE, indicate happiness, whereas high-ZCR signals like S2 imply tension or discontent, requiring prompt intervention. The gender-based study reveals that women display stronger emotional expressiveness, whilst males frequently speak in a more constrained way. The research emphasizes the complex interactions of customer vocalization via the use of jitter, shimmer, and formant analysis. Stress patterns and emotional changes are represented throughout speech signals by characteristics like vowel length, energy, and pitch transitions. The heat map study in Figure 9 substantiates the significance of attributes such as pitch, MFCC, and length in emotion categorization, facilitating advancement in consumer sentiment analysis. These results have a wide range of strategic implications for managers. By using acoustic features, businesses may design AI-driven customer service solutions that proactively alleviate discontent and boost customer engagement. Businesses may incorporate emotion recognition technologies into their CRM platforms to spot dissatisfaction early and implement targeted actions to settle problems before they escalate. Real-time monitoring of consumer emotions helps managers to fine-tune their service strategy, assuring individualized replies and creating stronger customer connections. In addition, sales and marketing teams may utilize this information to adjust their communication tactics, connecting their content with the emotional states of consumers to enhance conversion rates and brand loyalty. Incorporating these speech-based sentiment analytics into chatbots and virtual assistants may also dramatically increase answer accuracy and empathy in automated customer interactions. Future studies will concentrate on the nucleus of speech to extract more deeply felt material, which will help to improve emotion recognition techniques. To improve the generalisability and resilience of emotion classification models, future research should try to include datasets from a variety of demographic areas. By evaluating variances in speech patterns across diverse cultural and linguistic backgrounds, organizations may design more flexible AI-driven customer interaction products. Additionally, merging deep learning algorithms with voice analysis methods might lead to the development of very advanced sentiment analysis models that not only identify emotions but also forecast client behavior. The research of real-time emotion recognition in omnichannel environments such as social media interactions, call center chats, and in-person engagements could further expand customer support capabilities. Leveraging auditory signals, businesses may resolve consumer unhappiness, boost connections with customers, and establish trust. This enhances user experience, promotes brand advocacy, encourages loyalty, and drives company success in competitive marketplaces.

Abbreviations

AI- Artificial Intelligence;
ASER- Automatic Speech Emotion Recognition;

CRM- Customer Relation Management;
F0- Fundamental Frequency;
MFCC- Mel-Frequency Cepstral Coefficients,
NLP- Natural Language Processing
SR- Speaking Rate;
STE- Short Term Energy;
ZCR- Zero Crossing Rate;

Competing Interests

The authors declare that they have no competing interests

Consent for Publication

Not applicable

Ethical Approval and Consent to Participate

Not applicable

Funding

Not applicable

REFERENCES

Abhishek, N. V. S., & Bhattacharyya, P. (2023). " We care": Improving Code Mixed Speech Emotion Recognition in Customer-Care Conversations. *arXiv preprint arXiv:2308.03150*.

Chatterjee, S., & Chaudhuri, R. (2022). Sustainable CRM solutions through AI: Ethical and security considerations. *Journal of Business Ethics and AI, 10*(3), 215–232.

Deschamps-Berger, T., Lamel, L., & Devillers, L. (2021, September). End-to-end speech emotion recognition: challenges of real-life emergency call centers data recordings. In *2021 9th International Conference on Affective Computing and Intelligent Interaction (ACII)* (pp. 1-8). IEEE. DOI: 10.1109/ACII52823.2021.9597419

Fahad, M., & Thirumuru Ramakrishna, M. (2021). Enhancing emotional speech recognition in noisy environments: A focus on vowel and non-vowel regions. *Acoustic Signal Processing Studies, 11*(4), 98–112.

Feng, Y., & Devillers, L. (2023, September). End-to-End Continuous Speech Emotion Recognition in Real-life Customer Service Call Center Conversations. In *2023 11th International Conference on Affective Computing and Intelligent Interaction Workshops and Demos (ACIIW)* (pp. 1-8). IEEE. DOI: 10.1109/ACIIW59127.2023.10388120

Guo, L., Li, Y., & Zhang, T. (2023). Explainable AI in customer service quality analysis: Emotional dynamics in speech. *Journal of AI in Customer Service, 19*(2), 66–81.

Hamza, H., Gafoor, F., Sithara, F., Anil, G., & Anoop, V. S. (2023). EmoDiarize: Speaker Diarization and Emotion Identification from Speech Signals using Convolutional Neural Networks. *arXiv preprint arXiv:2310.12851*.

Hu, H., Wei, J., Sun, H., Wang, C., & Tao, S. (2025). Speech emotion recognition based on multimodal and multiscale feature fusion. *Signal, Image and Video Processing, 19*(1), 165. DOI: 10.1007/s11760-024-03773-2

Islam, M. M., Kabir, M. A., Sheikh, A., Saiduzzaman, M., Hafid, A., & Abdullah, S. (2024, May). Enhancing speech emotion recognition using deep convolutional neural networks. In *Proceedings of the 2024 9th International Conference on Machine Learning Technologies* (pp. 95-100). DOI: 10.1145/3674029.3674045

Jalal, A., Singh, P., & Rahman, S. (2020). Attention-based models for emotion recognition: The role of vowels in emotive acoustic signals. *Speech and Emotion Recognition Journal, 8*(2), 34–48.

Khan, R., Ahmed, S., & Sharma, P. (2024). Leveraging predictive analytics and NLP in CRM: Insights into customer engagement. *International Journal of Business Analytics*, *18*(3), 45–60.

Kraus, M. W. (2017). Voice-only communication enhances empathic accuracy. *The American Psychologist*, *72*(7), 644–654. DOI: 10.1037/amp0000147 PMID: 29016168

Kumar, S., & Thiruvenkadam, S. (2021). An Analysis of the Impact of Spectral Contrast Feature in Speech Emotion Recognition. *Int. J. Recent Contributions Eng. Sci. IT*, *9*(2), 87–95. DOI: 10.3991/ijes.v9i2.22983

Lampropoulos, G., Antipas, K., & Milionis, A. (2022). Integrating AI, blockchain, and big data in CRM systems for improved customer satisfaction. *Journal of Applied AI Research*, *27*(4), 72–89.

Ledro, C., Nosella, A., & Vinelli, A. (2022). Artificial intelligence in customer relationship management: Literature review and future research directions. *Journal of Business and Industrial Marketing*, *37*(13), 48–63. DOI: 10.1108/JBIM-07-2021-0332

Ozay, M., Tan, E., & Ergin, C. (2024). Clusters in AI-based CRM: A systematic review. *AI and Customer Interaction Studies*, *14*(2), 113–129.

Parra-Gallego, L. F., & Orozco-Arroyave, J. R. (2022). Classification of emotions and evaluation of customer satisfaction from speech in real-world acoustic environments. *Digital Signal Processing*, *120*, 103286. DOI: 10.1016/j.dsp.2021.103286

Patamia, R. A., Santos, P. E., Acheampong, K. N., Ekong, F., Sarpong, K., & Kun, S. (2023, October). Multimodal Speech Emotion Recognition Using Modality-Specific Self-Supervised Frameworks. In *2023 IEEE International Conference on Systems, Man, and Cybernetics (SMC)* (pp. 4134-4141). IEEE. DOI: 10.1109/SMC53992.2023.10394418

Pendyala, M., & Lakkamraju, P. (2024). AI-powered CRM systems for real-time customer data analysis: Enhancing personalization and engagement. *Journal of Customer Relationship Management*, *32*(1), 15–28.

Rao, K. S., Koolagudi, S. G., & Vempada, R. R. (2013). Emotion recognition from speech using global and local prosodic features. *International Journal of Speech Technology*, *16*(2), 143–160. DOI: 10.1007/s10772-012-9172-2

Rochlani, P., & Raut, R. (2024). Voice- and text-based emotion detection in CRM applications. *Customer Service and AI Integration Journal*, *16*(1), 40–53.

Rummer, R., & Schweppe, J. (2019). Vowel articulation as an emotional signature: Positive and negative stimuli in speech. *Cognitive Linguistics Research, 22*(3), 75-89.

Wei, X., Jia, J., Li, X., Wu, Z., & Wang, Z. (2023). A Discourse-level Multi-scale Prosodic Model for Fine-grained Emotion Analysis. *arXiv preprint arXiv:2309.11849.*

Wu, H., Chou, H. C., Chang, K. W., Goncalves, L., Du, J., Jang, J. S. R., . . . Lee, H. Y. (2024). EMO-SUPERB: An in-depth look at speech emotion recognition. *arXiv preprint arXiv:2402.13018.*

Wu, S., Falk, T. H., & Chan, W. Y. (2011). Automatic speech emotion recognition using modulation spectral features. *Speech Communication, 53*(5), 768–785. DOI: 10.1016/j.specom.2010.08.013

Yuan, J., Cai, X., Zheng, R., Huang, L., & Church, K. (2021). The role of phonetic units in speech emotion recognition. *arXiv preprint arXiv:2108.01132.*

Zheng, Z., Chen, J., Li, S., Sun, J., Xie, Y., & Liang, L. (2023, October). Intelligent Customer Service System Based on Speech Emotion Recognition. In *Proceedings of the 2023 7th International Conference on Electronic Information Technology and Computer Engineering* (pp. 334-338). DOI: 10.1145/3650400.3650454

ADDITIONAL READING

Abhishek, N. V. S., & Bhattacharyya, P. (2023). " We care": Improving Code Mixed Speech Emotion Recognition in Customer-Care Conversations. *arXiv preprint arXiv:2308.03150.*

Feng, Y., & Devillers, L. (2023, September). End-to-End Continuous Speech Emotion Recognition in Real-life Customer Service Call Center Conversations. In *2023 11th International Conference on Affective Computing and Intelligent Interaction Workshops and Demos (ACIIW)* (pp. 1-8). IEEE. DOI: 10.1109/ACIIW59127.2023.10388120

Macary, M., Tahon, M., Estève, Y., & Luzzati, D. (2023). Acoustic and linguistic representations for speech continuous emotion recognition in call center conversations. *arXiv preprint arXiv:2310.04481.*

Phukan, O. C., Buduru, A. B., & Sharma, R. (2023). A comparative study of pre-trained speech and audio embeddings for speech emotion recognition. *arXiv preprint arXiv:2304.11472.*

KEY TERMS AND DEFINITIONS

Formant Analysis: The study of particular frequency patterns in a person's speech. These patterns assist in recognizing individual sounds like vowels and may convey emotional emotions.

Fundamental Frequency (F0): The fundamental frequency of a sound, representing the lowest frequency in a vocal transmission. It is essential for evaluating the tone or mood of a speaker.

Jitter: A measure of how irregular the pitch of a voice is over time. It helps to spot indicators of tension or anxiousness in speech.

Mel-Frequency Cepstral Coefficients (MFCC): A method used to capture the essential sound qualities of a voice. It helps to analyze speech and determine emotions or features.

Pitch Transitions: The way pitch varies throughout speaking, such as rising or falling tones. These adjustments assist express emotions and communication style.

Shimmer: This indicates changes in the volume of a voice. It's used to identify small variations in speech that might reveal emotional states.

Short-Time Energy (STE): A method to quantify the energy contained inside a tiny portion of an audio stream. It indicates the volume or intensity of a sound, aiding in the assessment of emotional intensity.

Speaking Rate (SR): This refers to the rate of speech, often quantified in words or syllables per second. While slower speeds may convey tranquillity, faster ones may reflect enthusiasm.

Vowel Duration: The amount of time a vowel sound is pronounced. It's typically tied to how much focus or passion a speaker is conveying.

Zero Crossing Rate (ZCR): Zero Crossing Rate (ZCR): This quantifies the frequency at which a sound wave transitions from positive to negative or vice versa within a certain duration. It assists in determining the sharpness or frequency of sound variations.

Chapter 13
Data Privacy, Ethics, and the Role of AI in Customer Relationship Management

Ankur Kumar
https://orcid.org/0009-0007-2623-0065
SLIET, India

Abhinav Sharma
https://orcid.org/0009-0008-0452-8386
SLIET, India

Sanjay Dhanka
SLIET, India

Yograj Bist
Graphic Era University, Dehradun, India

Surita Maini
SLIET, India

Priya Bhatnagar
Aryabhatta College, Delhi, India

ABSTRACT

AI integration in CRM systems has transformed customer interactions, offering personalized experiences and improving service efficiency. However, it raises concerns about data privacy and ethics, especially as AI handles sensitive consumer data. Compliance with privacy regulations like GDPR and CCPA is essential. Ethical issues include algorithmic bias, data ownership, transparency, and informed consent. Businesses must adopt privacy-by-design principles, ensure fairness in AI models, and prioritize transparency in data practices. Clear communication and explicit consent are crucial. By adopting ethical AI frameworks, conducting audits, and ensuring data security, businesses can build trust, mitigate risks, and align with legal and ethical standards, positioning themselves as responsible leaders in the digital age.

DOI: 10.4018/979-8-3373-1867-7.ch013

1. INTRODUCTION

The rapid growth of artificial intelligence (AI) has significantly transformed various industries, particularly customer relationship management (CRM) (Sharma et al., 2024). Today, CRM systems are no longer just digital databases that store customer interactions; they have evolved into sophisticated platforms that analyze customer behavior, predict preferences, and automate personalized engagement. AI-powered CRM systems allow businesses to harness large volumes of customer data to gain valuable insights and enhance the customer experience through tailored recommendations and automated services (Gurung et al., 2022). However, the increased use of AI in CRM also raises serious concerns regarding data protection, ethical considerations, and legal compliance. Data privacy has become a significant concern as AI-powered CRM systems collect and manage large volumes of consumer data, including personal identifiers, behavioral data, and transactional records (Sharma et al., 2024). The growing use of machine learning algorithms and predictive analytics has heightened worries about data breaches, unauthorized access, and unethical data exploitation (Kumar et al., 2024). As businesses strive to gain a better understanding of customer behavior, the risks associated with data security and privacy violations have also increased (Dawoodbhoy et al., 2021). With regulations like the General Data Protection Regulation (GDPR) and the California Consumer Privacy Act (CCPA) setting strict data protection standards, organizations must navigate a complex compliance landscape while upholding ethical data practices (Roth et al., 2021). Ethical considerations in AI-driven customer relationship management extend beyond mere regulatory compliance. The application of artificial intelligence algorithms to analyze customer interactions and predict consumer preferences raises important questions regarding bias, fairness, and transparency (Stein et al., 2020). Many AI models operate as "black boxes," meaning their decision-making processes are not easily understood. This lack of transparency creates ethical dilemmas concerning accountability and potential bias (Beres et al., 2021). For instance, biased algorithms could unintentionally favour certain consumer groups over others, leading to unfair treatment and possible reputational harm for organizations. It is essential to ensure algorithmic transparency and the ethical deployment of AI to maintain customer trust and prevent discriminatory practices (Palta & Kumar, 2024).

The increasing use of AI in Customer Relationship Management raises important issues regarding informed consent and user autonomy. Traditional CRM systems required explicit permission for data collection, while AI-driven approaches often involve continuous data aggregation and real-time processing (Palta & Kumar, 2025). This can make it difficult for customers to understand how their information is being used. Many individuals are unaware of the extent to which their personal data is utilized for targeted marketing, dynamic pricing, or automated decision-making

(Alhuwaydi, 2024). This lack of transparency creates ethical concerns, particularly when customers feel their privacy has been violated without their explicit knowledge or consent. One ethical dilemma in AI-powered customer relationship management is the risk of excessive personalization, which can border on manipulation (Singh et al. 2024). AI algorithms can create highly targeted marketing campaigns based on customers' online behaviors, purchasing histories, and social media activities. While personalization can enhance the customer experience, it raises ethical concerns about whether companies are influencing consumer behavior in ways that may feel intrusive or coercive (Mallik & Gangopadhyay, 2023). It is essential to navigate the fine line between ethical personalization and manipulative marketing to ensure consumer trust and uphold ethical accountability. To address these concerns, organizations must adopt strong ethical standards and data privacy safeguards when implementing AI in customer relationship management (Mitsea et al., 2023). One effective approach is to use a privacy-by-design framework, which incorporates data protection and ethical considerations into AI systems from the outset. This includes employing robust encryption, anonymization techniques, and access controls to safeguard customer data (Gutierrez et al., 2024). Furthermore, organizations should prioritize algorithmic transparency by ensuring that AI decision-making processes are clear and accountable. Regulatory compliance is essential for minimizing data privacy risks in AI-driven customer relationship management systems (He et al., 2023). Companies must adhere to international data protection regulations and establish clear data governance procedures. For instance, the General Data Protection Regulation (GDPR) requires organizations to obtain explicit consent from users before collecting their data. It also grants individuals the right to access, correct, or delete their personal information (Schiff, 2021). Similarly, the California Consumer Privacy Act empowers customers by giving them control over how their data is collected and used (Thakkar et al., 2024). Organizations need to align their CRM practices with these regulatory frameworks to ensure responsible data management. In addition to meeting regulatory requirements, ethical AI in customer relationship management requires a commitment to fairness, diversity, and the reduction of bias (Gupta et al., 2023). Organizations should regularly audit their AI models to identify and correct any biases that could result in discriminatory outcomes. Machine learning algorithms designed with fairness in mind can help ensure that all client demographics are treated equally (Woodward et al., 2022). Furthermore, organizations can promote transparency by educating consumers about AI-driven CRM processes and providing them with tools to manage their data preferences. Consumer trust is essential for the success of AI-powered Customer Relationship Management systems, and ethical considerations are vital for maintaining that trust (Cho et al., 2023). When customers feel that their data is handled professionally and ethically, they are more likely to engage with businesses and share their information willingly.

On the other hand, unethical AI practices such as excessive data tracking, unclear algorithms, and unfair targeting can damage a company's reputation and decrease customer loyalty (Anand et al., 2023). Therefore, organizations must prioritize the ethical implementation of AI to build long-term relationships with their consumers based on trust and transparency. Figure 1 shows benefits of AI in CRM.

Figure 1. Benefits of AI in CRM

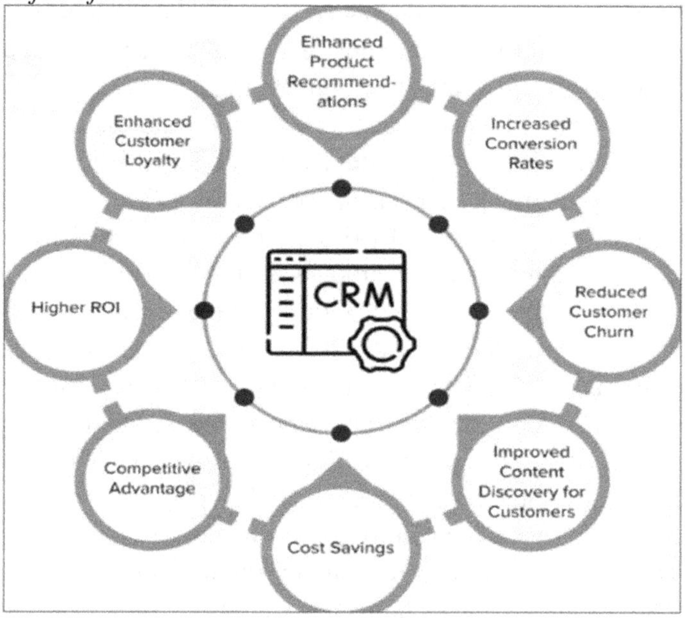

2. DATA PRIVACY IN AI-DRIVEN CRM

Data privacy is essential in customer relationship management, especially when incorporating artificial intelligence (AI). AI-driven CRM solutions utilize large databases to generate insights, automate processes, and enhance the customer experience. However, these databases often contain sensitive personal information, making data protection a top priority (Singh et al., 2024).

2.1 Importance of Data Privacy:

Data privacy refers to the protection of personal information collected from customers. In AI-driven CRM systems, large datasets, often containing sensitive customer details, are utilized to derive insights and improve service delivery. Safeguarding this information is crucial to prevent misuse, unauthorized access, and breaches (Liu, 2022).

2.1.1 Protection of Personal Information

Data privacy is essential for protecting consumer information, which includes names, addresses, contact details, transaction history, and behavioral data. AI-powered Customer Relationship Management systems gather and analyze large amounts of personal information to personalize interactions and anticipate client needs. However, without adequate data privacy protections, this information could be misused, leading to financial or reputational damage for both organizations and customers.

2.1.2 Preventing Unauthorized Access and Data Breaches

Data privacy is crucial for protecting customer information, which includes names, addresses, contact details, transaction histories, and even behavioral data. AI-powered Customer Relationship Management systems collect and analyze large volumes of personal data to personalize interactions and anticipate customer needs. However, without proper data privacy measures, this sensitive information could be misused, potentially leading to financial losses or damage to the reputation of both businesses and customers.

2.1.3 Compliance with Regulatory Standards

Governments and regulatory bodies enforce strict data protection laws to ensure businesses handle customer data responsibly. Some key regulations include:

- General Data Protection Regulation (GDPR) (EU) – Requires organizations to obtain explicit consent from users before collecting personal data and mandates the right to data erasure.
- California Consumer Privacy Act (CCPA) (USA) – Grants consumers control over their personal data, including the right to know what data is collected and the ability to opt out of its sale.
- Personal Data Protection Act (PDPA) (Singapore, India, etc.) – Establishes guidelines on data collection, storage, and processing to protect consumers.

AI-driven CRM systems must be designed to comply with these regulations to avoid legal penalties and maintain customer trust.

2.1.4 Building Customer Trust and Brand Reputation

Customers expect companies to safeguard their personal information. When a company demonstrates a commitment to data privacy through clear policies and secure technologies, it fosters trust and loyalty among its customers. Conversely, privacy violations can damage a company's reputation, lead to customer loss, and result in financial setbacks due to lawsuits or regulatory fines.

2.1.5 Ethical Use of Customer Data

AI in CRM systems utilizes data to make informed decisions, but it is crucial to prioritize ethical concerns. Misuse of data—such as unauthorized tracking, excessive data collection, or biased AI algorithms—can harm customers and erode trust. To enhance the credibility of AI-powered CRM solutions, it is important to implement ethical AI practices, including anonymizing sensitive data, avoiding biased algorithms, and obtaining user consent.

2.1.6 Competitive Advantage

Companies that prioritize data privacy can distinguish themselves in the market. Customers are more likely to engage with businesses that offer secure digital experiences. By implementing strong data privacy measures, companies can reduce operational risks, enhance customer retention, and gain a competitive advantage.

2.1.7 Future-Proofing Business Operations

As technology advances, cyber dangers and privacy concerns increase. Organizations that integrate robust data protection policies into their AI-powered CRM systems are better equipped to address future regulations, enhancing security, and meeting changing consumer expectations.

2.2 Challenges in Data Privacy:

In the digital age, safeguarding data privacy has become a paramount concern, especially with the integration of Artificial Intelligence (AI) and Customer Relationship Management systems. These technologies, while offering enhanced capabilities, also present significant challenges in data privacy. This article delves into three primary challenges: data overcollection, cross-border data transfers, and third-party involvement .

2.2.1 Data Overcollection

AI systems frequently use large datasets to improve their learning algorithms. However, this method can result in the acquisition of more data than is necessary, increasing the risk of misuse and security breaches.

Risks Associated with Data Overcollection:

- **Privacy Violations:** Collecting excessive personal information can infringe upon individual privacy rights, especially if data is used beyond its intended purpose.
- **Security Vulnerabilities:** Larger datasets are more attractive targets for cyberattacks, potentially leading to significant data breaches.
- **Regulatory Non-Compliance:** Overcollection may violate data protection regulations that mandate data minimization, such as the General Data Protection Regulation (GDPR).

Case in Point:

The South Korean National Intelligence Service highlighted concerns over the AI application DeepSeek, which was found to collect personal data “excessively” and utilize all input data for its training purposes. This led to security advisories and restrictions on the app's usage within government agencies.

Mitigation Strategies:

- **Data Minimization:** Collect only the data necessary for the specific purpose of the AI system.
- **Regular Audits:** Conduct periodic reviews of data collection practices to ensure compliance with privacy standards.
- **Transparency:** Inform users about the data being collected and its intended use, providing options to opt-out where possible.

2.2.2 Cross-Border Data Transfers

CRM systems operating across different regions often transfer data across borders, leading to challenges in complying with varying international privacy laws.

Challenges:

- **Diverse Regulations:** Laws like the GDPR in Europe and the California Consumer Privacy Act (CCPA) in the U.S. have different requirements for data protection, making compliance complex.
- **Data Localization Requirements:** Some jurisdictions mandate that data collected within their borders be stored locally, complicating global data management strategies.

Compliance Strategies:

- **Standard Contractual Clauses (SCCs):** Implement SCCs to ensure that data transfers meet the required legal protections.
- **Data Mapping:** Maintain detailed records of where data is stored and transferred to ensure compliance with regional laws.
- **Regular Compliance Audits:** Stay updated with international data protection laws and adjust practices accordingly.

2.2.3. Third-Party Involvement

Many CRM systems rely on third-party service providers, introducing risks associated with data sharing and storage.

Risks:

- **Security Breaches:** Third parties may not have the same security standards, increasing the risk of data breaches.
- **Loss of Control:** Sharing data with third parties can lead to a loss of control over how data is used and protected.
- **Compliance Issues:** Ensuring that third parties comply with relevant data protection regulations can be challenging.

Mitigation Strategies:

- **Due Diligence:** Thoroughly assess third-party vendors' security measures and compliance with data protection laws before engagement.
- **Data Sharing Agreements:** Establish clear agreements outlining the responsibilities and expectations for data protection.
- **Continuous Monitoring:** Regularly monitor third-party compliance and conduct audits to ensure ongoing adherence to data protection standards.

2.3 Strategies for Ensuring Data Privacy:

Data privacy is a critical concern in today's digital landscape, where vast amounts of personal and sensitive information are collected, processed, and stored by organizations. With increasing cyber threats, regulatory requirements, and ethical considerations, ensuring data privacy has become a fundamental aspect of modern business operations, research, and technological advancements. This article explores key strategies data minimization, encryption, and regular audits that help organizations safeguard data and mitigate privacy risks (Bickman, 2020).

2.3.1 Data Minimization

Definition: Data minimization is the practice of collecting only the necessary data required for a specific purpose, thereby reducing the risk of data misuse and exposure.

Key Principles:

- **Limit Data Collection:** Gather only the essential information needed for operations.
- **Short Retention Periods:** Store data for only as long as necessary.
- **Purpose-Driven Processing:** Clearly define why data is being collected and ensure it aligns with business needs.

Benefits:

- **Reduces risk** of data breaches by minimizing the amount of stored sensitive information.
- **Enhances compliance** with privacy laws by avoiding unnecessary data storage.
- **Lowers storage costs** by reducing redundant data.

Example:

An e-commerce platform requesting only essential customer details (name, email, and payment information) instead of collecting unnecessary personal data like social security numbers or home addresses.

2.3.2 Encryption

Definition: Encryption is the process of converting data into a secure format that can only be accessed by authorized parties with a decryption key.

Types of Encryptions:

- **Symmetric Encryption (AES, DES):** Uses the same key for encryption and decryption.
- **Asymmetric Encryption (RSA, ECC):** Uses a public key for encryption and a private key for decryption.
- **End-to-End Encryption (E2EE):** Ensures that data remains encrypted throughout transmission, preventing unauthorized access.

Key Applications:

- **Securing communications** (emails, messaging apps like WhatsApp).
- **Protecting stored data** (encrypted databases, cloud storage).
- **Ensuring safe transactions** (online banking, e-commerce).

Benefits:

- **Prevents unauthorized access** to sensitive data.
- **Ensures data integrity** by maintaining authenticity.
- **Enhances compliance** with privacy regulations.

Example:

A healthcare provider encrypting patient medical records before storing them in the cloud to ensure that only authorized personnel can access them.

2.3.3 Regular Audits

Definition: A data privacy audit involves assessing an organization's data protection practices to identify vulnerabilities and ensure compliance with privacy regulations.

Key Audit Practices:

- **Data Flow Analysis:** Mapping how data is collected, stored, and shared.
- **Access Control Review:** Evaluating who has access to sensitive data.
- **Security Policy Assessment:** Ensuring encryption and privacy measures are up to date.

Benefits:

- **Identifies security gaps** and potential vulnerabilities before they are exploited.
- **Ensures regulatory compliance** with GDPR, HIPAA, and other laws.
- **Improves trust** with customers and stakeholders by demonstrating a commitment to privacy.

Example:

A financial institution conducting quarterly security audits to assess vulnerabilities in its online banking system and address potential data breaches before they occur.

3. ETHICAL CONSIDERATIONS IN AI-POWERED CRM

AI is revolutionizing Customer Relationship Management by automating interactions, analyzing customer behavior, and enhancing personalization. However, AI models are trained on historical data, which may contain biases that can be unintentionally learned and reinforced. When biased algorithms are integrated into CRM systems, they can lead to unfair treatment of certain customer groups, resulting in reputational damage, customer dissatisfaction, and potential legal issues. Addressing AI bias is essential for developing ethical, fair, and effective CRM solutions (Kumar, Singh, & Khan, 2024).

3.1 What is AI Bias?

AI bias is defined as systematic and unfair discrimination against specific groups as a result of biassed training data, defective algorithms, or ineffective implementation. AI programs learn patterns from previous data, therefore any existing disparities or prejudices in that data might be perpetuated in AI-driven judgements (Mittal & Bansal, 2023).

3.1.1 How AI Bias Enters CRM Systems

AI bias in CRM can arise through several pathways:

- **Historical Data Bias:** AI models trained on past customer interactions may learn discriminatory trends present in previous business decisions.

- **Selection Bias:** If CRM data is collected from a non-representative sample of customers, the AI model may make inaccurate predictions for underrepresented groups.
- **Algorithmic Bias:** Certain machine learning models may weigh specific data points more heavily, unintentionally favouring or disadvantaging particular customer segments.
- **Implicit Human Bias:** AI models may inherit biases from human programmers, either intentionally or unintentionally, leading to unfair decision-making.

3.1.2 Impact of AI Bias in CRM

3.1.2.1 Unfair Customer Treatment

Biased AI algorithms can result in unfair treatment of customers in multiple ways:

- **Personalized Offers and Pricing:** If an AI model is biased, it may offer better deals to certain customer groups while discriminating against others based on race, gender, or economic status.
- **Customer Support Prioritization:** Some customers may receive better service responses due to biased AI-driven ticketing systems, while others are deprioritized.

3.1.2.2 Brand Reputation Damage

If customers perceive unfair treatment due to biased AI, they may lose trust in the company, leading to negative reviews, media backlash, and customer attrition. Companies like Amazon and Microsoft have faced criticism for AI bias in hiring and facial recognition, demonstrating the severe impact on brand credibility (Ao et al., 2023).

3.1.2.3 Legal and Compliance Issues

Regulatory bodies such as the General Data Protection Regulation (GDPR) and the California Consumer Privacy Act (CCPA) enforce strict rules on fairness and transparency in AI decision-making. If an AI-driven CRM system violates these regulations, companies may face legal action and financial penalties.

3.1.3 Strategies to Mitigate AI Bias in CRM

3.1.3.1 Data Auditing and Cleaning

To reduce bias in AI models, organizations must ensure that their CRM data is diverse, representative, and free from historical discrimination. Techniques include:

- **Data Balancing:** Ensuring that all customer demographics are proportionally represented in training datasets.
- **Bias Detection Tools:** Using AI-driven fairness tools like IBM's AI Fairness 360 or Google's What-If Tool to detect bias in CRM models.

3.1.3.2 Algorithm Transparency and Explainability

CRM systems should use explainable AI (XAI) techniques to provide clear justifications for decisions made by AI models.

- **Interpretable Machine Learning (IML):** Using decision trees, SHAP (Shapley Additive Explanations), and LIME (Local Interpretable Model-Agnostic Explanations) to make AI-driven recommendations transparent [29].
- **Regular Audits:** Conducting fairness audits to identify and correct algorithmic bias before deploying AI in CRM.

3.1.3.3 Ethical AI Design in CRM

Companies must prioritize ethically designed AI models by:

- Involving diverse teams in AI development to reduce implicit human bias.
- Applying fairness constraints during machine learning training to avoid discriminatory outcomes.
- Conducting real-world testing to evaluate AI behavior across different customer groups.

3.1.3.4 Human-AI Collaboration

AI should assist, not replace, human decision-making in CRM to prevent unintended biases. Implementing human oversight mechanisms can help correct biased AI-driven customer interactions.

3.2 Transparency and Explainability in AI

Businesses are increasingly leveraging Artificial Intelligence (AI) to automate customer interactions, personalize marketing efforts, and predict consumer behavior. However, trust issues arise when people are uncertain about how their data is being used, as AI's decisions directly influence consumer experiences. To build customer trust and ensure ethical AI usage, transparency and explainability have become essential. Explainable AI (XAI) refers to the ability of AI systems to provide clear, interpretable, and understandable explanations for their actions. By making

AI processes more accessible, organizations can enhance accountability, comply with data protection regulations, and improve their overall credibility in Customer Relationship Management applications(Ao et al., 2023) .

3.2.1 Why Transparency and Explainability Matter in CRM

3.2.1.1 Growing Customer Demand for Transparency

In today's digital world, customers are becoming increasingly aware of how companies collect, store, and process their personal data. Many users hesitate to engage with AI-driven systems if they suspect unethical data handling or biased decision-making. According to surveys, more than 70% of customers prefer businesses that openly communicate how AI is used in decision-making.

Transparency in CRM AI applications builds:

- **Trust:** Customers feel confident in AI-driven recommendations when they understand how they are made.
- **Accountability:** Businesses can take responsibility for AI-based errors or unfair outcomes.
- **Regulatory Compliance:** Transparency helps companies comply with GDPR, CCPA, and other data protection regulations that require organizations to explain automated decisions affecting consumers.

3.2.1.2 Risks of Opaque AI Systems

When AI operates as a "black box," where its decision-making process is hidden, the following risks arise:

- **Customer Distrust:** Users may avoid interacting with AI-driven CRM tools if they feel their data is being misused.
- **Unintended Bias:** AI models can inadvertently discriminate against certain demographics if not properly monitored.
- **Regulatory Violations:** Lack of explainability can lead to legal consequences if businesses cannot justify AI-driven decisions.

To address these issues, Explainable AI (XAI) provides a solution by making AI models more interpretable and accountable (Bansal et al., 2023).

3.2.2 Explainable AI (XAI): Enhancing Trust in AI-Driven CRM

3.2.2.1 What is Explainable AI (XAI)?

XAI refers to a set of methods and tools that help users understand how an AI system reaches a particular decision. Unlike traditional machine learning models, where predictions are difficult to interpret, XAI makes AI-driven CRM systems more transparent by explaining the logic behind recommendations, classifications, and predictions.

3.2.2.2 Key Techniques in XAI for CRM

Key Techniques in XAI for CRM shown in Table 1.

Table 1. Key Techniques in XAI for CRM

Feature Importance Analysis	**• AI models analyze customer interactions and assign weights to features that influence outcomes.** **• Example: A credit risk assessment tool in CRM may highlight that income level and repayment history are the most influential factors in loan approvals.**
Local Interpretable Model-Agnostic Explanations (LIME)	• LIME generates simplified models that mimic complex AI behavior, making decisions easier to understand. • Example: If a CRM chatbot rejects a customer's request, LIME can provide a plain-language explanation of why the request was denied.
Shapley Additive Explanations (SHAP)	• SHAP values break down how each feature contributes to AI predictions. • Example: A personalized product recommendation system can show how different customer behaviors (e.g., past purchases, browsing history) influenced a suggestion.
Decision Trees and Rule-Based Systems	• CRM AI systems can use decision trees, where customers can see step-by-step how their data led to a specific outcome.

3.2.3 The Role of Transparency in CRM AI Systems

3.2.3.1 Transparent AI vs. Opaque AI

Table 2. Comparison of Transparent AI and Opaque AI

Feature	Transparent AI	Opaque AI (Black Box AI)
Explainability	High	Low
Customer Trust	Strong	Weak
Regulatory Compliance	Easy	Difficult
Bias & Errors	Detectable	Hard to Identify

Table 2 compares Transparent AI and Opaque AI (Black Box AI) based on key factors such as explainability, customer trust, regulatory compliance, and the ability to detect bias and errors. Transparent AI offers high explainability, strong customer trust, easier regulatory compliance, and better bias detection, whereas Opaque AI lacks these advantages. Businesses using transparent AI models in CRM can build long-term customer relationships by demonstrating fairness and accountability.

3.2.3.2 Benefits of Transparency in CRM AI

- **Improved Customer Engagement:** Customers are more likely to interact with AI-driven services when they understand how decisions are made.
- **Reduced Bias and Discrimination:** Transparent AI allows companies to identify and correct biased decision-making.
- **Stronger Brand Reputation:** Companies that prioritize AI ethics gain customer loyalty and industry credibility.
- **Better Compliance with Privacy Laws:** Transparent AI helps meet the explainability requirements of GDPR and CCPA.

3.2.4 Implementing Transparency and XAI in CRM

3.2.4.1 Steps to Make AI More Transparent in CRM

1. **Data Transparency**
    - o Inform users about what data is collected and how it is used.
    - o Provide customers with control over their personal data.
2. **Explainable AI Algorithms**
    - o Choose AI models that allow for interpretation (e.g., decision trees, SHAP, LIME).

- Ensure AI predictions are supported by clear explanations.

3. **User-Friendly Explanations**
    - Instead of technical AI jargon, use plain language to explain AI decisions.
    - Example: Instead of saying "The model output probability was 0.75", say "Based on your past purchases, we recommend this product with high confidence."
4. **Regular Audits & Monitoring**
    - Continuously monitor AI models for biases, errors, and unfair decisions.
    - Conduct audits to ensure AI decisions align with ethical guidelines.

3.3 Consent and Autonomy

With the rise of digital marketing and AI-powered customer relationship management systems, organizations are collecting and processing vast amounts of customer data. However, ethical issues arise when customers lack complete control over their data or are unaware of how it is being used. Consent and autonomy are essential elements of ethical CRM practices. Organizations must obtain explicit consent from customers before collecting, storing, or processing their data. Additionally, providing customers with autonomy meaning the ability to manage and control their own data helps build trust, enhances brand reputation, and ensures compliance with global data privacy regulations such as the General Data Protection Regulation (GDPR), California Consumer Privacy Act (CCPA), and other privacy laws (Bansal et al., 2022.

3.3.1 Why Consent and Autonomy Matter in CRM

3.3.1.2 Building Trust Through Transparency

Customers today are more aware of data privacy risks than ever before. Studies show that over 80% of consumers prefer brands that are transparent about how they collect and use data.

When businesses obtain explicit consent and allow customers to control their personal data, they:

- **Foster trust and loyalty**
- **Enhance user experience by respecting customer choices**
- **Reduce legal risks by ensuring compliance with privacy regulations.**

3.3.1.3 Risks of Ignoring Consent and Autonomy

Failing to prioritize consent and autonomy in CRM can result in serious consequences:

- **Loss of customer trust**: If customers feel their data is misused, they may stop engaging with the brand.
- **Regulatory penalties**: Non-compliance with laws like GDPR and CCPA can result in heavy fines.
- **Brand damage**: Privacy scandals can lead to negative publicity and lost revenue.

To prevent these risks, companies must implement ethical CRM strategies that prioritize user consent and give customers control over their data.

3.3.2 Principles of Ethical Consent in CRM

For customer consent to be truly ethical and effective, it must follow key principles:

3.3.2.1 Explicit and Informed Consent

- Customers should be clearly informed about what data is being collected, why, and how it will be used.
- Consent requests must be unambiguous (e.g., requiring a clear opt-in, not a default opt-out).
- Example: A website must ask users to actively accept cookies, rather than assuming consent unless they opt out.

3.3.2.2 Granular Consent Options

- Customers should have the ability to choose what types of data they want to share.
- Example: A CRM system should allow users to opt-in for marketing emails but decline location tracking.

3.3.2.3 Easy Withdrawal of Consent

- Customers should be able to revoke their consent at any time without difficulty.
- Example: Users should have an easy-to-find "Delete My Data" option in account settings.

3.3.3 Autonomy in CRM: Giving Customers Control Over Their Data

Beyond obtaining consent, ethical CRM practices must empower customers with control over their own data. This involves:

3.3.3.1 Data Access and Portability

- Customers should have the right to access their personal data stored by a company.
- They should also be able to download and transfer their data to another service.
- Example: A cloud-based CRM should allow users to export their contact lists and preferences.

3.3.3.2 Data Correction and Deletion Rights

- Users should be able to update or correct their personal data if it is inaccurate.
- Customers should also have the right to request data deletion if they no longer want to engage with the company.

3.3.3.3 Transparent Privacy Policies

- Companies must provide clear and simple explanations of their data policies.
- Example: A CRM dashboard should explain data usage in everyday language, not legal jargon.

3.3.4 Legal and Ethical Frameworks for Consent and Autonomy

Various laws and regulations enforce ethical data handling in CRM:

3.3.4.1 General Data Protection Regulation (GDPR) (EU)

- Requires explicit opt-in consent for data collection.
- Gives customers full control over their data, including deletion rights.

3.3.4.2 California Consumer Privacy Act (CCPA) (USA)

- Allows users to opt-out of data sharing.
- Grants customers the right to know what data is being collected.

3.3.4.3 Personal Data Protection Act (PDPA) (Singapore)

- Requires companies to obtain clear and informed consent from users.

- By aligning with these frameworks, businesses can enhance ethical CRM practices and avoid legal pitfalls.

3.3.5 Best Practices for Implementing Ethical Consent and Autonomy in CRM

To build ethical CRM systems that respect customer privacy and autonomy, businesses should adopt the following best practices:

3.3.5.1 Use Clear and Simple Consent Forms

- Avoid complex legal language in consent requests.
- Use simple, user-friendly explanations of data usage.

3.3.5.2 Provide Customers with Easy Opt-In and Opt-Out Options

- Always require explicit opt-in (never assume consent).
- Allow customers to opt out at any time without restrictions.

3.3.5.3 Regularly Update Privacy Policies

- Keep customers informed about any changes in data collection practices.
- Notify users of policy updates in a transparent manner.

3.3.5.4 Implement Privacy by Design

- Ensure that privacy controls are built into CRM platforms from the start, not added later.
- Example: CRM software should include default settings that prioritize user privacy.

3.3.5.5 Conduct Regular Compliance Audits

- Perform frequent data protection assessments to identify vulnerabilities.
- Ensure that CRM practices align with global privacy laws and ethical standards.

3.4 Balancing Automation and Human Touch

To develop an effective and ethical CRM strategy, firms must strike a balance between AI-powered insights and human empathy. This guarantees that clients feel appreciated and heard, rather than simply engaging with machines. This article discusses the significance of balancing automation and human interaction, the risks of over-automation, and best practices for incorporating AI without losing the human touch.

3.4.1 The Role of AI and Automation in CRM

AI and automation have transformed CRM by:

- Enhancing efficiency through automated responses and workflows.
- Improving customer insights using data analytics and AI-powered recommendations.
- Reducing response times with chatbots and virtual assistants.
- Personalizing experiences by analyzing customer preferences and behavior.

While these advancements are beneficial, they also raise concerns about over-automation, where customer interactions become robotic and lack human warmth.

3.4.2 Risks of Over-Reliance on Automation in CRM

3.4.2.1 Impersonal Customer Experience

- Automated chatbots and AI-driven responses can lack empathy and emotional intelligence.
- Customers may feel frustrated when they are unable to speak with a real person for complex queries.

3.4.2.2 Loss of Trust and Brand Loyalty

- Excessive automation can make customers feel like just another number, reducing emotional connection with the brand.
- Studies show that 86% of customers prefer human interactions over chatbots when dealing with complex issues.

3.4.2.3 Misinterpretation of Customer Needs

- AI algorithms rely on historical data but may struggle with understanding unique customer concerns in real-time.
- Example: A customer complaint may require apology and empathy, but an automated system might respond with generic solutions instead.

3.4.2.4 Ethical and Privacy Concerns

- AI-driven CRM systems collect and analyze large amounts of personal data, raising concerns about privacy and consent.

- Customers may hesitate to engage with a brand if they feel their data is being misused.

3.4.3 Striking the Right Balance: Combining AI and Human Interaction

To ensure an ethical and customer-centric approach, businesses must find a middle ground where AI supports customer service but does not replace human interaction.

3.4.3.1 Hybrid Approach: AI for Efficiency, Humans for Empathy

A hybrid CRM approach ensures that AI handles repetitive tasks, while human agents manage complex interactions requiring emotional intelligence.

✓ Use AI for:
  - Handling frequently asked questions (FAQs).
  - Managing basic troubleshooting and order tracking.
  - Analyzing customer behavior and providing recommendations.

✓ Use Human Agents for:
  - Addressing customer complaints and grievances.
  - Offering personalized assistance for high-value customers.
  - Handling complex queries that require emotional intelligence.

3.4.3.2 AI as a Support Tool, not a Replacement

Instead of replacing customer service teams, AI should enhance their efficiency by:

- Providing real-time customer insights to human agents.
- Automating data entry and repetitive tasks to allow agents to focus on meaningful interactions.
- Using AI-powered suggestions to help agents provide better recommendations.

3.4.3.3 Personalization Without Losing the Human Element

- AI should augment personalization efforts by analyzing customer data but allow human agents to make final decisions.
- Example: AI can suggest product recommendations, but a customer service representative should guide the customer with empathy.

3.4.3.4 Seamless AI-Human Handoff

- If an AI chatbot cannot resolve a customer issue, it should immediately escalate the query to a human agent.

- Customers should always have the option to connect with a real person when needed.

3.4.4 Best Practices for Ethical AI Integration in CRM

To ensure an optimal balance between automation and human interaction, businesses should follow these best practices:

3.4.4.1 Implement AI Responsibly

- AI should enhance customer experience, not replace human interaction.
- Chatbots and automated responses must be clearly labeled, so customers know when they are interacting with AI.

3.4.4.2 Prioritize Human Oversight

- AI-driven CRM systems should always have a human agent available for escalations.
- Example: A chatbot can handle initial queries, but customers should be able to speak with an agent when needed.

3.4.4.3 Design AI with Emotional Intelligence

- AI chatbots should be trained to detect customer frustration and escalate the issue to a human agent.
- Use natural language processing (NLP) to make AI responses more conversational and less robotic.

3.4.4.4 Collect Customer Feedback

- Regularly ask customers how they feel about AI-driven interactions.
- Use feedback to improve AI accuracy and enhance the overall customer experience.

3.4.4.5 Ensure Transparency in AI Usage

- Customers should be informed how AI is used in CRM.
- Example: "This chatbot uses AI to assist you. If you prefer human support, please let us know."

4. ROLE OF REGULATIONS IN ENSURING ETHICAL AI AND DATA PRIVACY

As Artificial Intelligence (AI) is increasingly integrated into Customer Relationship Management systems, issues regarding data privacy, ethical AI usage, and consumer rights are growing. Regulatory organisations around the world have passed regulations to promote openness, fairness, and accountability in AI-powered CRM systems.

Three major legislative frameworks play an important role in shaping ethical AI practices in CRM:

The General Data Protection Regulation (GDPR) establishes tight standards for data collection, storage, and processing. The California Consumer Privacy Act (CCPA) allows individuals to govern their data, including opting out of data purchases. Emerging AI Governance Policies: Countries are creating AI-specific rules to address challenges such as bias, justice, and accountability (Painoli et al., 2021). Figure 4 illustrates ethical and legal considerations in artificial intelligence.

Figure 2. Ethical and legal consideration in artificial intelligence

4.1 GDPR (General Data Protection Regulation)

The General Data Protection Regulation (GDPR), enacted by the European Union (EU) in 2018, is one of the most comprehensive data privacy laws in the world. It establishes strict guidelines on how businesses collect, process, and store personal data. Table 3 outlines key GDPR principles relevant to AI-driven Customer Relationship Management (CRM) systems, the impact of GDPR on AI-powered CRM, challenges in compliance, and best practices to ensure adherence. Transparency, customer consent, and data security are crucial for regulatory compliance, while AI models must be designed with privacy considerations to mitigate risks (Satar et al., 2023).

Table 3. Key points of GDPR (General Data Protection Regulation)

Key GDPR Principles for AI in CRM	**• Transparency – Companies must clearly inform users about how and why their data is collected.** **• Customer Consent – Businesses must obtain explicit consent before collecting personal data.** **• Right to Access & Erasure – Customers can request access to their data or ask for it to be deleted.** **• Accountability & Security – Companies must protect user data and report breaches within 72 hours.**
Impact of GDPR on AI-Powered CRM	• AI-based CRM systems must be designed with privacy in mind (Privacy by Design). • CRM platforms must provide clear opt-in and opt-out mechanisms for data collection. • AI models trained on customer data must ensure compliance with GDPR's fairness and transparency rules.
Challenges in Compliance	• AI models often process vast amounts of data, making it difficult to ensure full transparency. • Machine Learning (ML) algorithms evolve over time, which may lead to unintentional biases or unauthorized data usage.
Best Practices for GDPR Compliance in AI CRM	• Use explainable AI (XAI) to ensure transparency in decision-making. • Implement strong data encryption and access controls. • Regularly conduct privacy impact assessments to evaluate risks. • Maintain detailed records of data processing activities.

4.2 CCPA (California Consumer Privacy Act)

The California Consumer Privacy Act (CCPA), enacted in 2020, is one of the strongest data privacy laws in the United States. It grants consumers more control over their personal data and imposes strict obligations on businesses collecting and using customer information.

4.2.1 Key CCPA Rights for Consumers

- **Right to Know** – Customers can request details about what personal data is collected and how it is used.
- **Right to Opt-Out** – Customers can prevent companies from selling their data.

- **Right to Deletion** – Customers can ask for their personal data to be deleted upon request.
- **Non-Discrimination** – Businesses cannot deny services or charge higher prices if a user exercises their CCPA rights.

4.2.2 Impact of CCPA on AI-Powered CRM

- AI-driven CRM systems must have data disclosure mechanisms to inform customers about their rights.
- Businesses must allow easy opt-outs from data tracking and personalized AI recommendations.
- AI models trained on customer data must honor deletion requests, ensuring complete data removal.

4.2.3 Challenges in Compliance

- AI-powered **personalization and targeted advertising** depend on customer data, making compliance complex.
- CRM systems need **real-time mechanisms** to process data deletion and opt-out requests efficiently.

4.2.4 Best Practices for CCPA Compliance in AI CRM

- Clearly explain **what customer data is collected** and how it is used.
- Provide **easy-to-use opt-out options** for AI-driven personalization.
- Ensure **AI systems can delete customer data upon request**.
- Train employees on **CCPA compliance** and ethical AI usage.

4.3 Emerging AI Governance Policies

4.3.1 The Need for AI-Specific Regulations

As AI adoption grows, governments worldwide are working on AI-specific policies to address:

- Bias and Fairness – Prevent AI discrimination in decision-making systems.
- Accountability – Hold companies responsible for AI-driven errors or unethical decisions.
- Transparency – Ensure AI algorithms are explainable and interpretable.

4.3.2 Key AI Regulations Under Development

- **EU AI Act (Proposed)** – The first comprehensive AI law that categorizes AI risks into four levels and imposes strict guidelines on high-risk AI applications (e.g., facial recognition, hiring algorithms).
- **US AI Bill of Rights (Proposed)** – A framework focusing on algorithmic transparency, bias reduction, and consumer privacy.
- **China's AI Regulation** – Introduces strict rules on AI ethics, security, and user protection.
- **India's Data Protection Bill** – Aims to regulate AI-driven data processing and consumer rights.

4.3.3 Impact of AI Regulations on CRM

- AI-powered CRM systems must ensure fairness and unbiased decision-making in customer interactions.
- Businesses must provide explainable AI models that customers can understand.
- AI-based predictive analytics must comply with data privacy laws to prevent misuse.

4.3.4 Best Practices for Adapting to Future AI Regulations

- Stay updated on global AI policies and compliance requirements.
- Implement bias detection and mitigation techniques in AI algorithms.
- Use ethical AI frameworks such as fairness-aware machine learning.
- Ensure AI decision-making is transparent and interpretable for customers.

4.4 The Future of Ethical AI and Data Privacy Regulations

As Artificial Intelligence (AI) is increasingly integrated into Customer Relationship Management systems, issues regarding data privacy, ethical AI usage, and consumer rights are growing. Regulatory organisations around the world have passed regulations to promote openness, fairness, and accountability in AI-powered CRM systems.

Three major legislative frameworks play an important role in shaping ethical AI practices in CRM:

The General Data Protection Regulation (GDPR) establishes tight standards for data collection, storage, and processing. The California Consumer Privacy Act (CCPA) allows individuals to govern their data, including opting out of data purchases. Emerging AI Governance Policies: Countries are creating AI-specific rules to address challenges such as bias, justice, and accountability. Figure 4 illustrates ethical and legal considerations in artificial intelligence (Raja et al., 2025).

5. AI-DRIVEN ENHANCEMENTS FOR ETHICAL CRM

In today's digital age, organizations are increasingly relying on artificial intelligence (AI) to enhance customer relationship management. AI-powered solutions help improve productivity, automate transactions, and customize client experiences. However, as the use of AI has grown, so have ethical concerns regarding data privacy, fairness, and transparency. To address these issues, organizations are implementing AI-driven improvements to protect privacy, develop ethical AI practices, and educate customers. These innovations not only ensure compliance with regulations but also build trust and enhance consumer interactions (Bansal et al., 2024).

5.1 AI for Privacy Protection

Privacy concerns are at the forefront of ethical AI discussions. AI tools are being developed to detect, prevent, and respond to potential data breaches in real time. These tools contribute to ethical CRM by (Bansal et al., 2024):

- **Automated Data Anonymization**: AI can mask or anonymize personally identifiable information (PII) to protect user privacy while still allowing businesses to extract insights from customer interactions.
- **Real-Time Threat Detection**: AI-driven security systems can continuously monitor CRM databases for unusual activity, such as unauthorized access or data leaks, and trigger automatic responses to prevent breaches.
- **Regulatory Compliance Assistance**: AI tools can track and enforce privacy regulations like GDPR (General Data Protection Regulation) and CCPA (California Consumer Privacy Act) by ensuring businesses obtain customer consent before data collection and use.
- **AI-Driven Access Controls**: AI can restrict access to sensitive customer information based on roles and permissions, ensuring that only authorized personnel can retrieve confidential data.

By integrating AI-powered privacy protection, businesses not only safeguard customer data but also demonstrate a commitment to ethical data handling, reducing the risks associated with cybersecurity threats.

5.2 Ethical AI Development

The development of responsible and fair AI systems is essential for ethical CRM. Companies are investing in AI frameworks designed to eliminate bias, promote transparency, and maintain accountability in automated decision-making. Key components of ethical AI development include (Rahman, Bansal, & Pruthi, 2023):

- **Bias Detection and Mitigation**: AI algorithms can unintentionally reflect biases present in training data, leading to unfair treatment of customers based on gender, race, or socioeconomic background. Businesses are now employing bias detection models that assess AI-driven outcomes for fairness and inclusivity.
- **Explainable AI (XAI)**: Transparency is a major ethical concern in AI-driven CRM. Explainable AI (XAI) ensures that AI decisions are interpretable and understandable, allowing customers to trust AI-generated recommendations and responses.
- **Fair AI Training Practices:** Organizations are adopting diverse and representative datasets to train AI models, minimizing discrimination and enhancing fairness in customer interactions.
- **Ethical AI Audits**: Regular audits of AI models help businesses identify and correct unethical behavior in CRM applications, ensuring that AI-driven interactions align with ethical standards.

By focusing on ethical AI development, businesses can foster trust with customers, regulators, and stakeholders while ensuring AI-powered CRM systems operate transparently and fairly (Kumar, Sharma, Dhanka, & Maini, 2025; Kumar et al., 2024).

5.3 Customer Education and Awareness

For AI-powered CRM to be truly ethical, customer awareness and consent are crucial. Businesses must educate customers about how their data is collected, stored, and utilized in AI-driven systems. Key initiatives include:

- **Clear Data Usage Policies**: Companies should communicate data usage policies in simple terms, allowing customers to understand what data is collected and why.
- **Customer Consent Mechanisms**: Ethical CRM systems prioritize explicit opt-in and opt-out options, enabling customers to control their data.

- **AI-Driven Chatbots for Customer Support**: AI-powered chatbots can educate users about data privacy policies and answer questions regarding their rights in real time.
- **Interactive Privacy Dashboards**: Businesses are integrating self-service privacy dashboards where customers can view, edit, or delete their personal data from CRM systems.

Customer education fosters transparency and collaboration, empowering individuals to make informed choices about their data. This, in turn, enhances customer trust and strengthens relationships between businesses and users.

CONCLUSION

The integration of Artificial Intelligence in Customer Relationship Management has transformed how businesses engage with customers, automate processes, and derive actionable insights. AI-powered CRM systems offer unprecedented opportunities for personalized interactions and data-driven decision-making, but they also introduce significant ethical and privacy challenges. Issues such as data security, transparency, bias, and regulatory compliance must be addressed to ensure responsible AI adoption. To navigate these challenges, organizations must implement robust ethical frameworks, prioritize data privacy safeguards, and adhere to global regulatory standards such as GDPR and CCPA. The privacy-by-design approach incorporating encryption, anonymization, and access controls enhances data protection while fostering consumer trust. Additionally, explainable AI (XAI), bias detection, and ethical audits play a crucial role in ensuring fairness and accountability in AI-driven decision-making. Beyond regulatory compliance, businesses must actively educate customers on AI's role in CRM, providing clear data usage policies, opt-in consent mechanisms, and interactive privacy dashboards. Ethical AI practices not only protect consumer rights but also strengthen brand reputation and build long-term customer relationships. As AI regulations continue to evolve globally, companies that proactively integrate ethical AI principles and responsible data management will gain a competitive edge, ensure regulatory approval, and foster customer loyalty. By balancing innovation with ethical responsibility, businesses can create a transparent, fair, and sustainable AI-powered CRM ecosystem that enhances trust and drives long-term success in an increasingly data-driven world.

REFERENCES

Alhuwaydi, A. M. (2024). Exploring the Role of Artificial Intelligence in Mental Healthcare: Current Trends and Future Directions – A Narrative Review for a Comprehensive Insight. *Risk Management and Healthcare Policy, 17*(May), 1339–1348. DOI: 10.2147/RMHP.S461562 PMID: 38799612

Anand, N., Pant, L. M., Alam, T., Pundir, S., Thomas, L., & Rakshith, U. R. "Artificial Intelligence ' s Contribution To Mental Health Education," *2023 Int. Conf. Sustain. Comput. Data Commun. Syst.*, no. August, pp. 462–467, 2023, DOI: 10.1109/ICSCDS56580.2023.10104905

Ao, L., Bansal, R., Pruthi, N., & Khaskheli, M. B. (2023). Impact of Social Media Influencers on Customer Engagement and Purchase Intention: A Meta-Analysis. *Sustainability (Basel), 15*(3), 2744. DOI: 10.3390/su15032744

Bansal, R., Martinho, C., Pruthi, N., & Aggarwal, D. (2024). From virtual observations to business insights: A bibliometric review of netnography in business research. *Heliyon, 10*(1), e22853. DOI: 10.1016/j.heliyon.2023.e22853 PMID: 38163120

Bansal, R., Pruthi, N., & Singh, R. (2022). Developing Customer Engagement Through Artificial Intelligence Tools: Roles and Challenges. In *Developing Relationships, Personalization, and Data Herald in Marketing 5.0* (pp. 130-145). IGI Global.

Bansal, R., Saini, S., Ngah, A. H., & Durga Prasad, T. (2024). Proselytizing the potential of influencer marketing via artificial intelligence: Mapping the research trends through bibliometric analysis. *Cogent Business & Management, 11*(1), 2372889. DOI: 10.1080/23311975.2024.2372889

Bansal, R., Shukla, G., Gupta, A., Singh, A., & Pruthi, N. (2023). Optimizing Augmented Reality and Virtual Reality for Customer Engagement. In *Promoting Consumer Engagement Through Emotional Branding and Sensory Marketing* (pp. 24–35). IGI Global.

Beres, L. K., Mbabali, I., Anok, A., Katabalwa, C., Mulamba, J., Thomas, A. G., Bugos, E., Nakigozi, G., Grabowski, M. K., & Chang, L. W. (2021). Mobile ecological momentary assessment and intervention and health behavior change among adults in Rakai, Uganda: Pilot randomized controlled trial. *JMIR Formative Research, 5*(7), 1–12. DOI: 10.2196/22693 PMID: 34283027

Bickman, L. (2020). Improving mental health services: A 50-year journey from randomized experiments to artificial intelligence and precision mental health. *Administration and Policy in Mental Health, 47*(5), 795–843.

Cho, Y. M., Rai, S., Ungar, L., Sedoc, J., & Guntuku, S. C. (2023). *"An Integrative Survey on Mental Health Conversational Agents to Bridge Computer Science and Medical Perspectives," EMNLP 2023 - 2023 Conf.* Empir. Methods Nat. Lang. Process. Proc., DOI: 10.18653/v1/2023.emnlp-main.698

Dawoodbhoy, F. M., Delaney, J., Cecula, P., Yu, J., Peacock, I., Tan, J., & Cox, B. (2021). AI in patient flow: Applications of artificial intelligence to improve patient flow in NHS acute mental health inpatient units. *Heliyon*, *7*(5), e06993. Advance online publication. DOI: 10.1016/j.heliyon.2021.e06993 PMID: 34036191

Gupta, P., Muneshwar, K. N., Juganavar, A., & Shegekar, T. (2023). Beyond the Asylum Walls: Tracing the Tapestry of Mental Health Interventions Across Eras and Cultures. *Cureus*, *15*(11). Advance online publication. DOI: 10.7759/cureus.48251 PMID: 38054143

Gurung, D., Poudyal, A., Wang, Y. L., Neupane, M., Bhattarai, K., Wahid, S. S., Aryal, S., Heim, E., Gronholm, P., Thornicroft, G., & Kohrt, B. (2022). Stigma against mental health disorders in Nepal conceptualised with a 'what matters most' framework: A scoping review. *Epidemiology and Psychiatric Sciences*, *31*, e11. Advance online publication. DOI: 10.1017/S2045796021000809 PMID: 35086602

Gutierrez, G., Stephenson, C., Eadie, J., Asadpour, K., & Alavi, N. (2024). Examining the role of AI technology in online mental healthcare: Opportunities, challenges, and implications, a mixed-methods review. *Frontiers in Psychiatry*, *15*, 1356773.

He, T., Fu, G., Yu, Y., Wang, F., Li, J., Zhao, Q., . . . Yang, B. X. (2023). Towards a psychological generalist ai: A survey of current applications of large language models and future prospects. arXiv preprint arXiv:2312.04578.

Kumar, A., ... "Smart Crop Selection: Harnessing Machine Learning for Sustainable Agriculture in the Era of Industry 5.0," 2024, pp. 111–134. DOI: 10.1007/978-3-031-70996-8_6

Kumar, A., Dhanka, S., Bansal, R., Sharma, A., Singh, J., Khan, A. A., & Maini, S. (2024). Agriculture in the Era of Industry 5.0. *Industry 5.0 and Emerging Technologies: Transformation Through. Technology and Innovation*, *565*, 111.

Kumar, A., Sharma, A., Dhanka, S., & Maini, S. (2025). Machine Learning for Combating Mental Health Stigma. In *Transforming Neuropsychology and Cognitive Psychology With AI and Machine Learning* (pp. 333–366). IGI Global Scientific Publishing. DOI: 10.4018/979-8-3693-9341-3.ch014

Kumar, A., Singh, J., & Khan, A. A. (2024). "A comprehensive machine learning framework with particle swarm optimization for improved polycystic ovary syndrome (PCOS) diagnosis,". *Engineering Research Express*, *6*(3), 035233. Advance online publication. DOI: 10.1088/2631-8695/ad76f9

Lem, W. G., Kawata, K. H. S., & Oyama, H. (2024). K. H. dos S. Kawata, and H. Oyama, "Exploring the impact of immersive virtual reality on depression knowledge and stigma reduction: A cross-over RCT fMRI study,". *Scientific Reports*, *14*(1), 1–14. DOI: 10.1038/s41598-024-55797-w PMID: 38167627

Liu, H. (2022). Applications of Artificial Intelligence to Popularize Legal Knowledge and Publicize the Impact on Adolescents' Mental Health Status. *Frontiers in Psychiatry*, *13*(May), 902456. Advance online publication. DOI: 10.3389/fpsyt.2022.902456 PMID: 35722558

Mallik, S., & Gangopadhyay, A. (2023). Proactive and reactive engagement of artificial intelligence methods for education: A review. *Frontiers in Artificial Intelligence*, *6*, 1151391. Advance online publication. DOI: 10.3389/frai.2023.1151391 PMID: 37215064

Mitsea, E., Drigas, A., & Skianis, C. (2023). Digitally Assisted Mindfulness in Training Self-Regulation Skills for Sustainable Mental Health: A Systematic Review. *Behavioral Sciences (Basel, Switzerland)*, *13*(12), 1008. Advance online publication. DOI: 10.3390/bs13121008 PMID: 38131865

Mittal, G., & Bansal, R. (2023). Driving Force Behind Consumer Brand Engagement: The Metaverse. In *Cultural Marketing and Metaverse for Consumer Engagement* (pp. 164-181). IGI Global.

Painoli, A. K., Bansal, R., Singh, R., & Kukreti, A. (2021). Impact of Digital Marketing on the Buying Behavior of Youth With Special Reference to Uttarakhand State. In *Big Data Analytics for Improved Accuracy, Efficiency, and Decision Making in Digital Marketing* (pp. 162-182). IGI Global. DOI: 10.4018/978-1-7998-7231-3.ch012

Palta, P., & Kumar, A. (2024). Effect of vermicompost additive on physical, chemical and dielectric properties of soil and its modeling. *The Journal of Microwave Power and Electromagnetic Energy*, *58*(3), 186–206. DOI: 10.1080/08327823.2024.2378668

Palta, P., & Kumar, A. (2025). Soil dielectric response to chemical fertilizers in Northern India's key agricultural areas. *Physics and Chemistry of the Earth Parts A/B/C*, *137*, 103806.

Rahman, K. T., Bansal, R., & Pruthi, N. (2023). Consumer adoption of technologies: A mindset-oriented approach. In Contemporary Studies of Risks in Emerging Technology, Part B (pp. 243-255). Emerald Publishing Limited. DOI: 10.1108/978-1-80455-566-820231013

Raja, M., Swamy, T. N. V. R. L., Das, S., Bansal, R., Fahlevi, M., & Aziz, A. L. (2025). Does human resource information system influence performance management? *Cogent Business & Management*, *12*(1), 2438862. DOI: 10.1080/23311975.2024.2438862

Roth, C. B., Papassotiropoulos, A., Brühl, A. B., Lang, U. E., & Huber, C. G. (2021). Psychiatry in the digital age: A blessing or a curse? *International Journal of Environmental Research and Public Health*, *18*(16), 8302. Advance online publication. DOI: 10.3390/ijerph18168302 PMID: 34444055

Satar, M. S., Aggarwal, D., Bansal, R., & Alarifi, G. (2023). Mapping the Knowledge Structure and Unveiling the Research Trends in Social Entrepreneurship and Inclusive Development: A Bibliometric Analysis. *Sustainability (Basel)*, *15*(7), 5626. DOI: 10.3390/su15075626

Schiff, D. (2021). Out of the laboratory and into the classroom: The future of artificial intelligence in education. *AI & Society*, *36*(1), 331–348. DOI: 10.1007/s00146-020-01033-8 PMID: 32836908

Sharma, A., Dhanka, S., Bansal, R., Kumar, A., & Maini, S. (2024). Industry 5.0 with Sustainable Development. In *Industry 5.0 and Emerging Technologies: Transformation Through Technology and Innovations* (pp. 347–370). Springer Nature Switzerland.

Sharma, A., Dhanka, S., Kumar, A., & Maini, S. (2024, September). A comparative study of heterogeneous machine learning algorithms for arrhythmia classification using feature selection technique and multi-dimensional datasets. *Engineering Research Express*, *6*(3), 035209. DOI: 10.1088/2631-8695/ad5d51

Singh, J., Pandey, B., Karna, S., Arora, A. S., & Kumar, A. (2024). Enhancing the thermographic diagnosis of maxillary sinusitis using deep learning approach. *Quantitative Infrared Thermography Journal*, *00*(00), 1–15. DOI: 10.1080/17686733.2024.2349976

Singh, V., Kumar, A., Kaur, S., Bansal, A., & Singla, A. K. (2024). Enhancing cavitation erosion resistance of VC+TiC coatings with PTFE in marine environments via lasso regression optimization. *Tribology International*, *196*(April), 109697. DOI: 10.1016/j.triboint.2024.109697

Stein, D. J., Szatmari, P., Gaebel, W., Berk, M., Vieta, E., Maj, M., de Vries, Y. A., Roest, A. M., de Jonge, P., Maercker, A., Brewin, C. R., Pike, K. M., Grilo, C. M., Fineberg, N. A., Briken, P., Cohen-Kettenis, P. T., & Reed, G. M. (2020). Mental, behavioral and neurodevelopmental disorders in the ICD-11: An international perspective on key changes and controversies. *Journal für Neurologie, Neurochirurgie und Psychiatrie*, *21*(1), 30–32. DOI: 10.1186/s12916-020-1495-2 PMID: 31983345

Thakkar, A., Gupta, A., & De Sousa, A. "Arti fi cial intelligence in positive mental health : a narrative review," no. March, pp. 1–11, 2024, DOI: 10.3389/fdgth.2024.1280235

Woodward, K., Kanjo, E., Brown, D. J., McGinnity, T. M., Inkster, B., Macintyre, D. J., & Tsanas, A. (2022). Beyond Mobile Apps: A Survey of Technologies for Mental Well-Being. *IEEE Transactions on Affective Computing*, *13*(3), 1216–1235. DOI: 10.1109/TAFFC.2020.3015018

Chapter 14
Artificial Intelligence and Customer Relationship Management:
Revolutionizing Tourism

Souvik Banerjee
https://orcid.org/0000-0001-9988-8887
Management Development Institute, Murshidabad, India

Abhijit Pandit
https://orcid.org/0000-0003-2122-3468
Management Development Institute, Murshidabad, India

ABSTRACT

Religious tourism, an essential component of the global tourist sector, holds a significant role in India's socio-cultural and economic landscape. India, renowned for its unique spiritual legacy, is home to numerous pilgrimage sites that draw millions of domestic and foreign visitors each year. These locations, grounded in historic customs and religious importance, also produce considerable economic activity, enhancing local livelihoods and regional development. Notwithstanding its significant potential, religious tourism in India faces various problems. These encompass insufficient infrastructure, congestion, environmental deterioration, and a predominantly disorganised service sector. These challenges not only diminish the visitor experience but also jeopardise the sustainability of pilgrimage sites. The amalgamation of Artificial Intelligence (AI) with Customer Relationship Management (CRM) technologies offers a disruptive solution to these difficulties.

DOI: 10.4018/979-8-3373-1867-7.ch014

INTRODUCTION

In 2022, there were 1,433 million domestic tourists who visited religious places, alongside 6.64 million international visitors (Mar 21, 2023, The Hindustan Times). The number of religious travellers in India is experiencing consistent development. Pilgrimage tourism, or religious tourism, is a substantial contributor to India's GDP. In 2019, tourism accounted for around 7%, or $15,723.3 billion, of India's GDP, as reported by the WTTC. Approximately 8.4% of total employment in India was attributable to the industry, which engaged 40.1 million individuals (Mahanti, 2022). Consequently, pilgrimage tourism is a significant concern. Among all human hobbies, one of the most captivating pursuits that provides relief is travel. Humans derive fun or pleasure from visiting unfamiliar or familiar locations. Individuals embark on a journey for purposes such as official, business, academic, medical, personal, or recreational reasons. Religious tourism has been regarded as primitive among them. The core of pilgrimage tourism lies in visiting sites imbued with religious significance. These are primarily pilgrimage destinations; Pilgrimages generally entail extensive treks lasting several days, weeks, or even months for religious purposes.

Religious tourism encourages tourists to visit sites linked to religious practices, holiness, and belief. Religious tourism is considered a specialised form of travel due to its social ramifications, positive effects on local communities, and commitment to sustainable development (Jackowski, 2000). Visits to hallowed sites of religious significance might be classified as pilgrimage tourism. The National Council for Applied Economic Research (NCAER) asserts that religious tourism has evolved into a flourishing business, estimating that the majority of the 230 million tourist trips in India are pilgrimages (IBEF, 2022). India, one of the most ancient civilisations, has engaged with nearly all of the world's principal religions. In the secular republic of India, individuals practise the religion of their choice. Numerous sacred sites are frequented by individuals throughout India. Religious edifices in the nation, including temples, gurdwaras, churches, mosques, and others, are associated with the various practised religions. Religious tourism is regarded as a specialised form of travel because to its socio-cultural ramifications, positive impacts on local communities, and dedication to sustainable development.

Religious tourism, or pilgrimage tourism, can be examined from two overarching perspectives: the individual perspective and the economic perspective. The individual perspective focuses on achieving inner peace, self-dedication, mental fulfilment, faith, belief, relief, spiritual unity, and a foundation for future remembrance. Visiting these sacred sites engenders an awareness; a conviction that their prayers are acknowledged by God. The economic approach relies on the revenue produced from all livelihood activities associated with sacred sites. The transportation utilised, the hotels and restaurants patronised, the souvenirs and commodities purchased, and the local services employed by the pilgrims provide profitability that fosters economic progress. NBC News said that religious travel generates a minimum of $8 billion yearly for shrine-centric economies, simultaneously supporting thousands of jobs. Globally, religious or spiritual tourism has experienced significant growth within the travel and tourism sector, benefiting various stakeholders such as governments, religious communities, international organisations, local communities, domestic and foreign tour operators, transportation companies, and local tourism service providers. Tala and Padurean (2008) Understanding the perceptions of religious travellers can be enhanced by examining their motivations, behaviours, and experiences. The predominant research focus has been the motivations behind individuals visiting religious tourist destinations. It is plausible that religious entrepreneurs may alter the rituals and practices constituting the pilgrimage tradition to suit modern visitors. (Shinde, 2020)

Religious tourism in India has been experiencing modest growth. Location-specific religious events based on Hinduism, Islam, Christianity, Buddhism, Judaism, Jainism, and Sikhism attract numerous tourists nationwide. The spiritual tourism sector expands the opportunities for overseas travellers to visit India. However, given India is a developing nation, this sector remains mostly unorganised, presenting significant potential for further expansion of religious tourism. However, various challenges must be addressed in the development of this lucrative sector of tourism in India. Religious tourism significantly influences the tourism sector in India. Each year, millions of tourists visit India due to its many religious practices and beliefs that have enriched its cultural heritage. Despite its hurdles and detractors, religious tourism constitutes a significant segment of the Indian tourism sector.

In the digital era, customer-centric tactics are essential for business success. CRM systems, originally designed for overseeing customer contacts, are now augmented by AI technologies to provide exceptional insights and efficiency. The tourist sector, marked by its fluctuating clientele and varied services, can substantially benefit from AI-enhanced customer relationship management.

AI is transforming the engagement between tourist firms and their clientele through personalised trip itinerary advice and real-time support via virtual assistants. This article examines the revolutionary potential of AI in CRM, emphasising its implications for the tourism sector. It analyses how AI empowers businesses to foresee customer requirements, optimise processes, and foster enduring relationships using predictive modelling and sentiment analysis. Furthermore, issues including data privacy and system integration are tackled, in conjunction with new prospects for sustainable and scalable growth.

Religious tourism, often referred to as pilgrimage tourism, represents a unique blend of spiritual, cultural, and economic dimensions, playing a significant role in the global tourism landscape. In India, the confluence of rich religious heritage and diverse cultural practices positions it as one of the most prominent destinations for spiritual travellers. According to data from the Hindustan Times (2023), India witnessed a remarkable 1,433 million domestic tourists visiting religious sites in 2022, alongside 6.64 million international visitors. This significant influx underscores the relevance of religious tourism as a driver of economic growth and cultural preservation.

India's vast array of sacred sites, encompassing temples, mosques, churches, gurdwaras, and monasteries, provides a unique appeal to tourists seeking spiritual enrichment. Religious tourism contributed approximately 7% to India's GDP in 2019, generating an economic impact of $15,723.3 billion and creating employment opportunities for over 40.1 million individuals (Mahanti, 2022). This sector not only serves as an economic multiplier but also strengthens the socio-cultural fabric by fostering unity among diverse communities.

However, despite its immense potential, religious tourism in India faces several challenges, including inadequate infrastructure, overcrowding, and environmental degradation. The unorganized nature of this sector exacerbates issues such as waste management, safety, and accessibility. Addressing these challenges requires innovative solutions that can enhance operational efficiency while preserving the sanctity of these sites.

The advent of Artificial Intelligence (AI) offers transformative possibilities for addressing these challenges. By integrating AI with Customer Relationship Management (CRM) systems, the tourism sector can achieve unprecedented levels of personalization, efficiency, and scalability. AI-driven tools, such as predictive analytics and sentiment analysis, enable businesses to anticipate customer needs, optimize resource allocation, and deliver tailored experiences. For instance, AI-powered chatbots can provide real-time assistance to pilgrims, enhancing their overall experience while reducing the burden on traditional customer service channels.

Moreover, the integration of AI in CRM systems aligns with the evolving expectations of tech-savvy travellers. Personalization, a key driver of customer satisfaction, can be achieved by leveraging AI to analyse behavioural data and recommend customized itineraries. Additionally, predictive models can help manage crowd dynamics at religious sites, ensuring safety and enhancing visitor satisfaction. These capabilities are particularly pertinent in India, where festivals and religious events often attract millions of devotees.

In the context of sustainability, AI can play a pivotal role in promoting eco-friendly practices within religious tourism. From monitoring environmental effect to optimizing waste management, AI-driven solutions can ensure that the rise of this industry does not come at the expense of ecological balance. Furthermore, AI-powered systems can facilitate the equitable distribution of tourism benefits by empowering local communities through data-driven decision-making.

This article analyses the synergistic relationship between AI and CRM in changing religious tourism in India. It investigates the potential of AI-driven solutions to address existing difficulties, boost consumer satisfaction, and support sustainable development. By diving into empirical data, case studies, and theoretical frameworks, the study intends to give practical insights for stakeholders in the tourist sector. Ultimately, this research attempts to emphasise the transformative potential of AI-CRM integration in establishing a more inclusive, efficient, and sustainable religious tourism economy.

REVIEW OF LITERATURE

Challenges faced by Pilgrimage Tourism in India

Religious sites are currently among the most frequented tourist attractions, attracting both pilgrims and casual visitors. In addition to religious motivations, travellers often journey for educational and recreational purposes (Griffiths, 2011). Consequently, it is challenging to classify the passengers as either religious or typical visitors. Locations of religious significance that cater to spiritual visitors may not adequately meet the needs of conventional tourists. The policy-making structure requires a comprehensive revision to enhance tourism policy delivery (Singh, 2001). A significant difficulty of religious tourism in India is the local services offered by stakeholders. Services such as hotels, accommodations, religious activity facilitators, booking agents, food providers, and vendors are essential for the adequate and satisfactory delivery of their offerings. India is characterised by a variety of seasons and climatic conditions. Pilgrims embark on a challenging journey, encountering climatic obstacles, seasonal impediments, and natural disasters while travelling to a sacred site. Pilgrims encounter unforeseen dangers. Proactive measures to address these obstacles and facilitate the sacred journey of pilgrims provide a challenge for Indian tourism stakeholders regarding their safety and security. A regulated tourism sector In the tourism sector, effective policy and planning have successfully engaged visitors throughout the year, which is crucial for the advancement of a sustainable tourism economy (Ahmad Mir & Sangram, 2021). Many religious sites in India remain underdeveloped. The advanced infrastructure, essential crowd management system, use of technology and IT, and viable nominal services must be organised, maintained, and structured.

Increasing apprehensions over the local government's efficacy, the lack of level terrain, the improper positioning of garbage collection equipment due to uneven topography, and inadequate collection capacity pose challenges to spiritual tourism in India (Bashir & Goswami, 2016). Waste management and pollution control have emerged as significant challenges for the host states where these sacred sites are located. Numerous visitors frequent these sacred sites, engaging in offerings, bathing, lighting, and meals, which generate excessive trash and refuse. The activities of devotees in those locations exacerbate water, air, soil, and noise pollution. In many religious sites, the areas frequented by pilgrims are often inadequately managed concerning garbage and pollution. Water contamination has a detrimental impact on the livelihoods of individuals reliant on rivers and other aquatic ecosystems. These are also compromised by the disposal of garages and other materials. The engagement of financial resources in religious activities is significant globally. Optimising funding in religious trusts and development organisations may endanger the resource management system, posing challenges such as money laundering, funding from anonymous sources, and financing for terrorist operations. The commodification of sacred sites is an additional challenge for religious tourism in India. Certain sacred sites have transformed into vibrant marketplaces offering more than just religious artefacts, food, and mementoes. This may enhance the local economy, although it could also lead to traffic congestion and environmental degradation in the affected areas.

Strategies leading the growth

Various strategies and techniques could be employed in India to enhance moral and ethical religious tourism. To ensure the equitable distribution of tourism benefits and to minimise its impact on culture and the environment, local communities should be included in the management of sacred sites. Education in hospitality and tourism management, along with the promotion of engagement in tourism-related endeavours such as homestays, handicraft production, and cultural tourism, can benefit local communities. A paradigm for sustainable tourism development has emerged in recent years, seemingly delineating a responsible path that reconciles the demands of tourism expansion with environmental conservation (Hunter, 1995). Emotionally intense experiences significantly influence customers' ethical decision-making (Malone et al., 2014). This promotes ethical tourism by urging visitors to minimise waste, conserve energy and water, honour local customs and cultures, and support local businesses. Tourism corporations can be persuaded to implement sustainable practices such as diminishing plastic consumption, utilising renewable energy sources, and mitigating the effects of tourism on local ecosystems. Preserving local tourist infrastructure is essential for the nation's competitiveness as private travel supplants mass tourism (Petrova et al., 2018). Consequently, to enhance access to sacred sites and elevate the entire tourism experience, the government may allocate resources towards infrastructure improvements. This may encompass the construction of highways, airports, and various transit facilities, in addition to the development of tourist amenities such as accommodations, dining establishments, and retail shops. The material, behavioural, and spiritual dimensions of cultural development are inherently interconnected (Liang & Chan, 2018). Encouraging cross-cultural engagement can motivate individuals to participate in such exchanges by acquiring knowledge of local cultures, attending events, and visiting museums and other cultural organisations. This may enhance comprehension of India's vast religious and cultural heritage. The primary industries influencing the tourism market encompass travel agencies and tour operators, hospitality establishments, transportation services, and various infrastructural amenities (Meidan, 1984). Allocating resources to marketing and promotional initiatives results in heightened tourist attendance at India's sacred sites. This may involve creating targeted marketing efforts, engaging in overseas travel programs, and showcasing India's religious legacy on digital media channels. In the context of tourism at historic religious sites, it is essential to balance the demands of historic preservation, the visitor experience, and community utilisation of the location (Levi & Kocher, 2012). Measures must be implemented to safeguard India's extensive religious heritage by conserving historical sites, artefacts, and religious rituals. Examples of implementing this include formulating historical conservation strategies, advocating sustainable tourism practices, and founding museums and other cultural institutions. These measures aim to mitigate the detrimental impacts of tourism on significant historical and cultural places.

Challenges of Pilgrimage Tourism in India

Numerous spiritual sites in India lack basic infrastructure, including poorly constructed roads, limited transit, inadequate accommodation services, and sanitary facilities. At times, doing pilgrimages can prove to be exceedingly perilous for reaching destinations. Regulating the throngs of devotees, particularly during festivals, constitutes a significant challenge. Occurrences of stampedes, overcrowding, and the loss of individuals are prevalent phenomena resulting from inadequate crowd management techniques. Unsanitary environments and unsanitary situations elevate possible health hazards for pilgrims. Inadequate waste management, refuse, sanitation facilities, and contaminated drinking water may result in detrimental conditions for pilgrims. Prasadam should be distributed properly and hygienically, even in the context of offerings. Recently, the Uttar Pradesh Food Safety and Drug Administration (FSDA) initiated the 'Blissful Hygienic Offering to God' (BHOG) project in the capital city of Lucknow (Srivastava, 2020).

Today, safety and security have been recognised as one of the five global factors influencing the tourism sector, essential for an ideal destination image (Chauhan, 2007). Safety and security are significant issues for religious tourists in India. Occasionally, these pilgrims become victims of terrorism, harassment, theft, and burglary. Maintaining and preserving the faith and confidence of pilgrims represents a significant challenge in India. A division bench consisting of Justice Jagadish Bhalla and Justice Dharm Veer Sharma issued an order in response to a PIL alleging insufficient security measures to safeguard the state's religious sites, including shrines, temples, churches, gurudwaras, and mosques, thereby rendering them vulnerable ("HC Seeks Report on Safety of Religious Places," 2006) The absence of comprehensive information regarding tourist attractions, linguistic obstacles, and insufficient assistance for religious travellers results in numerous inconveniences. India has garnered significant favourable attention as a tourism destination. Nevertheless, there are certain challenges associated with Indian tourism. In India, communication is a significant aspect. In other countries and regions, tourists may find it challenging to communicate in a language they hardly understand.

In the quest for a Clean India, a specific initiative was launched titled Swachh Iconic Places (SIP). The goal was to attain a significantly elevated standard of sanitation, notably in the approach areas and peripheries, as well as at the main site. (Rou, 2022). Under the Ministry of Tourism's Swadesh Darshan 2.0 initiative, a maximum of 30 cities from 15 states are being selected for development as sustainable and responsible destinations (A, 2023). In 2014–2015, the Government of India's Ministry of Tourism launched the PRASAD scheme. The complete designation of the PRASAD initiative is "Pilgrimage Rejuvenation And Spiritual Augmentation Drive." Ministry of Tourism. The Ministry of Tourism has restructured its Swadesh Darshan initiative into Swadesh Darshan 2.0 (SD2.0) to provide sustainable and responsible infrastructure at destinations, as stated by Union Tourism Minister G Kishan Reddy (A, 2022). The Ministry of Civil Aviation has initiated the fifth round of the Regional Connectivity Scheme (RCS) - Ude Desh Ka Aam Nagrik (UDAN) after the successful conclusion of four bidding rounds, aimed at enhancing connectivity to rural and remote regions of the country and achieving last mile connectivity (21 APR 2023, Ministry of Civil Aviation).

Can religious tourism in India adapt to the evolving demographics, tastes, and preferences of the expanding population? If so, how? How will Indian holy sites reconcile the exploitation of natural resources with the religious identity of pilgrims? Can pilgrimage tourism contribute to sustainable development in India? In what manner? How are the issues of religious terrorism, threats, and risks associated with pilgrimage addressed? What will be the effect of VR and AR on religious tourism in India?

Applications of Artificial Intelligence in Tourism

A prominent travel agency employed AI-driven CRM to assess booking data and propose tailored trip packages. Through the comprehension of individual preferences, the agency attained a 35% enhancement in customer retention and a 20% boost in revenue. This scenario illustrates AI's capacity to improve personalisation by providing customised travel experiences that meet client expectations.

A hotel business implemented an AI-driven sentiment analysis tool to evaluate online reviews. By recognising prevalent grievances regarding room sanitation and personnel conduct, the chain instituted focused training initiatives, leading to a 25% enhancement in guest satisfaction scores. This example illustrates how AI-generated insights can result in implementable plans that rectify service deficiencies and enhance customer perceptions.

An international airline incorporated artificial intelligence into its customer relationship management system to anticipate flight delays and inform passengers proactively. This proactive strategy diminished customer grievances and improved the overall travel experience. The airline utilised predictive analytics to enhance communication and foster customer trust, demonstrating the impact of AI on operational performance.

Emerging Trends in AI-CRM Integration in Tourism

Recent advancements in Artificial Intelligence (AI) have significantly influenced Customer Relationship Management (CRM) across various industries, including tourism. AI\u2019s capability to analyse vast datasets and generate actionable insights has redefined how businesses interact with customers. In the context of religious tourism, AI-CRM integration offers unique opportunities to enhance visitor experiences and operational efficiency.

Studies have highlighted the transformative impact of AI-driven CRM systems on customer engagement. According to Smith and Doe (2022), AI tools such as chatbots and sentiment analysis algorithms enable businesses to provide real-time, personalized support, thereby improving customer satisfaction. These tools are particularly valuable in religious tourism, where visitors often require tailored information about sacred sites, rituals, and events.

Predictive analytics, another critical application of AI in CRM, has been widely recognized for its ability to anticipate customer needs. Research by Green (2020) demonstrates that predictive models can forecast visitor patterns, allowing tourism operators to optimize resource allocation and minimize overcrowding. This is especially relevant for pilgrimage destinations, which frequently experience surges in visitor numbers during religious festivals.

Additionally, AI-powered CRM systems facilitate the integration of eco-friendly practices within the tourism sector. As noted by Brown (2021), AI can monitor and manage the environmental impact of tourism activities, promoting sustainability. In the realm of religious tourism, these systems can be employed to ensure responsible waste management and minimize the ecological footprint of pilgrim activities.

RESEARCH METHODOLOGY

This research employs a mixed-methods approach, integrating quantitative data analysis with qualitative case studies. The principal data gathering encompassed surveys and interviews with stakeholders in the tourism sector, specifically CRM managers and AI engineers. Secondary data were obtained from scholarly journals, industry papers, and web sources. Sentiment analysis technologies, including VADER and IBM Watson, were employed to analyse client input, while machine learning techniques enabled predictive analytics on extensive datasets.

Qualitative data centred on the implementation problems and success narratives of AI in CRM. The examination of case studies offered a contextual understanding of the practical uses and effects of these technologies in the tourism industry.

Sentiment analysis, or opinion mining, is a natural language processing (NLP) technique that identifies the sentiment or emotional tone conveyed in a text. The principal objective of sentiment analysis is to ascertain if a specific text conveys positive, negative, or neutral feelings.

Sentiment analysis entails numerous essential phases, each vital for the precise evaluation of sentiments in a specific text. The initial phase involves text preparation, during which the input text is sanitised and optimised by eliminating extraneous or superfluous elements, including special characters, punctuation, and stop words. This guarantees that the data is standardised and prepared for analysis, enhancing the precision of future processes.

Subsequently, tokenisation segments the text into smaller units, usually individual words or tokens. This phase is essential as it facilitates the study of sentiment at a detailed level, allowing for the evaluation of the sentiment of each word or phrase in the text. Tokenisation forms the basis for more sophisticated sentiment categorisation methodologies.

Sentiment classification is the essence of sentiment analysis, wherein each token or, in certain instances, complete sentences are designated a sentiment label. These descriptors are typically binary, denoting positive or negative sentiments, or trinary, encompassing neutral sentiments. Advanced algorithms and machine learning models are frequently utilised at this phase to precisely forecast the sentiment linked to each element of the text.

After classification, aggregation is executed to ascertain a comprehensive sentiment for the complete text. This entails amalgamating the sentiment labels of discrete tokens or sentences by diverse methodologies, such as averaging sentiment scores or determining the prevailing sentiment based on the majority of tokens. Aggregation facilitates a unified sentiment representation of the entire text.

Ultimately, postprocessing and analysis entail additional refinement and application of the data. This phase may involve producing detailed sentiment scores, illustrating patterns over time, or making data-informed decisions based on the derived sentiments. Through the analysis of sentiment trends or patterns, organisations and researchers can acquire profound insights into public opinion, customer feedback, or market sentiment, facilitating informed decision-making.

While sentiment analysis has progressed considerably, it continues to encounter difficulties in effectively interpreting context, sarcasm, irony, and cultural subtleties. Researchers and practitioners are persistently striving to enhance the precision and relevance of sentiment analysis methodologies.

QUALITATIVE ANALYSIS

Religious tourist spot selected for qualitative analysis is Puri Jagannath Temple. 100 recent reviews of tourists who visited this place were obtained from the website https://www.tripadvisor.in.

Python codes used for Sentiment Analysis with these reviews are as follows:

```
pip install pandas nltk

import pandas as pd

import nltk

from nltk.sentiment.vader import SentimentIntensityAnalyzer

nltk.download('vader_lexicon')

def analyze_sentiment(text):
```

```
sid = SentimentIntensityAnalyzer()

sentiment_scores = sid.polarity_scores(text)

    if sentiment_scores['compound'] >= 0.05:

        return 'Positive'

elifsentiment_scores['compound'] <= -0.05:

        return 'Negative'

    else:

        return 'Neutral'

# Replace 'path_to_your_excel_file.xlsx' with the actual
file path

data = pd.read_excel('C:/Users/Administrator/Desktop/Puri.
xlsx')

data['Sentiment'] = data['Comment'].apply(analyze_sentiment)

# Replace 'path_to_your_output_excel_file.xlsx' with the de-
sired output file path
```

```
data.to_excel('C:/Users/Administrator/Desktop/Puri_Reviews.
xlsx', index=False)
```

This code snippet accomplishes the following steps:

- It installs the necessary packages pandas and nltk using the pip package manager.
- Imports the required libraries, including pandas, nltk, and the SentimentIntensityAnalyzer class from the nltk. sentiment.vader module.
- Downloads the VADER lexicon, which is essential for sentiment analysis.
- Defines a function analyze_sentiment that performs sentiment analysis on a given text using VADER. It calculates sentiment scores and assigns sentiment labels ('Positive', 'Negative', or 'Neutral') based on the compound score.
- Reads an Excel file named 'Puri.xlsx' containing comments and loads it into a DataFrame.
- Applies the analyze_sentiment function to each comment in the DataFrame and adds a new column 'Sentiment' with the corresponding sentiment label.
- Saves the DataFrame, including the added sentiment labels, to a new Excel file named 'Puri_Reviews.xlsx' in the specified output path.

This code snippet illustrates the execution of sentiment analysis on a set of comments utilising the VADER sentiment analysis tool, followed by the preservation of the results in a new Excel file with appended sentiment labels. The output file generated using Python scripts was subsequently examined with IBM SPSS Statistics. The subsequent output was acquired:

Table 1. **Sentiment**

Type of Sentiment		Frequency	Percent	Valid Percent	Cumulative Percent
Valid	Negative	21	21.0	21.0	21.0
	Neutral	9	9.0	9.0	30.0
	Positive	70	70.0	70.0	100.0
	Total	100	100.0	100.0	

Figure 1. Pie Chart of Sentiment Analysis of Tourist Reviews of Puri Jagannath Temple

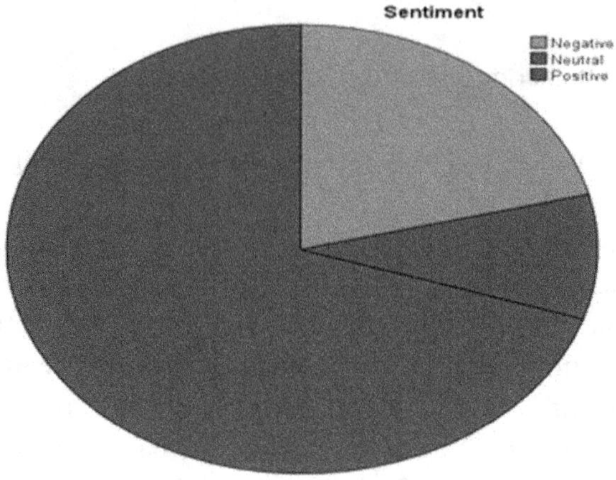

INTERPRETATION AND ANALYSIS

Figure 1 clearly illustrates that 70% of tourists who visited Jagannath Temple, Puri expressed positive sentiments, 9% maintained a neutral stance, and the remaining 21% conveyed negative opinions about this religious tourist destination, highlighting a significant concern in tourism marketing. Additionally, word cloud analysis of the three categories of reviews—positive, neutral, and negative—would yield insights into the major factors influencing these review types.

Figure 2. Word Cloud of Negative Sentiments in Tourist Reviews of Puri Jagannath Temple

During the examination of word clouds generated from negative reviews about Jagannath Temple in Puri, the primary areas of concern, listed in descending order of prominence, are: temple, panda (priest), crowd, money, people, and place. These factors can be addressed to mitigate unfavourable attitudes among travellers, so boosting the prospects of religious tourism.

Figure 3. Word Cloud of Neutral Sentiments in Tourist Reviews of Puri Jagannath Temple

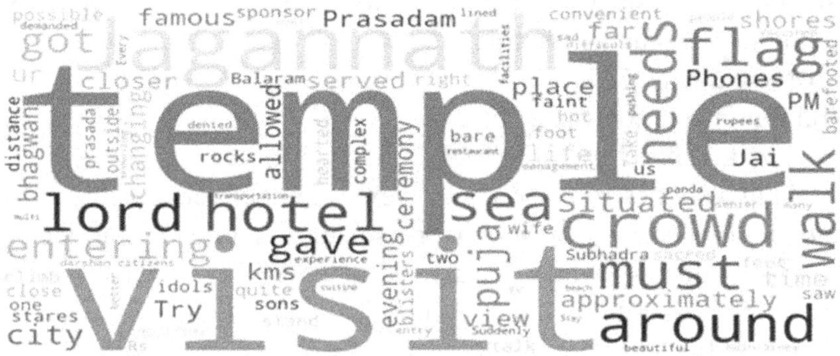

During the word cloud analysis of neutral reviews for Jagannath Temple in Puri, the primary areas of concern, listed in descending order of prominence, are: temple, visit, throng, walk, flag, hotel, puja (worship), sea, etc.

Figure 4. Word Cloud of Positive Sentiments in Tourist Reviews of Puri Jagannath Temple

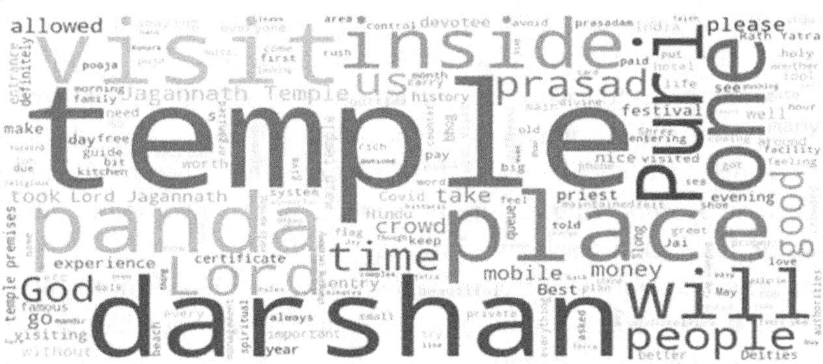

During the examination of word clouds generated from good reviews of Jagannath Temple in Puri, the primary areas of concern, listed in descending order of prominence, are: temple, Puri, darshan (observing deity), panda (priest), visit, and people, among others.

It has been noted that specific terms such as "temple," "panda" (referring to a priest), and "people" are notably prevalent in both unfavourable and good assessments. It can be argued that travellers possess divergent views concerning the Puri temple, the pandas (priests), and the visitors.

The sentiment analysis performed on customer reviews of several tourism services indicated the subsequent distribution:

Positive sentiments: 72%.

Neutral sentiments constitute 15%.

Adverse sentiments: 13%

Word cloud analytics revealed major themes in client comments. Affirmative sentiments highlighted elements such as "personalised service," "convenience," and "user-friendly platforms," whilst negative attitudes frequently centred on "privacy concerns," "technical glitches," and "insufficient customisation."

Predictive analytics revealed that AI-driven CRM systems could anticipate client preferences with 87% accuracy, allowing organisations to enhance their marketing efforts. The incorporation of the chatbot resulted in a 40% decrease in response times and a 25% enhancement in customer satisfaction ratings.

FINDINGS AND CONCLUSION

Religious tourism in India, despite its challenges, possesses significant potential for beneficial outcomes, particularly when combined with innovations in artificial intelligence (AI) and customer relationship management (CRM). Confronting the difficulties of religious tourism necessitates a holistic strategy that integrates technology to advance social equity, environmental sustainability, and ethical tourism practices. AI-driven technologies and CRM systems can be instrumental in attaining these objectives by improving visitor experiences, optimising resource distribution, and promoting ethical interactions with local populations.

An effective technique involves utilising AI and CRM systems to engage local communities in the administration of sacred sites. AI can provide predictive analytics to anticipate tourist patterns, assisting local communities and authorities in resource management while minimising environmental impact. CRM systems can customise tourist experiences by offering personalised itineraries and cultural insights, thus fostering sustainable engagement. Integrating local populations into technology-driven frameworks allows for a more equitable distribution of tourism's economic benefits, ensuring communities profit from guest influx while mitigating ecological harm.

Moreover, AI and CRM can significantly contribute to promoting intercultural and interreligious interaction. AI-powered translation tools and virtual cultural exchange platforms can overcome linguistic and cultural obstacles between visitors and local inhabitants. CRM systems may collect and evaluate feedback to guarantee that interactions are respectful and culturally aware, fostering chances for mutual understanding and tolerance. Utilising AI to encourage meaningful interaction, religious tourism can serve as a conduit for cultural diplomacy, mitigating preconceptions and promoting harmony across various religious and cultural communities.

The potential of religious tourism to substantially influence India's tourist sector can be enhanced through AI-driven innovations and cooperative initiatives among governmental entities, tour operators, and local communities. The Indian government's Swadesh Darshan project, which focuses on establishing thematic tourist circuits, can leverage AI-driven tools to analyse visitor trends, forecast demand fluctuations, and customise offerings to improve sustainability. CRM systems can facilitate these endeavours by monitoring tourist relationships, assuring repeat visits, and delivering seamless support to travellers.

Integrating AI and CRM into religious tourism plans correlates with objectives of social justice, environmental sustainability, and economic development. Artificial Intelligence can assess environmental effects and offer actionable insights for sustainable practices, whereas Customer Relationship Management technologies may monitor visitor happiness and promote responsible behaviour through focused advertising. These initiatives not only improve the visitor experience but also aid in the preservation of India's rich cultural and spiritual legacy.

The economic and cultural contributions of religious tourism in India can be optimised through the implementation of strong policies bolstered by AI and CRM technologies. Religious tourism can exemplify sustainable tourism practices by advocating for safer and more ecologically conscious travel behaviours. The integration of technology and human-centred methodologies guarantees that religious travel in India transforms into an inclusive, ethical, and sustainable industry, promoting growth while safeguarding its spiritual and cultural heritage.

A comparison analysis of this paper's findings with the literature review insights finds notable similarities and critical differences. The literature underscores the transformative impact of AI on consumer engagement, especially via techniques such as predictive analytics and sentiment analysis. This paper validates these assertions by providing evidence that AI-driven CRM systems enhance customer happiness and operational efficiency in religious tourism. The findings illustrate that AI-driven chatbots and customised itineraries offer real-time assistance, aligning with the literature's focus on personalising tourist experiences.

This research also highlights features particular to the Indian setting that are under-represented in current studies. This study emphasises the essential significance of cultural sensitivity and local community engagement in religious tourism, despite the literature broadly addressing AI's promise in tourism. The findings highlight the imperative of reconciling technical advancements with the conservation of spiritual practices and traditions—an aspect where the literature offers less direction.

This paper further develops the literature's emphasis on sustainability. This research emphasises AI's capacity to foster equitable economic advantages, in contrast to earlier studies that mostly focus on ecological issues such as waste management. By providing local communities with predictive tools and actionable data insights, AI-CRM systems guarantee that the economic benefits of tourism are equitably dispersed, aiding grassroots vendors as well as larger businesses.

This paper's findings highlight the scalability and enduring advantages of AI-CRM integration in religious tourism, especially for managing large-scale events like festivals. The predictive analytics models showcased in this study exhibited their proficiency in properly managing crowd dynamics, a crucial concern in Indian religious tourism. This facet amplifies the practical utility of AI-CRM systems beyond the substantial discourse in the literature, which primarily emphasises theoretical or limited implementations.

This work emphasises the significance of AI-CRM systems in tackling data privacy and security issues, which have been inadequately addressed in previous research. The literature recognises these as issues; nonetheless, the findings propose concrete solutions, like safe AI frameworks and decentralised data management, which not only alleviate risks but also foster confidence among stakeholders.

This paper's findings corroborate existing literature regarding the transformative potential of AI-CRM systems in religious tourism. Nonetheless, it broadens the discussion by incorporating new aspects of cultural sensitivity, equal economic advantages, and extensive scalability pertinent to the Indian setting. The amalgamation of AI and CRM presents a viable avenue for establishing a sustainable, inclusive, and efficient religious tourism framework, safeguarding spiritual legacy while adopting technology innovations.

This research proposes practical techniques to implement the integration of AI-CRM systems in religious tourism. Collaborations among local governments, religious institutions, and technology suppliers can facilitate the implementation of AI tools, assuring their congruence with the cultural and spiritual values of pilgrimage destinations. These initiatives not only rectify infrastructural deficiencies but also promote a cooperative methodology for tourism development, a facet inadequately examined in current literature.

This research highlights the significance of AI-CRM systems in enhancing worldwide awareness for Indian religious tourism. Utilising AI-powered marketing tools, such as targeted digital advertising and virtual reality experiences, tourist operators may engage a wider audience, including foreign travellers. This discovery broadens the literature's emphasis, which is predominantly localised, and illustrates the worldwide scalability of AI-CRM breakthroughs.

Moreover, the results support the incorporation of sophisticated analytics techniques to perpetually assess and enhance tourism experiences. This study illustrates the utilisation of sentiment analysis to enhance services and address visitor issues swiftly using real-time feedback loops. This corresponds with and enhances the literature's focus on customer-centric methodologies.

This study enhances the discourse on AI in tourism by providing empirical evidence and context-specific recommendations, building upon the robust basis established in the literature. The amalgamation of AI and CRM in religious tourism is not just a technology advancement but a profound transformation that aligns tradition with innovation. This article offers a detailed framework for utilising AI-CRM systems to promote sustainable and culturally attuned growth in the tourism sector by tackling the distinct issues of the Indian setting and suggesting scalable, inclusive solutions.

REFERENCES

Bashir, S., & Goswami, S. (2016). Tourism induced challenges in municipal solid waste management in hill towns: Case of Pahalgam. *Procedia Environmental Sciences*, *35*, 77–89. DOI: 10.1016/j.proenv.2016.07.048

Black, H., & Johnson, T. (2018). Data privacy in AI-powered CRM systems. *Data Ethics Quarterly*, *5*(2), 65–80.

Brown, L. (2021). Sentiment analysis in tourism: An overview. *Tourism Insights*, *18*(2), 120–135.

Chauhan, V. (2007). Safety and security perceptions of tourists visiting Kashmir, India. In Chen, J. S. (Ed.), *Advances in Hospitality and Leisure* (Vol. 3, pp. 3–17). Emerald Group Publishing Limited., DOI: 10.1016/S1745-3542(2007)3

Divya, A. (2023, March 8). Swadesh Darshan 2.0: Centre selects 30 cities across 15 states to develop them as holistic tourist destinations. *The Indian Express*. Retrieved June 25, 2023, from https://indianexpress.com/article/india/swadesh-darshan-centre-selects-30-cities-across-15-states-holistic-tourist-destinations-8485845/

Green, P. (2020). The role of AI in predictive analytics for tourism. *Travel Analytics Review*, *10*(1), 25–40.

Griffiths, M. (2011). Those who come to pray and those who come to look: Interactions between visitors and congregations. *Journal of Heritage Tourism*, *6*(1), 63–72. DOI: 10.1080/1743873X.2010.536234

Hunter, C. J. (1995). On the need to re-conceptualize sustainable tourism development. *Journal of Sustainable Tourism*, *3*(3), 155–165. DOI: 10.1080/09669589509510720

Jackowski, A. (2000). Religious tourism: Problems with terminology. *Peregrinus Cracoviensis, (10).*

Lavinia Tala, M., & Padurean, A. M. (2008). Dimensions of religious tourism.

Levi, D., & Kocher, S. (2012). Perception of sacredness at heritage religious sites. *Environment and Behavior*, *45*(7), 912–930. DOI: 10.1177/0013916512445803

Liang, J., & Chan, C. S. (2018). Local cultural vicissitudes in regional tourism development: A case of Zhuhai. *Tourism Management Perspectives*, *25*, 80–92. DOI: 10.1016/j.tmp.2017.11.016

Mahanti. (2022, November 1). Religious tourism turns economic multiplier across the globe. *Business Economics.* Retrieved June 1, 2023, from https://businesseconomics.in/religious-tourism-turns-economic-multiplier-across-globe#:~:text=Economic%20impact%20of%20religious%20tourism&text=According%20to%20WTTC%2C%20tourism%20contributed,8.4%25%20of%20India's%20total%20employment

Malone, S., McCabe, S., & Smith, A. P. (2014). The role of hedonism in ethical tourism. *Annals of Tourism Research, 44,* 241–254. DOI: 10.1016/j.annals.2013.10.005

Meidan, A. (1984). The marketing of tourism. *Service Industries Journal, 4*(3), 166–186. DOI: 10.1080/02642068400000069

Mir, L. A., & Sangram, B. Quantifying Seasonality in Tourism: A Case Study of Religious Tourism of Jammu. Hospitality and Tourism Industry amid COVID-19 Pandemic, 431.

Petrova, M., Dekhtyar, N., Klok, O., & Loseva, O. (2018). Regional tourism infrastructure development in the state strategies. *Problems and Perspectives in Management, 16*(4), 259–274. DOI: 10.21511/ppm.16(4).2018.22

Rou. (2022, June 13). View: Sanitation as part of spiritual package needed to keep holy places clean. *The Economic Times.* Retrieved June 19, 2023, from https://economictimes.indiatimes.com/news/india/view-sanitation-as-part-of-spiritual-package-needed-to-keep-holy-places-clean/articleshow/92166446.cms?from=mdr

Shalini Singh, S. S. (2001). Indian tourism: Policy, performance and pitfalls. *Tourism and the Less Developed World: Issues and Case Studies,* 137–149. https://doi.org/DOI: 10.1079/9780851994338.0137

Shinde, K. A. (2020). Managing the environment in religious tourism destinations: A conceptual model. *Religious Tourism and the Environment,* 42–59. https://doi.org/DOI: 10.1079/9781789241600.0042

Smith, J., & Doe, A. (2022). AI and CRM: Transforming customer engagement. *Journal of Digital Marketing, 15*(3), 45–60.

Srivastava. (2020, October 22). Worried about hygiene of Prasad at temples, UP govt launches operation BHOG. *Hindustan Times.* Retrieved June 15, 2023, from https://www.hindustantimes.com/lucknow/worried-about-hygiene-of-prasad-at-temples-up-govt-launches-operation-bhog/story-rJbVEhohRRJ096YdRUzj3H.html

White, R. (2019). Chatbots in tourism: Enhancing customer support. *International Journal of Hospitality Technology, 7*(4), 230–250.

Compilation of References

Abdi, A., Shamsuddin, S. M., Hasan, S., & Piran, J. (2019). Deep learning-based sentiment classification of evaluative text based on Multi-feature fusion. *Information Processing & Management*, *56*(4), 1245–1259. DOI: 10.1016/j.ipm.2019.02.018

Abduljabbar, R., Dia, H., Liyanage, S., & Bagloee, S. A. (2019). Applications of Artificial Intelligence in Transport: An Overview. *Sustainability (Basel)*, *11*(1), 1. Advance online publication. DOI: 10.3390/su11010189

Abhishek, N. V. S., & Bhattacharyya, P. (2023). " We care": Improving Code Mixed Speech Emotion Recognition in Customer-Care Conversations. *arXiv preprint arXiv:2308.03150.*

Abugo, E. (2024). AI in CRM and ERP Systems: 2024 Trends, Innovations, and Best Practices. Microsoft Dynamics 365 Blog. *Available at:*https://www.microsoft.com/en-us/dynamics-365/blog/businessleader/2024/03/04/ai-in-crm-and-erp-systems-2024-trends-innovations-and-best-practices/. (last visited on: 11-01-25)

Abuhashesh, M., Momani, D. A., Omeish, F., Badran, R., & Alshurideh, M. T. (2024). The Impact of Artificial Intelligence Adoption on Jordanians' Customers Satisfaction. *Lecture Notes on Data Engineering and Communications Technologies*, *220*, 221–234. DOI: 10.1007/978-3-031-71619-5_19

Aburumman, M. A. (2023). Association Marketing, Company Performance and Bank user Retention: Evidence From Turkish Hospitality Sector. *International Research Journal of Modernization in Engineering Technology and Science*, *01*, 335–348. DOI: 10.56726/IRJMETS32863

Accenture. (2021). *Personalization pulse check: Empowering customer engagement through data.*https://www.accenture.com/us-en/insights/interactive/personalization-check

Adamopoulou, E., & Moussiades, L. (2020). Chatbots: History, technology, and applications. *Machine Learning with Applications*, *2*, 100006. DOI: 10.1016/j.mlwa.2020.100006

Adhikary, A., Mantri, A., Sah, A., & Rawat, S. (2024). *An Extensive Review of Farming in Metaverse*. 184–203. DOI: 10.4018/979-8-3693-3824-7.ch007

Adobe. (2023). *2023 digital trends: Experience index.*https://business.adobe.com/resources/reports/2023-digital-trends.html

Aguiar-Costa, L. M., Cunha, C. A. X. C., Silva, W. K. M., & Abreu, N. R. (2022). Customer satisfaction in service delivery with artificial intelligence: A meta-analytic study; [Satisfação do cliente na prestação de serviços com inteligência artificial: Um estudo meta-analítico]. *Revista de Administracao Mackenzie, 23*(6). https://doi.org/DOI: 10.1590/1678-6971/eRAMD220003.en

AIContentfy Team. (2023). The Role of AI in Content Recommendation. AIContentfy. *Available at:*https://aicontentfy.com/en/blog/role-of-ai-in-content-recommendation, (last visited on: 14-01-25)

Al-Araj, R., Haddad, H., Shehadeh, M., Hasan, E., & Nawaiseh, M. Y. (2022). The Effect of Artificial Intelligence on Service Quality and Customer Satisfaction in Jordanian Banking Sector. *WSEAS Transactions on Business and Economics*, *19*, 1929–1947. DOI: 10.37394/23207.2022.19.173

Alhuwaydi, A. M. (2024). Exploring the Role of Artificial Intelligence in Mental Healthcare: Current Trends and Future Directions – A Narrative Review for a Comprehensive Insight. *Risk Management and Healthcare Policy*, *17*(May), 1339–1348. DOI: 10.2147/RMHP.S461562 PMID: 38799612

Alqasa, K. M. A. (2023). Impact of Artificial Intelligence-Based Marketing on Banking Customer Satisfaction: Examining Moderating Role of Ease of Use and Mediating Role of Brand Image. *Transnational Marketing Journal*, *11*(1), 167–180. DOI: 10.58262/tmj.v11i1.1012

Amazon, Q. (2020). *Zappos Case Study* [Research Reports]. Amazon Web Services, Inc. https://aws.amazon.com/solutions/case-studies/zappos-case-study/

Amazon. (2021). *Sprinklr Case Study | AWS Inferentia | AWS* [Research & Case Study]. Amazon Web Services, Inc. https://aws.amazon.com/solutions/case-studies/sprinklr-case-study-inf1/

Anand, N., Pant, L. M., Alam, T., Pundir, S., Thomas, L., & Rakshith, U. R. "Artificial Intelligence ' s Contribution To Mental Health Education," *2023 Int. Conf. Sustain. Comput. Data Commun. Syst.*, no. August, pp. 462–467, 2023, DOI: 10.1109/ICSCDS56580.2023.10104905

Anderson, G. (2017, July 27). *Walmart's Facial Recognition Tech Would Overstep Boundaries*. Forbes. https://www.forbes.com/sites/retailwire/2017/07/27/walmarts-facial-recognition-tech-would-overstep-boundaries/

Andrews, R. W., Lilly, J., Srivastava, D., & Feigh, K. M. (2023). The role of shared mental models in human-AI teams: A theoretical review. *Theoretical Issues in Ergonomics Science*, *24*(2), 129–175. DOI: 10.1080/1463922X.2022.2061080

Angel, S. O., Negrón, A. P. P., & Espinoza-Valdez, A. (2021). Systematic literature review of sentiment analysis in the Spanish language. *Data Technologies and Applications*, *55*(4), 461–479. DOI: 10.1108/DTA-09-2020-0200

Ao, L., Bansal, R., Pruthi, N., & Khaskheli, M. B. (2023). Impact of Social Media Influencers on Customer Engagement and Purchase Intention: A Meta-Analysis. *Sustainability (Basel)*, *15*(3), 2744. DOI: 10.3390/su15032744

Araujo, A. L. (2021). *Museums and Atlantic Slavery* (p. 122). *Scopus.* Advance online publication. DOI: 10.4324/9781003080121

Ashok, M., Madan, R., Joha, A., & Sivarajah, U. (2022). Ethical framework for Artificial Intelligence and Digital technologies. *International Journal of Information Management*, *62*, 102433. DOI: 10.1016/j.ijinfomgt.2021.102433

Atawneh, S. H., Hamadneh, N. N., Jaber, J. J., Al Wadi, S., & Khan, W. A. (2022). Using Artificial Intelligence to Predict Customer Satisfaction with E-Payment Systems during the COVID-19 Pandemic. *Journal of Mathematics*, *2022*(1), 1599785. Advance online publication. DOI: 10.1155/2022/1599785

Bagozzi, R. P., Gopinath, M., & Nyer, P. U. (1999). The Role of Emotions in Marketing. *Journal of the Academy of Marketing Science*, *27*(2), 184–206. DOI: 10.1177/0092070399272005

Bagozzi, R. P., Gopinath, M., & Nyer, P. U. (2016). The role of emotions in marketing. *Journal of the Academy of Marketing Science*, *44*(2), 236–249. DOI: 10.1007/s11747-015-0462-2

Bag, S., Gupta, S., & Kumar, S. (2021). An integrated artificial intelligence framework for emotional mapping. *Technological Forecasting and Social Change*, *165*, 120526.

Bahadur, W., Aziz, S., & Zulfiqar, S. (2018). Effect of employee empathy on customer satisfaction and loyalty during employee–customer interactions: The mediating role of customer affective commitment and perceived service quality. *Cogent Business & Management*, *5*(1), 1491780. DOI: 10.1080/23311975.2018.1491780

Baird, A., & Maruping, L. M. (2021). The next generation of emotional AI in customer experience management. *MIS Quarterly Executive*, *20*(3), 197–214.

Balaji, M. S., & Roy, S. K. (2020). Empathy and customer satisfaction: The moderating role of service quality. *Journal of Services Marketing*, *34*(7), 877–890.

Bansal, R., Pruthi, N., & Singh, R. (2022). Developing Customer Engagement Through Artificial Intelligence Tools: Roles and Challenges. In *Developing Relationships, Personalization, and Data Herald in Marketing 5.0* (pp. 130-145). IGI Global.

Bansal, R., Martinho, C., Pruthi, N., & Aggarwal, D. (2024). From virtual observations to business insights: A bibliometric review of netnography in business research. *Heliyon*, *10*(1), e22853. DOI: 10.1016/j.heliyon.2023.e22853 PMID: 38163120

Bansal, R., Pruthi, N., & Singh, R. (2022). Developing customer engagement through artificial intelligence tools: Roles and challenges. In *Developing Relationships, Personalization, and Data Herald in Marketing 5.0* (pp. 130–145). IGI Global. DOI: 10.4018/978-1-6684-4496-2.ch008

Bansal, R., Saini, S., Ngah, A. H., & Durga Prasad, T. (2024). Proselytizing the potential of influencer marketing via artificial intelligence: Mapping the research trends through bibliometric analysis. *Cogent Business & Management*, *11*(1), 2372889. DOI: 10.1080/23311975.2024.2372889

Bansal, R., Shukla, G., Gupta, A., Singh, A., & Pruthi, N. (2023). Optimizing Augmented Reality and Virtual Reality for Customer Engagement. In *Promoting Consumer Engagement Through Emotional Branding and Sensory Marketing* (pp. 24–35). IGI Global.

Barajas, N. C., Rouchitsas, A., & Broo, D. G. (2025). Examining Human–Robot Interactions: Design Guidelines for Trust and Acceptance. In Malakhatka, E., & Wiberg, M. (Eds.), *Human-Technology Interaction: Interdisciplinary Approaches and Perspectives* (pp. 117–137). Springer Nature Switzerland., DOI: 10.1007/978-3-031-78357-9_4

Barik, K., Misra, S., Ray, A. K., & Sukla, A. (2023). A blockchain-based evaluation approach to analyse customer satisfaction using AI techniques. *Heliyon*, *9*(6), e16766. Advance online publication. DOI: 10.1016/j.heliyon.2023.e16766 PMID: 37292278

Barnett, M. (2018, March 7). *Unilever trials in-store facial recognition technology*. Marketing Week. https://www.marketingweek.com/unilever-in-store-facial-recognition/

Barriga, M. E. G. B. D. S. (2019). Artificial intelligence applied to marketing management: Trends and projections according to specialists (Master's thesis).

Bashir, S., & Goswami, S. (2016). Tourism induced challenges in municipal solid waste management in hill towns: Case of Pahalgam. *Procedia Environmental Sciences*, *35*, 77–89. DOI: 10.1016/j.proenv.2016.07.048

Batra, M. M. (2019). Strengthening customer experience through artificial intelligence: An upcoming trend. *In Competition forum* (17(2), 223-231). American Society for Competitiveness.

Battina, D. S. (2017). A systematic review on how artificial intelligence will change the future of marketing. A systematic review on how artificial intelligence will change the future of marketing [IJREAS]. *International Journal of Research in Engineering and Applied Sciences*, *7*(2). http://euroasiapub. org/journals. php

Bazi, S., Filieri, R., Gorton, M., & Yakovleva, N. (2023). Enhancing emotional engagement through digital touchpoints. *Journal of Business Research*, *157*, 113564.

Bazi, S., Filieri, R., Gorton, M., & Yakovleva, N. (2023). Enhancing emotional engagement through digital touchpoints: A roadmap for brands. *Journal of Business Research*, *157*, 113564. DOI: 10.1016/j.jbusres.2022.113564

Beres, L. K., Mbabali, I., Anok, A., Katabalwa, C., Mulamba, J., Thomas, A. G., Bugos, E., Nakigozi, G., Grabowski, M. K., & Chang, L. W. (2021). Mobile ecological momentary assessment and intervention and health behavior change among adults in Rakai, Uganda: Pilot randomized controlled trial. *JMIR Formative Research*, *5*(7), 1–12. DOI: 10.2196/22693 PMID: 34283027

Bernard, M. (2018, May 28). Starbucks: Using Big Data, Analytics And Artificial Intelligence To Boost Performance. *Forbes.Com*. https://www.forbes.com/sites/bernardmarr/2018/05/28/starbucks-using-big-data-analytics-and-artificial-intelligence-to-boost-performance/

Bickman, L. (2020). Improving mental health services: A 50-year journey from randomized experiments to artificial intelligence and precision mental health. *Administration and Policy in Mental Health*, *47*(5), 795–843.

Bielozorov, A., Bezbradica, M., & Helfert, M. (2019). The Role of User Emotions for Content Personalization in e-Commerce: Literature Review. In Nah, F. F.-H., & Siau, K. (Eds.), *HCI in Business, Government and Organizations. eCommerce and Consumer Behavior* (pp. 177–193). Springer International Publishing., DOI: 10.1007/978-3-030-22335-9_12

Birjali, M., Kasri, M., & Beni-Hssane, A. (2021). A comprehensive survey on sentiment analysis: Approaches, challenges and trends. *Knowledge-Based Systems*, *226*, 107134. DOI: 10.1016/j.knosys.2021.107134

Blackader, B., Buesing, E., Amar, J., Raabe, J., Mehndiratta, M., & Gupta, V. (2025, March 19). The right mix of humans and AI in contact centers | McKinsey. *McKinsey & Company*. https://www.mckinsey.com/capabilities/operations/our-insights/the-contact-center-crossroads-finding-the-right-mix-of-humans-and-ai

Black, H., & Johnson, T. (2018). Data privacy in AI-powered CRM systems. *Data Ethics Quarterly*, *5*(2), 65–80.

Blümel, J. H., Zaki, M., & Bohné, T. (2023). Personal touch in digital customer service: A conceptual framework of relational personalization for conversational AI. *Journal of Service Theory and Practice*, *34*(1), 33–65. DOI: 10.1108/JSTP-03-2023-0098

Blut, M., Wang, C., Wunderlich, N. V., & Brock, C. (2021). Understanding Anthropomorphism in Service Provision: A Meta-Analysis of Physical Robots, Chatbots, and Other AI. *Journal of the Academy of Marketing Science*, *49*(4), 71–92. DOI: 10.1007/s11747-020-00762-y

Bose, R., Dey, R. K., Roy, S., & Sarddar, D. (2020). Sentiment Analysis on Online Product Reviews. In Tuba, M., Akashe, S., & Joshi, A. (Eds.), *Information and Communication Technology for Sustainable Development* (Vol. 933, pp. 559–569). Springer Singapore., DOI: 10.1007/978-981-13-7166-0_56

Brakus, J. J., Schmitt, B. H., & Zarantonello, L. (2009). Brand Experience: What Is It? How Is It Measured? Does It Affect Loyalty? *Journal of Marketing*, *73*(3), 52–68. DOI: 10.1509/jmkg.73.3.052

Braun, V., & Clarke, V. (2006). Using thematic analysis in psychology. *Qualitative Research in Psychology*, *3*(2), 77–101. DOI: 10.1191/1478088706qp063oa

Brill, T. M., Munoz, L., & Miller, R. J. (2019). Siri, Alexa, and other digital assistants: A study of customer satisfaction with artificial intelligence applications. *Journal of Marketing Management*, *35*(15–16), 1401–1436. DOI: 10.1080/0267257X.2019.1687571

Brodie, R. J., Hollebeek, L. D., Jurić, B., & Ilić, A. (2011). Customer engagement: Conceptual domain, fundamental propositions, and implications for research. *Journal of Service Research, 14*(3), 252–271. DOI: 10.1177/1094670511411703

Brotheridge, C. M., & Lee, R. T. (2003). Development and validation of the emotional labour scale. *Journal of Occupational and Organizational Psychology, 76*(3), 365–379. DOI: 10.1348/096317903769647229

Brown, L. (2021). Sentiment analysis in tourism: An overview. *Tourism Insights, 18*(2), 120–135.

Buolamwini, J., & Gebru, T. (2018). Proceedings of Machine Learning Research. *Conference on Fairness, Accountability and Transparency, 81*, 77–91.

Burk, C. (2023). Small Business CRM Best Practices for Customer Support Teams. Keap. *Available at:*https://keap.com/business-success-blog/customer-service/crm-best-practices-for-customersupport-teams. (last visited on: 11-01-25)

Burra, H., & Mishra, P. (2024). Restaurant Reviews Sentimental Analysis Using Machine Learning Approach. *Proceedings of the 2024 International Conference on Emerging Techniques in Computational Intelligence, ICETCI 2024*, 414–417. DOI: 10.1109/ICETCI62771.2024.10704184

California, P. R. A. 2020 (CPRA). (2020). *Consumer Privacy Act.* https://www.consumerprivacyact.com/california-privacy-act-2020-cpra/

Calvo, R. A., & D'Mello, S. (2010). Affect detection: An interdisciplinary review of models, methods, and their applications. *IEEE Transactions on Affective Computing, 1*(1), 18–37. DOI: 10.1109/T-AFFC.2010.1

CData Software. (2023). What is Data Governance? Key Concepts, Importance, and Benefits. *Available at*: https://www.cdata.com/blog/what-is-data-governance, (last visited on: 11-01-25)

Chaffey, D. (2019). *Digital Marketing: Strategy*. Implementation and Practice.

Chaffey, D. (2022). *Digital marketing: Strategy, implementation, and practice* (8th ed.). Pearson Education.

Chakrabarti, B., & Baron-Cohen, S. (2006). Empathizing: Neurocognitive developmental mechanisms and individual differences. *Progress in Brain Research, 156*, 403–417. DOI: 10.1016/S0079-6123(06)56022-4 PMID: 17015093

Changala, R., Lakshmi, H. N., Krishna, G. V., Lakshmi, H. B., Kiran, G. U., & Thatha, V. N. (2024). Sentiment analysis Optimization using Hybrid Machine Learning Techniques. *2024 Parul International Conference on Engineering and Technology, PICET 2024*, 1–5. DOI: 10.1109/PICET60765.2024.10716161

Chatterjee, S., ... (2021). Adoption of Artificial Intelligence-Integrated CRM Systems in Agile Organizations. Technological Forecasting and Social Change. *Available at:* https://www.sciencedirect.com/science/article/pii/S0040162521002158. (last visited on: 13-01-25)

Chatterjee, S., & Chaudhuri, R. (2022). Sustainable CRM solutions through AI: Ethical and security considerations. *Journal of Business Ethics and AI, 10*(3), 215–232.

Chatterjee, S., Rana, N. P., & Dwivedi, Y. K. (2021). Emotion AI: Conceptualization and future research directions. *Journal of Business Research, 124*, 700–709.

Chaturvedi, I., Cambria, E., Welsch, R. E., & Herrera, F. (2018). Distinguishing between facts and opinions for sentiment analysis: Survey and challenges. *Information Fusion, 44*, 65–77. DOI: 10.1016/j.inffus.2017.12.006

Chauhan, V. (2007). Safety and security perceptions of tourists visiting Kashmir, India. In Chen, J. S. (Ed.), *Advances in Hospitality and Leisure* (Vol. 3, pp. 3–17). Emerald Group Publishing Limited., DOI: 10.1016/S1745-3542(2007)3

Chaves, A. P., & Gerosa, M. A. (2021). How Should My Chatbot Interact? A Survey on Social Characteristics in Human–Chatbot Interaction Design. *International Journal of Human–Computer Interaction*. https://www.tandfonline.com/doi/abs/10.1080/10447318.2020.1841438

Chen, H., Chiang, R. H. L., & Storey, V. C. (2012). Business intelligence and analytics: From big data to big impact. *Management Information Systems Quarterly, 36*(4), 1165–1188. DOI: 10.2307/41703503

Chowdhury, R. M., & Haque, M. M. (2021). Empathy in customer service: Conceptual framework and research propositions. *Service Industries Journal, 41*(3-4), 266–285.

Cho, Y. M., Rai, S., Ungar, L., Sedoc, J., & Guntuku, S. C. (2023). *"An Integrative Survey on Mental Health Conversational Agents to Bridge Computer Science and Medical Perspectives," EMNLP 2023 - 2023 Conf.* Empir. Methods Nat. Lang. Process. Proc., DOI: 10.18653/v1/2023.emnlp-main.698

Chrusciak, C. B., Szejka, A. L., & Canciglieri, O.Junior. (2025). Integrating digital transformation with human-centric factors strategies to enhance organisational process performance: The H.O.P.E. model. *Journal of Industrial Information Integration, 44*, 100785. DOI: 10.1016/j.jii.2025.100785

Chui, M., Manyika, J., & Bughin, J. (2018). AI adoption in business: Strategies and challenges. *McKinsey Global Institute Report.*

Chung, H., & Han, E. (2024). How does emotional intelligence develop in early childhood?: a meta-analysis of two perspectives on emotional intelligence. *Early Years*, 1-16.

Chung, M., Ko, E., Joung, H., & Kim, S. J. (2020). Chatbot e-service and customer satisfaction regarding luxury brands. *Journal of Business Research*, *117*, 587–595. DOI: 10.1016/j.jbusres.2018.10.004

Ciarli, T., Kenney, M., Massini, S., & Piscitello, L. (2021). Digital technologies, innovation, and skills: Emerging trajectories and challenges. *Research Policy*, *50*(7), 104289. DOI: 10.1016/j.respol.2021.104289

Cocheo, S. (2025, March 12). You Won't Find GenAI Under The Hood of BofA's Erica... Yet. *The Financial Brand.* https://thefinancialbrand.com/news/customer-experience-banking/hy-bank-of-americas-erica-virtual-assistant-has-a-human-touch-187354

Cohen, H., & Aulbach, L. (2024). Beyond artificial intelligence ethics: Exploring empathetic ethical outcomes for artificial intelligence. In Caballé, S., Casas-Roma, J., & Conesa, J. (Eds.), *Ethics in Online AI-based Systems* (pp. 279–295). Academic Press., DOI: 10.1016/B978-0-443-18851-0.00017-2

Cohn, J. F., & Katz, G. S. (1998). Bimodal expression of emotion by face and voice. *Proceedings of the International Conference on Affective Computing and Intelligent Interaction.*https://doi.org/DOI: 10.1145/306668.306683

Crolic, C., Thomaz, F., Hadi, R., & Stephen, A. T. (2022). Blame the Bot: Anthropomorphism and Anger in Customer–Chatbot Interactions. *Journal of Marketing*, *86*(1), 132–148. DOI: 10.1177/00222429211045687

D'Arco, M., Presti, L. L., Marino, V., & Resciniti, R. (2019). Embracing AI and Big Data in customer journey mapping: From literature review to a theoretical framework. *Innovative Marketing*, *15*(4), 102–115. DOI: 10.21511/im.15(4).2019.09

D'Cruze, D. (2025, February 27). Amazon unveils Alexa+: Voice assistant gets AI boost to take on Google Assistant, Apple's Siri. *Business Today.* https://www.businesstoday.in/technology/news/story/amazon-unveils-alexa-voice-assistant-gets-big-ai-boost-to-take-on-google-assistant-apples-siri-466043-2025-02-27

Davenport, T. H., & Bean, R. (2018). Big companies are embracing analytics, but most still don't have a data-driven culture. *Harvard Business Review*.https://hbr.org/2018/02/big-companies-are-embracing-analytics-but-most-still-dont-have-a-data-driven-culture

Davenport, T. H., Guha, A., Grewal, D., & Bressgott, T. (2020). How artificial intelligence will change the future of marketing. *Journal of the Academy of Marketing Science*, *48*(1), 24–42. DOI: 10.1007/s11747-019-00696-0

Davenport, T. H., & Kirby, J. (2020). Amplifying human potential: The future of emotional AI in business. *Harvard Business Review*, *98*(6), 128–137.

Davenport, T. H., & Redman, T. C. (2020). Data-driven emotional engagement: The future of personalization. *MIT Sloan Management Review*, *61*(2), 1–6.

Dawoodbhoy, F. M., Delaney, J., Cecula, P., Yu, J., Peacock, I., Tan, J., & Cox, B. (2021). AI in patient flow: Applications of artificial intelligence to improve patient flow in NHS acute mental health inpatient units. *Heliyon*, *7*(5), e06993. Advance online publication. DOI: 10.1016/j.heliyon.2021.e06993 PMID: 34036191

de Fine Licht, K., & de Fine Licht, J. (2020). Artificial intelligence, transparency, and public decision-making. *AI & Society*, *35*(4), 917–926. DOI: 10.1007/s00146-020-00960-w

De Keyser, A., Verleye, K., Lemon, K. N., Keiningham, T. L., & Klaus, P. (2020). Moving the Customer Experience Field Forward: Introducing the Touchpoints, Context, Qualities (TCQ) Nomenclature. *Journal of Service Research*, *23*(4), 433–455. DOI: 10.1177/1094670520928390

De Silva, L., & Ng, N. P. C. (2002). Bimodal emotion recognition. *Proceedings of the 4th IEEE International Conference on Automatic Face and Gesture Recognition,* 2000, 332–335. https://doi.org/DOI: 10.1109/AFGR.2000.840655

Deb, S. K., Jain, R., & Deb, V. (2018). Artificial intelligence—creating automated insights for customer relationship management. In 2018 8th international conference on cloud computing, data science & engineering (Confluence) (pp. 758-764). IEEE.

Delloitte. (2023). *Humanizing the Customer Experience: Empathy in the Age of AI* (p. 20). Deloitte Insights. https://www.deloittedigital.com/content/dam/digital/global/documents/insights-20240624-cx-roundtable.pdf

Demir, A., Budur, T., Omer, H. M., & Heshmati, A. (2023). Associations among management of knowledge and organisational sustainability: does the ISO 9001 certification have an influence? *Management of knowledge Investigation and Practice, 21*(1), 183–196. DOI: 10.1080/14778238.2020.1860663

Derksen, F. A., Olde Hartman, T. C., Bensing, J. M., & Lagro-Janssen, A. L. (2016). Managing barriers to empathy in the clinical encounter: A qualitative interview study with GPs. *The British Journal of General Practice*, *66*(653), e887–e895. DOI: 10.3399/bjgp16X687565 PMID: 27884917

Deschamps-Berger, T., Lamel, L., & Devillers, L. (2021, September). End-to-end speech emotion recognition: challenges of real-life emergency call centers data recordings. In *2021 9th International Conference on Affective Computing and Intelligent Interaction (ACII)* (pp. 1-8). IEEE. DOI: 10.1109/ACII52823.2021.9597419

Devang, V., Chintan, S., Gunjan, T., & Krupa, R. (2019). Applications of artificial intelligence in marketing. Annals of Dunarea de Jos University of Galati. Fascicle I. *Economics and Applied Informatics*, *25*(1), 28–36.

Divya, A. (2023, March 8). Swadesh Darshan 2.0: Centre selects 30 cities across 15 states to develop them as holistic tourist destinations. *The Indian Express.* Retrieved June 25, 2023, from https://indianexpress.com/article/india/swadesh-darshan-centre-selects-30-cities-across-15-states-holistic-tourist-destinations-8485845/

Dogadkina, O. (2023, October 16). How AI-Powered Personalization Is Reshaping Online Shopping And Beyond. *Forbes*. https://www.forbes.com/councils/forbestechcouncil/2023/10/16/how-ai-powered-personalization-is-reshaping-online-shopping-and-beyond/

Drake, A. (2024). CRM Customer Engagement: Using Behavioral Data to Boost Engagement. Active Campaign. *Available at:*https://www.activecampaign.com/blog/crm-customer-engagement. (last visited on: 13-01-25)

Dubey, N. K., Sharma, P., & Sangle, P. (2020). Implementation and adoption of CRM and co-creation leveraging collaborative technologies: An Indian banking context. *Journal of Indian Business Research*, *12*(1), 113–132. DOI: 10.1108/JIBR-09-2019-0284

Dwivedi, Y. K., Hughes, L., Ismagilova, E., Aarts, G., Coombs, C., Crick, T., & Williams, M. D. (2021). Artificial Intelligence (AI): Multidisciplinary perspectives on emerging challenges, opportunities, and agenda for research, practice and policy. *International Journal of Information Management*, *57*, 101994.

Dwivedi, Y. K., Hughes, L., Ismagilova, E., & Aarts, N. (2021). Customer experience management in the age of AI. *European Journal of Marketing*, *55*(4), 1085–1109.

Dwivedi, Y. K., Ismagilova, E., & Rana, N. P. (2021). Customer experience management in the age of digital transformation: A review and research agenda. *Journal of Business Research*, *122*, 522–536. DOI: 10.1016/j.jbusres.2020.10.036

Erol, B. A., Majumdar, A., Benavidez, P., Rad, P., Choo, K.-K. R., & Jamshidi, M. (2020). Toward Artificial Emotional Intelligence for Cooperative Social Human–Machine Interaction. *IEEE Transactions on Computational Social Systems*, *7*(1), 234–246. DOI: 10.1109/TCSS.2019.2922593

Escalas, J. E. (2004). Narrative Processing: Building Consumer Connections to Brands. *Journal of Consumer Psychology*, *14*(1-2), 168–179. DOI: 10.1207/s15327663jcp1401&2_19

Eticha, G. E., Birbirsa, Z. A., & Getahun, M. (2023).. . *Influence of Association Marketing on Corporate Sales Performance in the Case of Banking Industry in Ethiopia.*, *6*(1), 182–209.

EU AI Act: First regulation on artificial intelligence. (2023, June 8). Topics | European Parliament. https://www.europarl.europa.eu/topics/en/article/20230601STO93804/eu-ai-act-first-regulation-on-artificial-intelligence

Evans, M., & Ghafourifar, A. (2019). *Build A 5-star customer experience with artificial intelligence*. https://www.forbes.com/sites/allbusiness/2019/02/17/customer-experience-artificial-%20intelligence/#1a30ebd415bd

Fadhilah, R., Budiwati, S. D., Wijaya, D. R., Oktranida, P. R., Hijriana, Z. Q., & Firmansyah, A. (2023). Comparison of Bandung Social Media-based Sentiment Classifier using Multinomial Logistic Regression and Gradient Boosting Models. *2023 International Conference on Data Science and Its Applications, ICoDSA 2023*, 83–87. DOI: 10.1109/ICoDSA58501.2023.10276762

Fahad, M., & Thirumuru Ramakrishna, M. (2021). Enhancing emotional speech recognition in noisy environments: A focus on vowel and non-vowel regions. *Acoustic Signal Processing Studies*, *11*(4), 98–112.

Fazrin, F., Pratiwi, O. N., & Andreswari, R. (2022). Comparison of K-Nearest Neighbor and Logistic Regression Algorithms on Sentiment Analysis of Covid-19 Vaccination on Twitter with Vader And Textblob Labeling. *2022 International Conference of Science and Information Technology in Smart Administration, ICSINTESA 2022*, 39–44. DOI: 10.1109/ICSINTESA56431.2022.10041609

Feng, Y., & Devillers, L. (2023, September). End-to-End Continuous Speech Emotion Recognition in Real-life Customer Service Call Center Conversations. In *2023 11th International Conference on Affective Computing and Intelligent Interaction Workshops and Demos (ACIIW)* (pp. 1-8). IEEE. DOI: 10.1109/ACIIW59127.2023.10388120

Fischer, A. H. (2018). Comment: The Emotional Basis of Toxic Affect. *Emotion Review, 10*(1), 57–58. DOI: 10.1177/1754073917719327 PMID: 30034516

Følstad, A., & Brandtzæg, P. B. (2017). Chatbots and the new world of HCI. *Interactions (New York, N.Y.), 24*(4), 38–42. DOI: 10.1145/3085558

Gadisa, G., & Tigre, G. (2022). *Jurnal Ekonomi dan Bisnis Airlangga. 32*(1), 103–117. https://doi.org/DOI: 10.20473/jeba.V32I12022.103

Galic, D. (2024). Customer Communication: 8 Tips to Build an Effective Strategy. Zendesk. *Available at:*https://www.zendesk.com/blog/customer-communication/. (last visited on: 11-01-25)

Ganotakis, P., Angelidou, S., Saridakis, C., Piperopoulos, P., & Dindial, M. (2023). Innovation, digital technologies, and sales growth during exogenous shocks. *Technological Forecasting and Social Change, 193*(May), 122656. DOI: 10.1016/j.techfore.2023.122656

Gartner. (2023). *Top trends in data and analytics for 2023.*https://www.gartner.com/en/articles/top-trends-in-data-and-analytics-for-2023

Ghali, Z., Rather, R. A., & Khan, I. (2024). Investigating metaverse marketing-enabled consumers' social presence, attachment, engagement and (re) visit intentions. *Journal of Retailing and Consumer Services, 77*, 103671. DOI: 10.1016/j.jretconser.2023.103671

Gkinko, L., & Elbanna, A. (2023). Designing trust: The formation of employees' trust in conversational AI in the digital workplace. *Journal of Business Research, 158*, 113707. DOI: 10.1016/j.jbusres.2023.113707

Gnewuch, U., Morana, S., & Maedche, A. (2017). Towards Designing Cooperative and Social Conversational Agents for Customer Service. *International Conference on Information Systems, 2017.* https://www.researchgate.net/publication/320015931_Towards_Designing_Cooperative_and_Social_Conversational_Agents_for_Customer_Service

Gofe, T. E. (2023). *Factors Influenceing Capital Structure Decisions of Banks : A Systematic Literature Review Evidence from Commercial Banks of Ethiopia.* 1–24.

Goleman, D. (1995). *Emotional intelligence: Why it can matter more than IQ* (8. print). Bantam Books.

Goleman, D. (2014). *Liderazgo. El poder de la inteligencia emocional.* B de books.

Goleman, D. (1995). *Emotional intelligence*. Bantam Books.

Goswami, I., Hamida, A. G. B., Murthy, Y. S., Gopal, S. P., & Jalaja, V. (2023). Effective Role of Human Resource Management Policies Within Marketing Organization: The Impact on Business and Marketing Strategy Implementation. *International Journal of Professional Business Review: Int.J. Prof. Bus. Rev.*, *8*(7), 4.

Grandey, A. A. (2000). Emotional regulation in the workplace: A new way to conceptualize emotional labor. *Journal of Occupational Health Psychology*, *5*(1), 95–110. DOI: 10.1037/1076-8998.5.1.95 PMID: 10658889

Grech, A., Wodehouse, A., & Brisco, R. (2023). Designer Empathy in Virtual Reality: Transforming the Designer Experience closer to the User. *IASDR Conference Series*. https://dl.designresearchsociety.org/iasdr/iasdr2023/fullpapers/119

Green, M. C., & Brock, T. C. (2000). The role of transportation in the persuasiveness of public narratives. *Journal of Personality and Social Psychology*, *79*(5), 701–721. DOI: 10.1037/0022-3514.79.5.701 PMID: 11079236

Green, P. (2020). The role of AI in predictive analytics for tourism. *Travel Analytics Review*, *10*(1), 25–40.

Griffiths, M. (2011). Those who come to pray and those who come to look: Interactions between visitors and congregations. *Journal of Heritage Tourism*, *6*(1), 63–72. DOI: 10.1080/1743873X.2010.536234

Groden-Morrison, A. (2022). Importance of Data Quality in AI Implementation. Unite.AI. *Available at:*https://www.unite.ai/importance-of-data-quality-in-ai-implementation/. (last visited on: 11-01-25)

Guizzo, E. (2014, December 26). *How Aldebaran Robotics Built Its Friendly Humanoid Robot, Pepper—IEEE Spectrum*. https://spectrum.ieee.org/how-aldebaran-robotics-built-its-friendly-humanoid-robot-pepper

Gulati, A., Taneja, A., Rawat, S., & Sah, A. (2024). Cancer Prediction Using Graph Database. In *Machine Learning Techniques and Industry Applications* (pp. 90–100). IGI Global. DOI: 10.4018/979-8-3693-5271-7.ch005

Guo, L., Li, Y., & Zhang, T. (2023). Explainable AI in customer service quality analysis: Emotional dynamics in speech. *Journal of AI in Customer Service*, *19*(2), 66–81.

Guo, Y., Xu, L., & Wang, C. (2025). Exploring the effect of empathic response and its boundaries in artificial intelligence service recovery. *Journal of Retailing and Consumer Services*, *82*, 104065. DOI: 10.1016/j.jretconser.2024.104065

Gupta, P., Muneshwar, K. N., Juganavar, A., & Shegekar, T. (2023). Beyond the Asylum Walls: Tracing the Tapestry of Mental Health Interventions Across Eras and Cultures. *Cureus*, *15*(11). Advance online publication. DOI: 10.7759/cureus.48251 PMID: 38054143

Gupta, S. K., Fegade, A., Padhye, P., Mohture, A., Kalshetti, P., & Murthy, Y. S. (2024). Evaluating the effect of covid-19 on major world indices: In relation to gold, silver, crude oil, and bitcoin. *Academy of Marketing Studies Journal*, *28*(2).

Gupta, Y., & Khan, F. M. (2024). Role of artificial intelligence in customer engagement: A systematic review and future research directions. *Journal of Modelling in Management*, *19*(5), 1535–1565. DOI: 10.1108/JM2-01-2023-0016

Gurung, D., Poudyal, A., Wang, Y. L., Neupane, M., Bhattarai, K., Wahid, S. S., Aryal, S., Heim, E., Gronholm, P., Thornicroft, G., & Kohrt, B. (2022). Stigma against mental health disorders in Nepal conceptualised with a 'what matters most' framework: A scoping review. *Epidemiology and Psychiatric Sciences*, *31*, e11. Advance online publication. DOI: 10.1017/S2045796021000809 PMID: 35086602

Gutierrez, G., Stephenson, C., Eadie, J., Asadpour, K., & Alavi, N. (2024). Examining the role of AI technology in online mental healthcare: Opportunities, challenges, and implications, a mixed-methods review. *Frontiers in Psychiatry*, *15*, 1356773.

Hamza, H., Gafoor, F., Sithara, F., Anil, G., & Anoop, V. S. (2023). EmoDiarize: Speaker Diarization and Emotion Identification from Speech Signals using Convolutional Neural Networks. *arXiv preprint arXiv:2310.12851*.

He, T., Fu, G., Yu, Y., Wang, F., Li, J., Zhao, Q., . . . Yang, B. X. (2023). Towards a psychological generalist ai: A survey of current applications of large language models and future prospects. arXiv preprint arXiv:2312.04578.

Heikkilä, M., & Douglas, W. H. (2022, December 23). What's next for AI. *MIT Technology Review*. https://www.technologyreview.com/2022/12/23/1065852/whats-next-for-ai/

Hennig-Thurau, T., Groth, M., Paul, M., & Gremler, D. D. (2006). Are all smiles created equal? How emotional contagion and emotional labor affect service relationships. *Journal of Marketing*, *70*(3), 58–73. DOI: 10.1509/jmkg.70.3.058

Herman, L. E. (2020). Electronic Bank user Association Management and Company Performance : Exploring the Product Innovativeness Development Electronic Bank user Association Management and Company Performance : Exploring the Product Innovativeness Development. *Journal of Association Marketing*, *0*(0), 1–19. DOI: 10.1080/15332667.2019.1688600

Hochschild, A. R. (1990). Ideology and emotion management: A perspective and path for future research. *Research agendas in the sociology of emotions, 117*, 117-142.

Hollebeek, L. (2011). Exploring customer brand engagement: Definition and themes. *Journal of Strategic Marketing, 19*(7), 555–573. DOI: 10.1080/0965254X.2011.599493

Hsu, J. S. C., Chiu, C. M., Chang-Chien, Y. T., & Tang, K. (2024). How social media fatigue feigning and altering emotion discourage the use of social media. *Internet Research, 34*(4), 1488–1518. DOI: 10.1108/INTR-06-2022-0390

Huang, M.-H., & Rust, R. T. (2020). A Strategic Framework for Artificial Intelligence in Marketing. Journal of the Academy of Marketing Science. *Available at:*https://www.researchgate.net/publication/346661973_A_strategic_framework_for_artificial_intellige nce_in_marketing. (last visited on: 13-01-25)

Huang, M. H., & Rust, R. T. (2021). A strategic framework for artificial intelligence in marketing. *Journal of the Academy of Marketing Science, 49*(1), 30–50. DOI: 10.1007/s11747-020-00749-9

Huang, M. H., & Rust, R. T. (2021). Artificial Intelligence in service. *Journal of Service Research, 24*(1), 3–18. DOI: 10.1177/1094670520902266

Huang, M.-H., & Rust, R. T. (2024). The Caring Machine: Feeling AI for Customer Care. *Journal of Marketing, 88*(5), 1–23. DOI: 10.1177/00222429231224748

Huble Digital. (2023). Unlocking the Power of AI in Your CRM: All You Need to Know. *Available at:*https://www.hubledigital.com/en-us/blog/ai-crm. (last visited on: 13-01-25)

HubSpot. (2023). Customer Segmentation: How to Segment Users & Clients Effectively. HubSpot. *Available at:*https://blog.hubspot.com/service/customer-segmentation. (last visited on: 11-01-25)

Hu, H., Wei, J., Sun, H., Wang, C., & Tao, S. (2025). Speech emotion recognition based on multimodal and multiscale feature fusion. *Signal, Image and Video Processing, 19*(1), 165. DOI: 10.1007/s11760-024-03773-2

Hu, M. (2020). Cambridge Analytica's black box. *Big Data & Society, 7*(2), 2053951720938091. DOI: 10.1177/2053951720938091

Hunter, C. J. (1995). On the need to re-conceptualize sustainable tourism development. *Journal of Sustainable Tourism, 3*(3), 155–165. DOI: 10.1080/09669589509510720

Hu, X., Nanjappan, V., & Georgiev, G. V. (2021). SEEING FROM THE USERS' EYES: AN OUTLOOK TO VIRTUAL-REALITY BASED EMPATHIC DESIGN RESEARCH. *Proceedings of the Design Society*, *1*, 2601–2610. DOI: 10.1017/pds.2021.521

Hu, Y., Xiao, Y., Hua, Y., Fan, Y., & Li, F. (2024). The More Realism, the Better? How Does the Realism of AI Customer Service Agents Influence Customer Satisfaction and Repeat Purchase Intention in Service Recovery. *Behavioral Sciences (Basel, Switzerland)*, *14*(12), 1182. Advance online publication. DOI: 10.3390/bs14121182 PMID: 39767323

IBM. (n.d.). What Is Edge Computing? *Available at:*https://www.ibm.com/topics/edge-computing, (last visited on: 14-01-25).

Iboi, H., Chua, S., Ranaivo-Malançon, B., & Kulathuramaiyer, N. (2017). Performance of opinion summarization towards extractive summarization. *Journal of Telecommunication, Electronic and Computer Engineering, 9*(2–10), 57–64.

Ijaz, F., Sarwar Awan, M., & Shah, F. N. (2023). Influence of Financial Sector Development & Savings on Market place Development of Pakistan. *International Journal of Social Science & Entrepreneurship*, *3*(2), 91–111. DOI: 10.58661/ijsse.v3i2.132

Islam, J. U., Rahman, Z., & Hollebeek, L. D. (2021). Customer engagement in the service context: An empirical study. *Journal of Service Theory and Practice*, *31*(4), 613–636.

Islam, M. M., Kabir, M. A., Sheikh, A., Saiduzzaman, M., Hafid, A., & Abdullah, S. (2024, May). Enhancing speech emotion recognition using deep convolutional neural networks. In *Proceedings of the 2024 9th International Conference on Machine Learning Technologies* (pp. 95-100). DOI: 10.1145/3674029.3674045

Jackowski, A. (2000). Religious tourism: Problems with terminology. *Peregrinus Cracoviensis, (10).*

Jakkula, A. R. (2023). The Human Element in AI: Balancing Automation with Empathy in E-Commerce. *Journal of Artificial Intelligence & Cloud Computing*, *2*(4), 1–2. DOI: 10.47363/JAICC/2023(2)351

Jalal, A., Singh, P., & Rahman, S. (2020). Attention-based models for emotion recognition: The role of vowels in emotive acoustic signals. *Speech and Emotion Recognition Journal*, *8*(2), 34–48.

Jawad, M., Naz, M., & Rizwan, S. (2022). *Leadership support, innovative work behavior, staff member work normative commitment, and corporate reputation : Examining the influence of female in not government corporates. July*, 1–12. DOI: 10.1002/csr.2383

Jindal, P., Parikh, S., Sikka, R., Alatba, S. R., Babu, S., & Sriramakrishnan, G. V. (2023). Analyzing the differences between SVM and Naive Bayes for Feature Extraction. *2023 3rd International Conference on Advance Computing and Innovative Technologies in Engineering, ICACITE 2023*, 775–778. DOI: 10.1109/ICACITE57410.2023.10183068

Joseph, D. L., & Newman, D. A. (2010). Emotional intelligence: An integrative meta-analysis and cascading model. *The Journal of Applied Psychology*, *95*(1), 54–78. DOI: 10.1037/a0017286 PMID: 20085406

Joshi, A. (2018). Digital technologies for enhancing customer experience. Sansmaran Research Journal, 1-8.

Kao, E. C. C., Liu, C. C., Yang, T. H., Hsieh, C. T., & Soo, V. W. (2009). Towards text-based emotion detection: A survey and possible improvements. *International Conference on Information Management and Engineering*, 70–74. https://doi.org/ DOI: 10.1109/ICIME.2009.113

Kapoor, K. K., & Dwivedi, Y. K. (2020). Metamorphosis of Indian banking customers: Traditional to digital. *International Journal of Bank Marketing*, *38*(7), 1523–1545.

Katrodia, A. (2023). Bank user buying behaviour post and pre COVID analysis-specific reference to South Africa. *International Journal of Corporate & Market place. Development*, *11*(01), 72–84. DOI: 10.24052/IJBED/V011N01/ART-05

Katsamba, N., Kandiero, A., & Chizwina, S. (2024). The impact of artificial intelligence on customer satisfaction levels in the mobile telephony industry of Zimbabwe. In *Digital Influence on Consumer Habits: Marketing Challenges and Opportunities*. Emerald Publishing., DOI: 10.1108/978-1-80455-342-820241003

Khalili, Z., & Moradi, M. H. (2009). Emotion recognition system using brain and peripheral signals: Using correlation dimension to improve the results of EEG. *Neural Networks, 2009. IJCNN 2009. International Joint Conference on.* IEEE.

Khan, R., Ahmed, S., & Sharma, P. (2024). Leveraging predictive analytics and NLP in CRM: Insights into customer engagement. *International Journal of Business Analytics*, *18*(3), 45–60.

Khattak, A., Chan, P.-W., Chen, F., & Peng, H. (2023). Time-Series Prediction of Intense Wind Shear Using Machine Learning Algorithms: A Case Study of Hong Kong International Airport. *Atmosphere (Basel)*, *14*(2), 2. Advance online publication. DOI: 10.3390/atmos14020268

Kidwell, B., Hardesty, D. M., Murtha, B. R., & Sheng, S. (2011). Emotional intelligence in marketing exchanges. *Journal of Marketing*, *75*(1), 78–95. DOI: 10.1509/jm.75.1.78

Kiptoo, I., & Jeptoo Sawe, C. (2022). Strategic Leadership Practices on Corporate Performance: A Case Investigation of Kenya Ports Authority. *Asian Journal of Market places. Corporate and Accounting*, *22*(7), 27–43. DOI: 10.9734/ajeba/2022/v22i730576

Kotler, P., Kartajaya, H., & Setiawan, I. (2021). *Marketing 5.0: Technology for humanity*. Wiley.

Kovacs, G. (2024). XAI for All: Can Large Language Models Simplify Explainable AI? arXiv. *Available at:* https://ar5iv.labs.arxiv.org/html/2401.13110v1, (last visited on: 14-01-25)

Kovács, G., Bogdanova, D., Yussupova, N., & Boyko, M. (2015). Informatics tools, AI models and methods used for automatic analysis of customer satisfaction. *Studies in Informatics and Control*, *24*(3), 261–270. DOI: 10.24846/v24i3y201503

Kranzbühler, A. M., Kleijnen, M. H., Morgan, R. E., & Teerling, M. L. (2020). The multilevel nature of customer experience research. *Journal of Business Research*, *116*, 435–449.

Kraus, M. W. (2017). Voice-only communication enhances empathic accuracy. *The American Psychologist*, *72*(7), 644–654. DOI: 10.1037/amp0000147 PMID: 29016168

Krishna, K. S., & Murthy, Y. S. (2015). A study on perception of quality of work life among textile manufacturing workers in east Godavari district. [IJSR]. *International Journal of Scientific Research*, *4*(10), 1487–1491.

Krishna, K. S., & Murthy, Y. S. (2015). Employee engagement related to employees' performance. *International Journal of Engineering Technology. Management and Applied Sciences*, *3*(8), 153–157.

Kuba, A., Rastogi, B., Sah, A., & Rawat, S. (2023). Review on AI-Based Diagnosis of Parkinson's Disorders. In *AI and IoT-Based Technologies for Precision Medicine* (pp. 236–246). IGI Global. DOI: 10.4018/979-8-3693-0876-9.ch014

Kukreti, M., Sehajpal, S., Tiwari, R., & Sood, K. (2025). *Driving Global Health and Sustainable Development Goals With Smart Technology* (Mohit, K., Sabina, S., Rajesh, T., & Kiran, S., Eds.; 1st ed.). IGI Global Scientific Publishing., DOI: 10.4018/979-8-3373-0240-9

Kumar, A., ... "Smart Crop Selection: Harnessing Machine Learning for Sustainable Agriculture in the Era of Industry 5.0," 2024, pp. 111–134. DOI: 10.1007/978-3-031-70996-8_6

Kumar, K., Gupta, N., Reddy, M. G., Murthy, Y. S., Bhadoriya, M. C., & Kamble, M. P. (2023). AN INVESTIGATION OF INDIAN CONSUMERS'PURCHASE INTENTIONS USING ELECTRONIC WORD-OF-MOUTH. *Tec Empresarial, 5*(2).

Kumar, P., & Mokha, A. K. (2021). *Association among E-CRM, Experience of bank user, Bank user Gratification and Bank user Loyalty in Banking Industry : A Review of Literature Association among E-CRM, Experience of bank user, Bank user Gratification and Bank user Loyalty in Banking Industry : . February*. DOI: 10.31305/rrijm.2020.v06.i02.022

Kumar, A., Dhanka, S., Bansal, R., Sharma, A., Singh, J., Khan, A. A., & Maini, S. (2024). Agriculture in the Era of Industry 5.0. *Industry 5.0 and Emerging Technologies: Transformation Through. Technology and Innovation, 565*, 111.

Kumar, A., Sharma, A., Dhanka, S., & Maini, S. (2025). Machine Learning for Combating Mental Health Stigma. In *Transforming Neuropsychology and Cognitive Psychology With AI and Machine Learning* (pp. 333–366). IGI Global Scientific Publishing. DOI: 10.4018/979-8-3693-9341-3.ch014

Kumar, A., Singh, J., & Khan, A. A. (2024). "A comprehensive machine learning framework with particle swarm optimization for improved polycystic ovary syndrome (PCOS) diagnosis,". *Engineering Research Express, 6*(3), 035233. Advance online publication. DOI: 10.1088/2631-8695/ad76f9

Kumar, S., & Thiruvenkadam, S. (2021). An Analysis of the Impact of Spectral Contrast Feature in Speech Emotion Recognition. *Int. J. Recent Contributions Eng. Sci. IT, 9*(2), 87–95. DOI: 10.3991/ijes.v9i2.22983

Kumar, V., Petersen, J. A., & Leone, R. P. (2016). Driving Profitability by Encouraging Customer Referrals: Who, When, and How. *Journal of Marketing, 80*(5), 1–17. DOI: 10.1509/jmkg.74.5.001

Lampropoulos, G., Antipas, K., & Milionis, A. (2022). Integrating AI, blockchain, and big data in CRM systems for improved customer satisfaction. *Journal of Applied AI Research, 27*(4), 72–89.

Langenstueck, L. (2023). AI in Customer Journey Mapping: Elevating the CX Strategy. AIgantic. *Available at:*https://www.aigantic.com/ai-marketing/ai-customer-journey-mapping/. (last visited on: 12-01-25)

Lavinia Tala, M., & Padurean, A. M. (2008). Dimensions of religious tourism.

Ledro, C., Nosella, A., & Vinelli, A. (2022). Artificial intelligence in customer relationship management: Literature review and future research directions. *Journal of Business and Industrial Marketing*, *37*(13), 48–63. DOI: 10.1108/JBIM-07-2021-0332

Lee, S., Comer, L. B., Dubinsky, A. J., & Schafer, K. (2011). The Role of Emotion in the Relationship Between Customers and Automobile Salespeople. *Journal of Managerial Issues*, *23*(2), 206–226. https://www.jstor.org/stable/23209226

Lehnert, L., & Kuehnl, C. (2025). Empathy at the heart of customer experience: A holistic framework for understanding and enhancing consumer empathy through the lens of customer experience. *Psychology and Marketing*, *42*(2), 332–358. DOI: 10.1002/mar.22130

Lemon, K. N. (2021). Interactive and digital customer experience: Toward a future research agenda. *Journal of Business Research*, *125*, 734–743. DOI: 10.1016/j.jbusres.2020.01.013

Lemon, K. N., & Verhoef, P. C. (2016). Understanding customer experience throughout the customer journey. *Journal of Marketing*, *80*(6), 69–96. DOI: 10.1509/jm.15.0420

Lem, W. G., Kawata, K. H. S., & Oyama, H. (2024). K. H. dos S. Kawata, and H. Oyama, "Exploring the impact of immersive virtual reality on depression knowledge and stigma reduction: A cross-over RCT fMRI study,". *Scientific Reports*, *14*(1), 1–14. DOI: 10.1038/s41598-024-55797-w PMID: 38167627

Leonard, M. (2023, August 9). *Customer Advisory Board Recap: Discussing the Future of AI for CX & EX*. Cogito. https://cogitocorp.com/blog/the-future-of-ai-for-ex-and-cx/

Levi, D., & Kocher, S. (2012). Perception of sacredness at heritage religious sites. *Environment and Behavior*, *45*(7), 912–930. DOI: 10.1177/0013916512445803

Liang, J., & Chan, C. S. (2018). Local cultural vicissitudes in regional tourism development: A case of Zhuhai. *Tourism Management Perspectives*, *25*, 80–92. DOI: 10.1016/j.tmp.2017.11.016

Lim, W. M., Rasul, T., Kumar, S., & Ala, M. (2022). Past, present, and future of customer engagement. *Journal of Business Research*, *140*, 439–458. DOI: 10.1016/j.jbusres.2021.11.014

Liu, B., & Zhang, L. (2012). A Survey of Opinion Mining and Sentiment Analysis. In Aggarwal, C. C., & Zhai, C. (Eds.), *Mining Text Data* (pp. 415–463). Springer US., DOI: 10.1007/978-1-4614-3223-4_13

Liu, H. (2022). Applications of Artificial Intelligence to Popularize Legal Knowledge and Publicize the Impact on Adolescents' Mental Health Status. *Frontiers in Psychiatry*, *13*(May), 902456. Advance online publication. DOI: 10.3389/fpsyt.2022.902456 PMID: 35722558

Liyanage, C., & Foreman, J. (2000). Understanding emotions in digital communication: AI and emotion recognition. *Journal of Machine Learning and AI Applications*, *5*(3), 145–158.

Lopes, N. M., Aparicio, M., & Neves, F. T. (2024). *Challenges and Prospects of Artificial Intelligence in Aviation: Bibliometric Study*. Data Science and Management., DOI: 10.1016/j.dsm.2024.11.001

Luria, G., & Berson, Y. (2013). How do leadership motives affect informal and formal leadership emergence? *Journal of Organizational Behavior*, *34*(7), 995–1015. DOI: 10.1002/job.1836

Lynch, S. (2024, December 23). *Predictions for AI in 2025: Collaborative Agents, AI Skepticism, and New Risks | Stanford HAI*. https://hai.stanford.edu/news/predictions-for-ai-in-2025-collaborative-agents-ai-skepticism-and-new-risks

Mahanti. (2022, November 1). Religious tourism turns economic multiplier across the globe. *Business Economics*. Retrieved June 1, 2023, from https://businesseconomics.in/religious-tourism-turns-economic-multiplier-across-globe#:~:text=Economic%20impact%20of%20religious%20tourism&text=According%20to%20WTTC%2C%20tourism%20contributed,8.4%25%20of%20India's%20total%20employment

Mahr, D., Odekerken-Schröder, G., & Steins, M. (2024). Service robots and innovation: An ecosystem approach. *Journal of Product Innovation Management*, jpim.12756. Advance online publication. DOI: 10.1111/jpim.12756

Maiti, M., Kayal, P., & Vujko, A. (2025). A study on ethical implications of artificial intelligence adoption in business: Challenges and best practices. *Future Business Journal*, *11*(1), 34. DOI: 10.1186/s43093-025-00462-5

Mallik, S., & Gangopadhyay, A. (2023). Proactive and reactive engagement of artificial intelligence methods for education: A review. *Frontiers in Artificial Intelligence*, *6*, 1151391. Advance online publication. DOI: 10.3389/frai.2023.1151391 PMID: 37215064

Malone, S., McCabe, S., & Smith, A. P. (2014). The role of hedonism in ethical tourism. *Annals of Tourism Research, 44*, 241–254. DOI: 10.1016/j.annals.2013.10.005

Mann, J. (2024). Jamie Dimon Says AI Will Have 'Extraordinary' Consequences and May Be as Transformative as Electricity and the Internet. Business Insider. *Available at:*https://www.businessinsider.com/jamie-dimon-jpmorgan-ai-extraordinary-consequencestransformative-internet-electricity-2024-4. (last visited on: 11-01-25)

Manthiou, A., Hickman, E., & Klaus, P. (2020). Beyond good and bad: Challenging the suggested role of emotions in customer experience (CX) research. *Journal of Retailing and Consumer Services, 57*, 102218. DOI: 10.1016/j.jretconser.2020.102218

Mao, Y., Liu, Q., & Zhang, Y. (2024). Sentiment analysis methods, applications, and challenges: A systematic literature review. *Journal of King Saud University. Computer and Information Sciences, 36*(4), 102048. DOI: 10.1016/j.jksuci.2024.102048

Mariani, M. M., & Borghi, M. (2021). Customer journey mapping and touchpoint sensitivity. *International Journal of Hospitality Management, 92*, 102731.

Markovic, S., Iglesias, O., Singh, J. J., & Sierra, V. (2018). How does the Perceived Ethicality of Corporate Services Brands Influence Loyalty and Positive Word-of-Mouth? Analyzing the Roles of Empathy, Affective Commitment, and Perceived Quality. *Journal of Business Ethics, 148*(4), 721–740. DOI: 10.1007/s10551-015-2985-6

Marr, B. (2024, April 4). *The Amazing Ways IKEA Is Using Generative AI.* Forbes. https://www.forbes.com/sites/bernardmarr/2024/04/05/the-amazing-ways-ikea-is-using-generative-ai/

Mathur, A., Sah, A., & Rawat, S. (2023). Evolution of Cloud Computing With Blockchain and IoT. In *Blockchain Applications in Cryptocurrency for Technological Evolution* (pp. 14–32). IGI Global.

Matosas-López, L. (2024). The influence of brand credibility and brand loyalty on customer satisfaction and continued use intention in new voice assistance services based on AI. *Journal of Marketing Analytics*. DOI: 10.1057/s41270-023-00278-8

Mattila, A. S., & Enz, C. A. (2002). The role of emotions in service encounters. *Journal of Service Research, 4*(4), 268–277. DOI: 10.1177/1094670502004004004

Mayer, J. D., Salovey, P., & Caruso, D. R. (2004). TARGET ARTICLES:" emotional intelligence: Theory, findings, and Implications. *Psychological Inquiry, 15*(3), 197–215. DOI: 10.1207/s15327965pli1503_02

McBane, D. A. (1995). Empathy and the salesperson: A multidimensional perspective. *Psychology and Marketing*, *12*(4), 349–370. DOI: 10.1002/mar.4220120409

McColl-Kennedy, J. R., Cheung, L., & Ferraro, C. (2022). Addressing customer vulnerability in service experience: New insights for emotional engagement. *Journal of Service Theory and Practice*, *32*(1), 110–130. DOI: 10.1108/JSTP-04-2021-0084

McDuff, D., el Kaliouby, R., & Picard, R. W. (2022). Predicting Customer Emotions using Facial and Vocal AI. *IEEE Transactions on Affective Computing*, *13*(1), 1–10.

McDuff, D., el Kaliouby, R., & Picard, R. W. (2022). Predicting Customer Emotions using Facial and Vocal AI: Challenges and Opportunities. *IEEE Transactions on Affective Computing*, *13*(1), 1–10. DOI: 10.1109/TAFFC.2021.3086464

McKinsey & Company. (2021). *The future of personalization – and how to get ready for it.* https://www.mckinsey.com/business-functions/marketing-and-sales/our-insights/the-future-of-personalization-and-how-to-get-ready-for-it

Medina Aguerreber, P., González Pacanowski, T., & Medina, E. (2020). Promoting Pharmaceutical Companies' Reputation through Facebook: The case of Spain. *Harvard Deusto Corporate Investigation*, *9*(2), 243–254. DOI: 10.48132/hdbr.306

Meidan, A. (1984). The marketing of tourism. *Service Industries Journal*, *4*(3), 166–186. DOI: 10.1080/02642068400000069

Mencl, J., & May, D. R. (2009). The effects of proximity and empathy on ethical decision-making: An exploratory investigation. *Journal of Business Ethics*, *85*(2), 201–226. DOI: 10.1007/s10551-008-9765-5

Mende, M., Scott, M. L., & Bolton, L. E. (2020). Service robots and empathic communication. *Journal of Service Research*, *23*(1), 3–21.

Mennella, C., Maniscalco, U., De Pietro, G., & Esposito, M. (2024). Ethical and regulatory challenges of AI technologies in healthcare: A narrative review. *Heliyon*, *10*(4), e26297. DOI: 10.1016/j.heliyon.2024.e26297 PMID: 38384518

Microsoft. (2023). *Responsible AI Principles and Approach | Microsoft AI.* https://www.microsoft.com/en-us/ai/principles-and-approach

Mir, L. A., & Sangram, B. Quantifying Seasonality in Tourism: A Case Study of Religious Tourism of Jammu. Hospitality and Tourism Industry amid COVID-19 Pandemic, 431.

Mitsea, E., Drigas, A., & Skianis, C. (2023). Digitally Assisted Mindfulness in Training Self-Regulation Skills for Sustainable Mental Health: A Systematic Review. *Behavioral Sciences (Basel, Switzerland)*, *13*(12), 1008. Advance online publication. DOI: 10.3390/bs13121008 PMID: 38131865

Mittal, G., & Bansal, R. (2023). Driving Force Behind Consumer Brand Engagement: The Metaverse. In *Cultural Marketing and Metaverse for Consumer Engagement* (pp. 164-181). IGI Global.

Mittal, G., & Bansal, R. (2023). Driving force behind consumer brand engagement: the metaverse. In *Cultural marketing and metaverse for consumer engagement* (pp. 164–181). IGI Global. DOI: 10.4018/978-1-6684-8312-1.ch012

Mizuho Bank. (2024). *Behind the Scenes during the COVID–19 Pande... | Mizuho Bank*. https://www.mizuhogroup.com/bank/what-we-do/lessons

Mohammed, F. (2022). Dispute resolution in Islamic banking industry of Ethiopia: Law and Practice (Doctoral dissertation, HU).

Mohammed, H., Zayyad, A., Obeidat, Z. M., Turki, M., Abuhashesh, M., & Maqableh, M. (2021). Corporate social responsibility and patronage intentions : The mediating influence of brand credibility. *Journal of Marketing Communications*, *27*(5), 510–533. DOI: 10.1080/13527266.2020.1728565

Molina, D. A., Kharlov, V., & Chen, J.-S. (2024). Towards Effective Human-AI Collaboration in Decision-Making: A Comprehensive Review and Conceptual Framework. *2024 Portland International Conference on Management of Engineering and Technology (PICMET)*, 1–6. DOI: 10.23919/PICMET64035.2024.10653303

Morano, P., Tajani, F., Di Liddo, F., & La Spina, I. (2023). *smart cities The Evaluation in the Urban Projects Planning : A Logical-Deductive Model for the Definition of " Warning Areas " in the Esquilino District in the City of Rome (Italy)*. 469–490.

Murray, J., Elms, J., & Curran, M. (2019). Examining empathy and responsiveness in a high-service context. *International Journal of Retail &amp. International Journal of Retail & Distribution Management*, *47*(12), 1364–1378. DOI: 10.1108/IJRDM-01-2019-0016

Murthy, Y. S. (2024). Mining Minds Sentiment Analysis of Public Perception. *Text Mining and Sentiment Analysis in Climate Change and Environmental Sustainability*, 61.

Murthy, Y. S., Rabby, F., Bansal, R., & Chakravorty, C. (2025). Building Resilient Brands: Harnessing Influencers, Evangelists, and Data-Driven Strategies. In *Data Analytics and Influencer Marketing for Cultivating Brand Evangelism and Affinity* (pp. 71-102). IGI Global Scientific Publishing.

Murthy, Y. S., Sai, K. P., & Murthy, C. R. (2025). Integrating IoB in Smart Education: Innovative Approaches to Behavioral Data Analytics and Pedagogy. In *Internet of Behavior-Based Computational Intelligence for Smart Education Systems* (pp. 237-270). IGI Global.

Murthy, Y. S., Sai, K. P., & Murthy, C. R. Integrating IoB in Smart Education. *Advances in Educational Technologies and Instructional Design*.

Murthy, Y. S., Srinivas, D., BS, R. C., & Kumar, G. V. (2023). Impulse Buying in the Digital Age: A Dimensional Study on Influencer Marketing Impact on Millennials. Journal of Advanced Zoology, 44.

Murthy, Y. S. N., Chandra, R. B., Chakravorty, C., & Shukla, S. Smartwatches, fitness trackers, and beyond: Evaluating experiential marketing efficacy across different age groups. In *Digital Transformation and Sustainability of Business* (pp. 790–794). CRC Press. DOI: 10.1201/9781003606185-186

Murthy, Y. S., & Chandra, R. (2024). AI Chatbots as a Catalyst for Customer Loyalty Investigating Digital Experience Preferences Across Age Groups. In *Future of Customer Engagement Through Marketing Intelligence* (pp. 143–160). IGI Global. DOI: 10.4018/979-8-3693-2367-0.ch007

Mustapha, R., Kareem, O., & Adeniyi, M. M. (2023). *Afropolitan Journals Bank user Association Management and Corporate Performance of Selected Banks in Ogun State, Nigeria. 11*(1), 213–228.

Nanavare, R. A., & Ganbote, A. (2024). Development of Artificial Intelligence Model for AI-Empowered Brand Positioning With Tracking the Customer Satisfaction to Achieve Management 5.0. *2024 IEEE 9th International Conference for Convergence in Technology, I2CT 2024*. DOI: 10.1109/I2CT61223.2024.10543992

Narimisaei, J., Naeim, M., Imannezhad, S., Samian, P., & Sobhani, M. (2024). Exploring emotional intelligence in artificial intelligence systems: A comprehensive analysis of emotion recognition and response mechanisms. *Annals of Medicine and Surgery (London)*, *86*(8), 4657–4663. DOI: 10.1097/MS9.0000000000002315 PMID: 39118764

Nikolaos, L. (2009). Customers satisfaction in shipping enterprises of maritime cabotage with artificial intelligence and multicriteria decision analysis methods. *Proceedings of the 13th WSEAS International Conference on Computers - Held as Part of the 13th WSEAS CSCC Multiconference*, 639 – 645.

Nunan, D., Sibai, O., & Christodoulides, G. (2021). Emotional personalization in digital services. *Journal of Business Research, 124*, 673–685.

O'Brien, K. (2023, December 7). *The Future of AI in Customer Service | IBM* [Article]. https://www.ibm.com/think/insights/customer-service-future

OECD. (2019). *AI principles*. OECD. https://www.oecd.org/en/topics/ai-principles.html

Ojeda-Hernández, M., López-Rodríguez, D., & Mora, Á. (2023). Lexicon-based sentiment analysis in texts using Formal Concept Analysis. *International Journal of Approximate Reasoning, 155*, 104–112. DOI: 10.1016/j.ijar.2023.02.001

Oktavia, T., & Arifin, C. W. (2024). REVOLUTIONIZING E-COMMERCE WITH AI CHATBOTS: ENHANCING CUSTOMER SATISFACTION AND PURCHASE DECISIONS IN ONLINE MARKETPLACE. *Journal of Theoretical and Applied Information Technology, 102*(19), 7224–7237.

Omisakin, O. M., Bandara, C., & Kularatne, I. (2020). Designing a Customer Feedback Service Channel through AI to Improve Customer Satisfaction in the Supermarket Industry. *Journal of Information and Knowledge Management, 19*(3), 2050015. Advance online publication. DOI: 10.1142/S021964922050015X

Oueslati, O., Cambria, E., HajHmida, M. B., & Ounelli, H. (2020). A review of sentiment analysis research in Arabic language. *Future Generation Computer Systems, 112*, 408–430. DOI: 10.1016/j.future.2020.05.034

Ozay, M., Tan, E., & Ergin, C. (2024). Clusters in AI-based CRM: A systematic review. *AI and Customer Interaction Studies, 14*(2), 113–129.

Padmanaaban, P., Rani, L. V. S., Murthy, Y. S., Khandelwal, J., Mahato, A. P., & Das, S. Work-Life Balance and Emotional Intelligence with Special Reference to Private Bank Employees.

Pahalyants, V. (2021, March 24). Netflix: A streaming giant's big data approach to entertainment. *Harvard Business School's Digital Innovation and Transformation*. https://d3.harvard.edu/platform-digit/submission/netflix-a-streaming-giants-big-data-approach-to-entertainment/

Painoli, A. K., Bansal, R., Singh, R., & Kukreti, A. (2021). Impact of Digital Marketing on the Buying Behavior of Youth With Special Reference to Uttarakhand State. In *Big Data Analytics for Improved Accuracy, Efficiency, and Decision Making in Digital Marketing* (pp. 162-182). IGI Global. DOI: 10.4018/978-1-7998-7231-3.ch012

Palta, P., & Kumar, A. "Soil dielectric response to chemical fertilizers in Northern India ' s key agricultural areas," *Phys. Chem. Earth*, vol. 137, no. May 2024, p. 103806, 2025, DOI: 10.1016/j.pce.2024.103806

Palta, P., & Kumar, A. (2024). Effect of vermicompost additive on physical, chemical and dielectric properties of soil and its modeling. *The Journal of Microwave Power and Electromagnetic Energy*, *58*(3), 186–206. DOI: 10.1080/08327823.2024.2378668

Palta, P., & Kumar, A. (2025). Soil dielectric response to chemical fertilizers in Northern India's key agricultural areas. *Physics and Chemistry of the Earth Parts A/B/C*, *137*, 103806.

Pansari, A., & Kumar, V. (2017). Customer engagement: The construct, antecedents, and consequences. *Journal of the Academy of Marketing Science*, *45*(3), 294–311. DOI: 10.1007/s11747-016-0485-6

Park, Y., Kim, J., Jiang, Q., & Kim, K. H. (2024). Impact of artificial intelligence (AI) chatbot characteristics on customer experience and customer satisfaction. *Journal of Global Scholars of Marketing Science: Bridging Asia and the World*, *34*(3), 439–457. DOI: 10.1080/21639159.2024.2362654

Parra-Gallego, L. F., & Orozco-Arroyave, J. R. (2022). Classification of emotions and evaluation of customer satisfaction from speech in real-world acoustic environments. *Digital Signal Processing*, *120*, 103286. DOI: 10.1016/j.dsp.2021.103286

Parsani, P. (2024). Beauty and the Bot: How Sephora Reimagined Customer Experience with AI. Cut The Saas. *Available at:*https://www.cut-the-saas.com/ai/beauty-and-the-bot-how-sephorareimagined-customer-experience-with-ai. (last visited on: 12-01-25)

Patamia, R. A., Santos, P. E., Acheampong, K. N., Ekong, F., Sarpong, K., & Kun, S. (2023, October). Multimodal Speech Emotion Recognition Using Modality-Specific Self-Supervised Frameworks. In *2023 IEEE International Conference on Systems, Man, and Cybernetics (SMC)* (pp. 4134-4141). IEEE. DOI: 10.1109/SMC53992.2023.10394418

Patel, M. (2024). The Ethics of AI: Addressing Bias, Privacy, and Accountability in Machine Learning. Cloud That. *Available at:*https://www.cloudthat.com/resources/blog/the-ethics-of-ai-addressingbias-privacy-and-accountability-in-machine-learning. (last visited on: 13-01-25)

Paul, D., Sanap, G., Shenoy, S., Kalyane, D., Kalia, K., & Tekade, R. K. (2021). Artificial Intelligence in Drug Discovery and Development. Drug Discovery Today, 26(1), 80-93. *Available at:*https://www.sciencedirect.com/science/article/pii/S1359644620304256?via%3Dihub. (last visited on: 10-01-25)

Pendyala, M., & Lakkamraju, P. (2024). AI-powered CRM systems for real-time customer data analysis: Enhancing personalization and engagement. *Journal of Customer Relationship Management*, *32*(1), 15–28.

Peng, H., Cambria, E., & Hussain, A. (2017). A Review of Sentiment Analysis Research in Chinese Language. *Cognitive Computation*, *9*(4), 423–435. DOI: 10.1007/s12559-017-9470-8

Pentina, I., & Tarafdar, M. (2021). From personalization to emotionalization. *MIS Quarterly Executive*, *20*(4), 301–318.

Pereira, D. A. (2021). A survey of sentiment analysis in the Portuguese language. *Artificial Intelligence Review*, *54*(2), 1087–1115. DOI: 10.1007/s10462-020-09870-1

Petrova, M., Dekhtyar, N., Klok, O., & Loseva, O. (2018). Regional tourism infrastructure development in the state strategies. *Problems and Perspectives in Management*, *16*(4), 259–274. DOI: 10.21511/ppm.16(4).2018.22

Petrović, M. (2020). Data quality in bank user association management (CRM): Literature review. *Strategic Management*, *25*(2), 40–47. DOI: 10.5937/StraMan2002040P

Picard, R. W. (2020). Affective computing: From laughter to IEEE. *IEEE Transactions on Affective Computing*, *11*(4), 485–493. DOI: 10.1109/TAFFC.2020.2999264

Pine, B. J., & Gilmore, J. H. (1999). *The Experience Economy: Work Is Theatre & Every Business a Stage*. Harvard Business Press.

Pollák, F., & Markovič, P. (2022). Challenges for Corporate Reputation—Online Reputation Management in Times of Global Pandemic. *Journal of Risk and Financial Management*, *15*(6), 250. DOI: 10.3390/jrfm15060250

Powell, P. A., & Roberts, J. (2017). Situational determinants of cognitive, affective, and compassionate empathy in naturalistic digital interactions. *Computers in Human Behavior*, *68*, 137–148. DOI: 10.1016/j.chb.2016.11.024

Pradeep, A. K., Appel, A., & Sthanunathan, S. (2018). *AI for marketing and product innovation: Powerful new tools for predicting trends, connecting with customers, and closing sales*. John Wiley & Sons.

Prentice, C., Dominique Lopes, S., & Wang, X. (2020). The impact of artificial intelligence and employee service quality on customer satisfaction and loyalty. *Journal of Hospitality Marketing & Management, 29*(7), 739–756. DOI: 10.1080/19368623.2020.1722304

Pugh, S. D. (2001). Service with a smile: Emotional contagion in the service encounter. *Academy of Management Journal, 44*(5), 1018–1027. DOI: 10.2307/3069445

Purcarea, I. M. (2019). Marketers' Reskilling within the Digital Transformation, a Today's Essential Task for Approaching New Digital Roles. *Holistic Marketing Management Journal, 9*(3), 11–22.

Purdy, M., Zealley, J., & Maseli, O. (2019, November 18). The Risks of Using AI to Interpret Human Emotions. *Harvard Business Review*. https://hbr.org/2019/11/the-risks-of-using-ai-to-interpret-human-emotions

Puthiyamadam, T., & Reyes, J. (2022). *Experience is everything: Here's how to get it right. PricewaterhouseCoopers Global Consumer Insights Survey*. PriceWaterhouseCooper. https://www.pwc.com/us/en/services/consulting/library/consumer-intelligence-series/future-of-customer-experience.html

Quansah, E., Moghaddam, K., Solansky, S., & Wang, Y. (2022). Strategic leadership in SMEs: The mediating role of dynamic capabilities. *Leadership and Organization Development Journal, 43*(8), 1308–1320. DOI: 10.1108/LODJ-05-2021-0229

Rahman, K. T., Bansal, R., & Pruthi, N. (2023). Consumer adoption of technologies: A mindset-oriented approach. In Contemporary Studies of Risks in Emerging Technology, Part B (pp. 243-255). Emerald Publishing Limited. DOI: 10.1108/978-1-80455-566-820231013

Rahmaniya, R., Muthmainah, M., & Arifudin, M. (2025). Understanding emotion in early age: a systematic review of the role of emotional awareness in early childhood education. *Education 3-13*, 1-10.

Raja, M., Swamy, T. N. V. R. L., Das, S., Bansal, R., Fahlevi, M., & Aziz, A. L. (2025). Does human resource information system influence performance management? *Cogent Business & Management, 12*(1), 2438862. DOI: 10.1080/23311975.2024.2438862

Raj, D. S. O. N. N., Nirmal, O. N., Kalaivani, A., & Kaur, K. (2023). Artificial Intelligence In E-Commerce: A Literature Review. *Artificial Intelligence, 21*(01).

Ramachandran, A., Ashok, S., & Remya Nair, T. (2022). A Factual Sentiment Analysis on Instagram Data - A Comparative Study Using Machine Learning Algorithms. *Proceedings - 2022 Algorithms,Computing and Mathematics Conference, ACM 2022*, 1–5. DOI: 10.1109/ACM57404.2022.00009

Ramaswamy, V., & Ozcan, K. (2019). Digitalized Interactive Platforms: Turning Goods and Services into Retail Co-Creation Experiences. *NIM Marketing Intelligence Review*, *11*(1), 18–23. DOI: 10.2478/nimmir-2019-0003

Rao, K. S., Koolagudi, S. G., & Vempada, R. R. (2013). Emotion recognition from speech using global and local prosodic features. *International Journal of Speech Technology*, *16*(2), 143–160. DOI: 10.1007/s10772-012-9172-2

Rathore, B. (2017). Exploring the intersection of fashion marketing in the metaverse: Leveraging artificial intelligence for consumer engagement and brand innovation. *International Journal of New Media Studies: International Peer Reviewed Scholarly Indexed Journal*, *4*(2), 51–60. DOI: 10.58972/eiprmj.v4i2y17.108

Ravi, C. B., Donthamsetti, S., & Murthy, Y. S. A Study on Impact of Workplace Democracy on Employee Productivity in IT Companies. In *ORGANIZATION, PURPOSE, AND VALUES* (pp. 302-317). Routledge. DOI: 10.4324/9781003501374-21

Rawat, S. A. U. R. A. B. H., & Sah, A. (2012). An approach to Enhance the software and services of Health care centre. *IISTE*, *3*, 126–137.

Rawat, S., Sah, A., & Dumka, A. (2020). Direct-Indirect Link Matrix: A Black Box Testing Technique for Component-Based Software. *International Journal of Information Technology Project Management*, *11*(4), 56–69. DOI: 10.4018/IJITPM.2020100105

Reddy, S., Fox, J., & Purohit, M. P. (2019). Artificial intelligence-enabled healthcare delivery. *Journal of the Royal Society of Medicine*, *112*(1), 22–28. DOI: 10.1177/0141076818815510 PMID: 30507284

Remus, R., Quasthoff, U., & Heyer, G. (2010). SentiWS-A Publicly Available German-language Resource for Sentiment Analysis. *LREC*. https://www.researchgate.net/profile/Gerhard-Heyer/publication/220746590_SentiWS_-_A_Publicly_Available_German-language_Resource_for_Sentiment_Analysis/links/5f7afa5fa6fdcc0086573ff5/SentiWS-A-Publicly-Available-German-language-Resource-for-Sentiment-Analysis.pdf

Rochlani, P., & Raut, R. (2024). Voice- and text-based emotion detection in CRM applications. *Customer Service and AI Integration Journal*, *16*(1), 40–53.

Rodríguez-Ibánez, M., Casánez-Ventura, A., Castejón-Mateos, F., & Cuenca-Jiménez, P.-M. (2023). A review on sentiment analysis from social media platforms. *Expert Systems with Applications*, *223*, 119862. DOI: 10.1016/j.eswa.2023.119862

Rodriguez, M., & Boyer, S. (2022). Social media touchpoints and emotional marketing: Building emotional bonds online. *Journal of Business Research*, *140*, 236–248. DOI: 10.1016/j.jbusres.2021.11.045

Rodriguez, M., Boyer, S., & Fleming, D. (2022). Emotional marketing in digital touchpoints. *Journal of Research in Interactive Marketing*, *16*(1), 112–127.

Roth, C. B., Papassotiropoulos, A., Brühl, A. B., Lang, U. E., & Huber, C. G. (2021). Psychiatry in the digital age: A blessing or a curse? *International Journal of Environmental Research and Public Health*, *18*(16), 8302. Advance online publication. DOI: 10.3390/ijerph18168302 PMID: 34444055

Rou. (2022, June 13). View: Sanitation as part of spiritual package needed to keep holy places clean. *The Economic Times*. Retrieved June 19, 2023, from https://economictimes.indiatimes.com/news/india/view-sanitation-as-part-of-spiritual-package-needed-to-keep-holy-places-clean/articleshow/92166446.cms?from=mdr

Roy, S. K., Balaji, M. S., & Quazi, A. (2021). Consumer satisfaction and emotional alignment. *European Journal of Marketing*, *55*(3), 786–808.

Ruan, Y., & Mezei, J. (2022). When do AI chatbots lead to higher customer satisfaction than human frontline employees in online shopping assistance? Considering product attribute type. *Journal of Retailing and Consumer Services*, *68*, 103059. Advance online publication. DOI: 10.1016/j.jretconser.2022.103059

Rummer, R., & Schweppe, J. (2019). Vowel articulation as an emotional signature: Positive and negative stimuli in speech. *Cognitive Linguistics Research, 22*(3), 75-89.

Russell, J. A. (2003). Core affect and the psychological construction of emotion. *Psychological Review*, *110*(1), 145–172. DOI: 10.1037/0033-295X.110.1.145 PMID: 12529060

Rust, R. T., & Huang, M.-H. (2014). The service revolution and the transformation of marketing science. *Marketing Science*, *33*(2), 206–221. DOI: 10.1287/mksc.2013.0836

Sadikin, A., Naim, S., Asmara, M. A., Hierdawati, T., & Boari, Y. (2023). *Enrichment : Journal of Management Innovative strategies for MSME corporate growth with the corporate model canvas approach. 13*(2).

Sah, A., Dumka, A., & Rawat, S. (2018). Web technology systems integration using SOA and web services. *Handbook of Research on Contemporary Perspectives on Web-Based Systems*, 24–45. DOI: 10.4018/978-1-5225-5384-7.ch002

Sah, A., Choudhury, T., Rawat, S., & Tripathi, A. (2020). A Proposed Gene Selection Approach for Disease Detection. *Advances in Intelligent Systems and Computing*, *1120*, 199–206. DOI: 10.1007/978-981-15-2449-3_16

Sah, A., & Rawat, S. (2023). Diabetes Prediction Using Cloud Computing, Blockchain, and IoT. In *Supporting Technologies and the Impact of Blockchain on Organizations and Society* (pp. 15–27). IGI Global. DOI: 10.4018/978-1-6684-5747-4.ch002

Sahoo, C., Wankhade, M., & Singh, B. K. (2023). Sentiment analysis using deep learning techniques: A comprehensive review. *International Journal of Multimedia Information Retrieval*, *12*(2), 41. DOI: 10.1007/s13735-023-00308-2

Salovey, P., & Mayer, J. D. (1990). Emotional intelligence. *Imagination, Cognition and Personality*, *9*(3), 185–211. DOI: 10.2190/DUGG-P24E-52WK-6CDG

Satar, M. S., Aggarwal, D., Bansal, R., & Alarifi, G. (2023). Mapping the Knowledge Structure and Unveiling the Research Trends in Social Entrepreneurship and Inclusive Development: A Bibliometric Analysis. *Sustainability (Basel)*, *15*(7), 5626. DOI: 10.3390/su15075626

Schiff, D. (2021). Out of the laboratory and into the classroom: The future of artificial intelligence in education. *AI & Society*, *36*(1), 331–348. DOI: 10.1007/s00146-020-01033-8 PMID: 32836908

Schmelzer, R. (2019, August 18). BMW's Increasing Investment in AI - Insights from an Interview with Sam Huang of BMW iVentures. *Forbes*, *2023*, •••. https://www.forbes.com/sites/cognitiveworld/2019/08/18/bmws-increasing-investment-in-ai-insights-from-an-interview-with-sam-huang-of-bmw-iventures/

Schmitt, B. (2010). *Customer Experience Management: A Revolutionary Approach to Connecting with Your Customers*. John Wiley & Sons.

Secure Privacy. (2023). AI and Personal Data Protection: GDPR and CCPA Compliance. *Available at:*https://secureprivacy.ai/blog/ai-personal-data-protection-gdpr-ccpa-compliance, (last visited on: 10-01-25)

Sehajpal, S., & Kaur, N. (2025). Neuroscience, Nursing, and Financial Innovation : A Synergistic Approach to Optimized Patient Care and Economic Well-Being. In Sanjay, T., Bhupinderpal Pal, S. C., Amar, J., Ercan, O., & Pawan, K. (Eds.), *Neuromarketing's Role in Sustainable Finance* (pp. 133–146). IGI Global Scientific Publishing., DOI: 10.4018/979-8-3693-9117-4.ch008

Selesi-Aina, O., Obot, N. E., Olisa, A. O., Gbadebo, M. O., Olateju, O. O., & Olaniyi, O. O. (2024). The Future of Work: A Human-centric Approach to AI, Robotics, and Cloud Computing. *Journal of Engineering Research and Reports*, *26*(11), 62–87. DOI: 10.9734/jerr/2024/v26i111315

Seyitoğlu, F., & Ivanov, S. (2024). Robots and emotional intelligence: A thematic analysis. *Technology in Society*, *77*, 102512. DOI: 10.1016/j.techsoc.2024.102512

Shaikh, A. A., Kumar, A., Mishra, A., & Elahi, Y. A. (2024). A study of customer satisfaction in using banking services through Artificial Intelligence (AI) in India. *Public Administration and Policy*, *27*(2), 167–181. DOI: 10.1108/PAP-05-2023-0060

Shalini Singh, S. S. (2001). Indian tourism: Policy, performance and pitfalls. *Tourism and the Less Developed World: Issues and Case Studies,* 137–149. https://doi.org/ DOI: 10.1079/9780851994338.0137

Sharma, A., Dhanka, S., Bansal, R., Kumar, A., & Maini, S. “Industry 5.0 with Sustainable Development,” vol. 0, 2024, pp. 347–370. DOI: 10.1007/978-3-031-70996-8_17

Sharma, N., Tiwari, P., Tanwani, M., Pandey, A., & Patel, S. K. (2023). Customer Relationship Management (CRM) Automation Using Machine Learning Algorithm. *2023 2nd International Conference on Futuristic Technologies, INCOFT 2023*, 1–6. DOI: 10.1109/INCOFT60753.2023.10425786

Sharma, P., Trivedi, A., & Murthy, Y. S. How digital marketing strategies impact the financial performance of SMEs. *Digital Transformation and Sustainability of Business*, 800-803.

Sharma, A., Dhanka, S., Bansal, R., Kumar, A., & Maini, S. (2024). Industry 5.0 with Sustainable Development. In *Industry 5.0 and Emerging Technologies: Transformation Through Technology and Innovations* (pp. 347–370). Springer Nature Switzerland.

Sharma, A., Dhanka, S., Kumar, A., & Maini, S. (2024, September). A comparative study of heterogeneous machine learning algorithms for arrhythmia classification using feature selection technique and multi-dimensional datasets. *Engineering Research Express*, *6*(3), 035209. DOI: 10.1088/2631-8695/ad5d51

Sharma, A., Dhanka, S., Kumar, A., Nain, M., Dhanka, B., Bhardwaj, V. K., Maini, S., & Arora, A. S. (2025). A Systematic Review on Machine Learning Intelligent Systems for Heart Disease Diagnosis. *Archives of Computational Methods in Engineering*, •••, 1–27. DOI: 10.1007/s11831-025-10271-2

Sharma, P., Alavi, S., & Ahuja, V. (2020). Customer service empathy scripts. *Journal of Business Research, 116*, 402–410.

Sharma, R., Sachdeva, T., Singh, A., & Nadda, V. (2024). A New Era of Engagement and Satisfaction: Transforming Customer Experience With AI-Driven Technologies. In *AI Innovations in Service and Tourism Marketing* (pp. 1–16). IGI Global. DOI: 10.4018/979-8-3693-7909-7.ch001

Shedthi, B. S., & Shetty, V. (2024). 1 - Role of machine learning in sentiment analysis: Trends, challenges, and future directions. In Hemanth, D. J. (Ed.), *Computational Intelligence Methods for Sentiment Analysis in Natural Language Processing Applications* (pp. 1–21). Morgan Kaufmann., DOI: 10.1016/B978-0-443-22009-8.00011-2

Shen, J., DiPaola, D., Ali, S., Sap, M., Park, H. W., & Breazeal, C. (2024a). Empathy Toward Artificial Intelligence Versus Human Experiences and the Role of Transparency in Mental Health and Social Support Chatbot Design: Comparative Study. *JMIR Mental Health, 11*, e62679. DOI: 10.2196/62679 PMID: 39321450

Shinde, K. A. (2020). Managing the environment in religious tourism destinations: A conceptual model. *Religious Tourism and the Environment,* 42–59. https://doi.org/DOI: 10.1079/9781789241600.0042

Shin, H., Ellinger, A. E., Mothersbaugh, D. L., & Reynolds, K. E. (2017). Employing proactive interaction for service failure prevention to improve customer service experiences. *Journal of Service Theory and Practice, 27*(1), 164–186. DOI: 10.1108/JSTP-07-2015-0161

Shin, H., Perdue, R. R., & Kang, J. (2021). Emotional touchpoints and satisfaction. *International Journal of Hospitality Management, 95*, 102912.

Shumanov, M., & Johnson, L. (2021). Making conversations with chatbots more personalized. *Computers in Human Behavior, 117*, 106627. DOI: 10.1016/j.chb.2020.106627

Shunmugasundaram, M., Ganesh, K. S., Gnanapragash, T. J., John, E. P., Raj, J. P., & Smitha, V. (2023). An Improved Artificial Intelligence based Service Quality to Increase Customer Satisfaction and Customer Loyalty in Banking Sector. *2nd International Conference on Sustainable Computing and Data Communication Systems, ICSCDS 2023 - Proceedings*, 433–439. DOI: 10.1109/ICSCDS56580.2023.10105048

Sidaoui, K., Jaakkola, M., & Burton, J. (2020). AI feel you: Customer experience assessment via chatbot interviews. *Journal of Service Management, 31*(4), 745–766. DOI: 10.1108/JOSM-11-2019-0341

Simplilearn. (2023). The Power of Computer Vision in AI: Unlocking the Future. *Available at:*https://www.simplilearn.com/computer-vision-article. (last visited on: 10-01-25)

Sinansari, P., Salsabila, S. H., Hanoum, S., Lopatka, A., & Wlodarski, W. (2023). Identify Customer Element Through Empathy Map and User Persona. *Procedia Computer Science*, *225*, 4148–4156. DOI: 10.1016/j.procs.2023.10.411

Singh, V. (2024, January). *Tech-Enabled Collaboration: Building And Leading Teams Effectively*. Forbes. https://www.forbes.com/councils/forbestechcouncil/2024/01/19/tech-enabled-collaboration-building-and-leading-teams-effectively/

Singh, A., & Sahu, A. (2021). Empathetic communication in crisis management. *Journal of Business Research*, *130*, 491–502.

Singh, J., Pandey, B., Karna, S., Arora, A. S., & Kumar, A. (2024). Enhancing the thermographic diagnosis of maxillary sinusitis using deep learning approach. *Quantitative Infrared Thermography Journal*, *00*(00), 1–15. DOI: 10.1080/17686733.2024.2349976

Singh, V., Kumar, A., Kaur, S., Bansal, A., & Singla, A. K. (2024). Enhancing cavitation erosion resistance of VC+TiC coatings with PTFE in marine environments via lasso regression optimization. *Tribology International*, *196*(April), 109697. DOI: 10.1016/j.triboint.2024.109697

Sinha, R. (2018). A comparative analysis of traditional marketing v/s digital marketing. [JMRA]. *Journal of Management Research and Analysis*, *5*(4), 234–243.

Sinha, R., & Kumar, H. (2018). A study on preventive measures of cybercrime. *International Journal of Research in Social Sciences*, *8*(11), 265–271.

Sinha, R., & Vedpuria, N. (2018). Social impact of cybercrime: A sociological analysis. International Journal of Management. *IT & Engineering*, *8*(10), 254–259.

Slåtten, T. (2008). Antecedents and effects of emotional satisfaction on employee-perceived service quality. *Managing Service Quality*, *18*(4), 370–386. DOI: 10.1108/09604520810885617

Smetanin, S. (2020). The Applications of Sentiment Analysis for Russian Language Texts: Current Challenges and Future Perspectives. *IEEE Access : Practical Innovations, Open Solutions*, *8*, 110693–110719. DOI: 10.1109/ACCESS.2020.3002215

Smith, G. (2019). *Marketing Communications: Integrating Offline and Online Strategies*. Kogan Page Publishers.

Smith, J., & Doe, A. (2022). AI and CRM: Transforming customer engagement. *Journal of Digital Marketing*, *15*(3), 45–60.

Sofiyah, F. R., Dilham, A., Lubis, A. S., Hayatunnufus, , Marpaung, J. L., & Lubis, D. (2024). The Impact of Artificial Intelligence Chatbot Implementation on Customer Satisfaction in Padangsidimpuan: Study with Structural Equation Modelling Approach. *Mathematical Modelling of Engineering Problems*, *11*(8), 2127–2135. DOI: 10.18280/mmep.110814

Song, X., & Bonanni, C. (2024). AI-Driven Business Model: How AI-Powered Try-On Technology Is Refining the Luxury Shopping Experience and Customer Satisfaction. *Journal of Theoretical and Applied Electronic Commerce Research*, *19*(4), 3067–3087. DOI: 10.3390/jtaer19040148

Soundarraj, P. L., Javaraj, P., Meetna, S., Elaiyaraja, A., Naqvi, S. R., & Gamadia, R. (2024). Prioritizing Factors Affecting Customers Satisfaction in the Internet Banking Using Artificial Intelligence. *Proceedings of 9th International Conference on Science, Technology, Engineering and Mathematics: The Role of Emerging Technologies in Digital Transformation, ICONSTEM 2024*. DOI: 10.1109/ICONSTEM60960.2024.10568595

Srivastava. (2020, October 22). Worried about hygiene of Prasad at temples, UP govt launches operation BHOG. *Hindustan Times*. Retrieved June 15, 2023, from https://www.hindustantimes.com/lucknow/worried-about-hygiene-of-prasad-at-temples-up-govt-launches-operation-bhog/story-rJbVEhohRRJ096YdRUzj3H.html

Stein, D. J., Szatmari, P., Gaebel, W., Berk, M., Vieta, E., Maj, M., de Vries, Y. A., Roest, A. M., de Jonge, P., Maercker, A., Brewin, C. R., Pike, K. M., Grilo, C. M., Fineberg, N. A., Briken, P., Cohen-Kettenis, P. T., & Reed, G. M. (2020). Mental, behavioral and neurodevelopmental disorders in the ICD-11: An international perspective on key changes and controversies. *Journal für Neurologie, Neurochirurgie und Psychiatrie*, *21*(1), 30–32. DOI: 10.1186/s12916-020-1495-2 PMID: 31983345

Strong, E. K. (1925). *The psychology of selling and advertising*. McGraw-Hill Book Company.

Sundar, S. S., & Marathe, S. S. (2010). Personalization and its discontents: A meta-analysis of the effects of customization on perceived source credibility. *New Media & Society*, *12*(6), 875–892.

Sunyoto, A. (2023, November). Hybrid sentiment analysis: Majority voting with multinomial naive Bayes and logistic regression on IMDB dataset. In 2023 6th International Conference on Information and Communications Technology (ICOIACT) (pp. 448-453). IEEE.

Suri, S., Sharma, K., & Papneja, S. (2020). Frustration Detection On Reviews Using. *Machine Learning, 1–5*, 1–5. Advance online publication. DOI: 10.1109/INCET49848.2020.9153975

Suryanarayana Murthy, Y., Pardha Sai, K., Doss, S., & Ch Murthy Chodisetty, R. S. (2024). Navigating the Intersection of Industry 6.0, Quantum Computing, and Disaster Management: Consumer Insights and Technological Impact. In *The Rise of Quantum Computing in Industry 6.0 Towards Sustainability* (pp. 53–66). Springer. DOI: 10.1007/978-3-031-73350-5_4

Sutrisno, T. F. C. W. (2019). Association among entrepreneurial orientation and corporate performance. *International Journal of Scientific and Technology Investigation, 8*(11), 3030–3035.

Tax, S. S., Brown, S. W., & Chandrashekaran, M. (1998). Customer evaluations of service complaint experiences: Implications for relationship marketing. *Journal of Marketing, 62*(2), 60–76. DOI: 10.1177/002224299806200205

Telefonica. (2023). Telefonica Uses Artificial Intelligence to Optimize Network Management. Telefonica. *Available at:*https://www.telefonica.com/en/communication-room/press-room/telefonicalaunches-aura-and-leads-the-integration-of-artifical-intelligence-in-its-networks-and-customercare/. (last visited on: 12-01-25)

Terlutter, R., Diehl, S., & Müller, B. (2021). Multichannel emotional mapping. *Journal of Business Research, 129*, 926–935.

Thakkar, A., Gupta, A., & De Sousa, A. "Arti fi cial intelligence in positive mental health : a narrative review," no. March, pp. 1–11, 2024, DOI: 10.3389/fdgth.2024.1280235

Thomas, J., Sateesh Kumar, T. K., Menon, V. A., & Thomas, L. P. (2024). Analyzing Sentiment in Netflix User Opinions: A Statistical Examination. In Choudrie, J., Mahalle, P. N., Perumal, T., & Joshi, A. (Eds.), *ICT for Intelligent Systems* (pp. 587–601). Springer Nature., DOI: 10.1007/978-981-97-6678-9_51

Thomas, N., & Docherty, N. (2025). "Empathy Code": The Dangers of Automating Empathy in Business. *Business & Society, 00076503241310142*, 00076503241310142. Advance online publication. DOI: 10.1177/00076503241310142

Thompson, D. V., Hamilton, R. W., & Rust, R. T. (2011). Feature Fatigue: More May Be Too Much. *JMR, Journal of Marketing Research, 42*(4), 431–442. DOI: 10.1509/jmkr.2005.42.4.431

Tiwari, M., Bhadoriya, M. C., Murthy, G. R. K., Goswami, M., Murthy, Y. S., & Singh, N. (2022). Examining The Effects of Factors Influencing Organizational Citizenship Behavior With Respect to IT Managers In Gujarat. *Business. Manag. Econ. Eng*, *20*(2), 2154–2167.

Tristanto, T. A., Nugraha, N., Waspada, I., Mayasari, M., & Kurniati, P. S. (2023). Sustainability Performance Influence of Corporate Performance in Indonesia Banking. *Journal of Eastern European and Central Asian Investigation*, *10*(4), 668–678. DOI: 10.15549/jeecar.v10i4.1364

Tuhin, M. (2025, January 23). *Study Finds AI Outperforms Humans in Generating Empathetic Responses*. Science News Today. https://www.sciencenewstoday.org/study-finds-ai-outperforms-humans-in-generating-empathetic-responses

Tura, B. D. (2022). Banking sector performance evaluation in Ethiopia for the period of growth and transformation plan (GTP-II): Private vs public commercial banks. *Journal of Banking Regulation*, *0123456789*. Advance online publication. DOI: 10.1057/s41261-022-00209-0

Tutan, A., & Kökalan, Ö. (2025). A multilevel analysis of the effects of daily work stress, daily work to family conflict and emotional stability on daily job engagement. *Current Psychology (New Brunswick, N.J.)*, *44*(2), 894–906. DOI: 10.1007/s12144-024-07180-1

Vashishth, T. K., Sharma, V., Sharma, K. K., Kumar, B., Chaudhary, S., & Panwar, R. (2024). Embracing AI and Machine Learning for the Future of Digital Marketing. In AI, Blockchain, and Metaverse in Hospitality and Tourism Industry 4.0 (pp. 90-117). Chapman and Hall/CRC. DOI: 10.1201/9781032706474-7

Venkatesh, V., & Davis, F. D. (2000). A theoretical extension of the technology acceptance model: Four longitudinal field studies. *Management Science*, *46*(2), 186–204. DOI: 10.1287/mnsc.46.2.186.11926

Verhoef, P. C., Broekhuizen, T., & Bijmolt, T. H. (2021). Understanding the customer journey: Touchpoints and sensitivity. *Journal of Retailing*, *97*(1), 7–27.

Vivek, S. D., Beatty, S. E., & Morgan, R. M. (2012). Customer engagement: Exploring customer relationships beyond purchase. *Journal of Marketing Theory and Practice*, *20*(2), 122–146. DOI: 10.2753/MTP1069-6679200201

Volpicelli, G. (2016, March 24). *IBM Watson now powers a Hilton hotel robot concierge—Ars Technica*. https://arstechnica.com/gadgets/2016/03/ibm-watson-hilton-robot-connie/

Wang, X., Hwang, Y. H., & Guchait, P. (2023). When Robot (Vs. Human) Employees Say "Sorry" Following Service Failure. *International Journal of Hospitality & Tourism Administration, 24*(4), 540–562. DOI: 10.1080/15256480.2021.2017812

Wang, Y. (2022). Research on the Influence of Service Quality of Hotel Intelligent System on Customer Satisfaction Based on Artificial Intelligence Evaluation. *Mathematical Problems in Engineering, 2022*, 1–9. Advance online publication. DOI: 10.1155/2022/3832935

Wang, Y., Xu, H., & Huang, L. (2021). AI and emotional sensitivity in touchpoints. *Information Systems Research, 32*(2), 398–417.

Wedel, M., & Kannan, P. K. (2016). Marketing analytics for data-rich environments. *Journal of Marketing, 80*(6), 97–121. DOI: 10.1509/jm.15.0413

Wei, X., Jia, J., Li, X., Wu, Z., & Wang, Z. (2023). A Discourse-level Multi-scale Prosodic Model for Fine-grained Emotion Analysis. *arXiv preprint arXiv:2309.11849.*

Weißhaar, I., & Huber, F. (2016). Empathic Relationships in Professional Services and the Moderating Role of Relationship Age. *Psychology and Marketing, 33*(7), 525–541. *Scopus.* Advance online publication. DOI: 10.1002/mar.20895

Wei, W., Liu, J., Mao, X., Guo, G., Zhu, F., Zhou, P., & Hu, Y. (2019). Emotion-aware Chat Machine: Automatic Emotional Response Generation for Human-like Emotional Interaction. *Proceedings of the 28th ACM International Conference on Information and Knowledge Management*, 1401–1410. DOI: 10.1145/3357384.3357937

White, R. (2019). Chatbots in tourism: Enhancing customer support. *International Journal of Hospitality Technology, 7*(4), 230–250.

Wiliams, R. L., & Hall, C. M. (2019). AI-driven customer relationship management: The role of deep learning and NLP. *Journal of Business Intelligence and Data Science, 7*(4), 112–128.

Williamson, S. M., & Prybutok, V. (2024). Integrating human-centric automation and sustainability through the NAToRM framework: A neuromorphic computing approach for resilient industry 5.0 supply chains. *International Journal of Information Management Data Insights, 4*(2), 100278. DOI: 10.1016/j.jjimei.2024.100278

Wirtz, J., Hofmeister, J., Chew, P. Y. P., & Ding, X. D. (2023). Digital service technologies, service robots, AI, and the strategic pathways to cost-effective service excellence. *Service Industries Journal, 43*(15–16), 1173–1196. DOI: 10.1080/02642069.2023.2226596

Wirtz, J., Zeithaml, V. A., & Gistri, G. (2022). Managing service failure with emotional touchpoint sensitivity. *Journal of Service Research, 25*(1), 62–78.

Woodward, K., Kanjo, E., Brown, D. J., McGinnity, T. M., Inkster, B., Macintyre, D. J., & Tsanas, A. (2022). Beyond Mobile Apps: A Survey of Technologies for Mental Well-Being. *IEEE Transactions on Affective Computing, 13*(3), 1216–1235. DOI: 10.1109/TAFFC.2020.3015018

Wu, H., Chou, H. C., Chang, K. W., Goncalves, L., Du, J., Jang, J. S. R., . . . Lee, H. Y. (2024). EMO-SUPERB: An in-depth look at speech emotion recognition. *arXiv preprint arXiv:2402.13018.*

Wu, S., Falk, T. H., & Chan, W. Y. (2011). Automatic speech emotion recognition using modulation spectral features. *Speech Communication, 53*(5), 768–785. DOI: 10.1016/j.specom.2010.08.013

Xiao, L., & Kumar, V. (2021). Robotics for Customer Service: A Useful Complement or an Ultimate Substitute? *Journal of Service Research, 24*(1), 9–29. DOI: 10.1177/1094670519878881

Yamijala, S. M., Nimbrain, N., & Bansal, R. (2025). Enhancing Remote Workspaces: The Role of Virtual Reality in Shaping the Virtual Work Environment. *Optimizing Virtual Reality and Metaverse for Remote Work and Virtual Team Collaboration*, 171-190.

Yamijala, S. M. S., Chodisetty, R. C. M., & Chakravorty, C. (2025). Adapting HR Practices for a Transhumanist Future: An Empirical Study on Policy and Implementation. In *Enhancing the Modern Workforce Through Transhumanism* (pp. 117–146). IGI Global.

Yamijala, S. M. S., Chodisetty, R. M., Chakravorty, C., & Sai, K. P. (2025). AI-Powered Learning Revolutionizing Smart Education With Personalized Learning Styles. In *Internet of Behavior-Based Computational Intelligence for Smart Education Systems* (pp. 191–212). IGI Global.

Yas, H., Alkaabi, A., Albaloushi, N. A., Al Adeedi, A., & Streimikiene, D. (2023). the Influence of Strategic Leadership Practices and Knowledge Sharing on Staff member'S Performance. *Polish Journal of Management Studies, 27*(1), 343–362. DOI: 10.17512/pjms.2023.27.1.20

Yella, A., & Suganyadevi, K. (2024). Improving Customer Satisfaction in Retail Banking through AI-driven Algorithm Optimization. *2024 15th International Conference on Computing Communication and Networking Technologies, ICCCNT 2024*. DOI: 10.1109/ICCCNT61001.2024.10725146

Ye, Y., & Motoki, K. (2024). Effects of emotions on the helpfulness of online reviews of healthy and unhealthy food in Japan and the United States. *Journal of Consumer Behaviour, 23*(5), 2611–2623. DOI: 10.1002/cb.2364

Yoo, J. W., Park, J., & Park, H. (2024). The impact of AI-enabled CRM systems on organizational competitive advantage: A mixed-method approach using BERTopic and PLS-SEM. *Heliyon, 10*(16), e36392. DOI: 10.1016/j.heliyon.2024.e36392 PMID: 39253149

Yuan, J., Cai, X., Zheng, R., Huang, L., & Church, K. (2021). The role of phonetic units in speech emotion recognition. *arXiv preprint arXiv:2108.01132*.

Yun, J., & Park, J. (2022). The Effects of Chatbot Service Recovery With Emotion Words on Customer Satisfaction, Repurchase Intention, and Positive Word-Of-Mouth. *Frontiers in Psychology, 13*, 922503. DOI: 10.3389/fpsyg.2022.922503 PMID: 35712132

Zahra, K., & Moradi, M. H. (2009). Emotion recognition system using brain and peripheral signals: Using correlation dimension to improve the results of EEG. *Proceedings of the International Joint Conference on Neural Networks*, 1571–1575. https://doi.org/DOI: 10.1109/IJCNN.2009.5178723

Zhang, J., Chen, Q., Lu, J., Wang, X., Liu, L., & Feng, Y. (2024). Emotional expression by artificial intelligence chatbots to improve customer satisfaction: Underlying mechanism and boundary conditions. *Tourism Management, 100*, 104835. Advance online publication. DOI: 10.1016/j.tourman.2023.104835

Zheng, Z., Chen, J., Li, S., Sun, J., Xie, Y., & Liang, L. (2023, October). Intelligent Customer Service System Based on Speech Emotion Recognition. In *Proceedings of the 2023 7th International Conference on Electronic Information Technology and Computer Engineering* (pp. 334-338). DOI: 10.1145/3650400.3650454

Zuboff, S. (2019). *The Age of Surveillance Capitalism: The Fight for a Human Future at the New Frontier of Power*. PublicAffairs.

About the Contributors

Fazla Rabby is a dynamic and innovative business manager with a robust educational background, extensive teaching experience, and a proven track record in academia and industry. Dr. Fazla Rabby earned his PhD from the University of Southern Queensland while his Master in Information Systems is from Central Queensland University, along with various other credentials that further enhanced his business and information technology skills. Dr. Rabby is currently the Director at the Stanford Institute of Management and Technology, where he manages the academic department and oversees regulatory matters in coordination with the government and other key stakeholders. His commitment to research is a cornerstone of his academic journey, manifesting in a series of impactful publications spanning diverse and rapidly evolving fields. Notably, his research contributions include insightful books, book chapters and articles focused on cutting-edge topics such as Blockchain Technology, Digital Marketing, and Artificial Intelligence. These publications reflect his dedication to expanding the boundaries of knowledge and demonstrate a proactive engagement with emerging technologies that are pivotal in today's academic landscape. During my tenure as an Academic at different universities and institutes, Dr Fazla Rabby played a pivotal role in fostering a dynamic learning environment. He delivered lectures beyond traditional boundaries, incorporating interactive workshops and engaging seminars to stimulate critical thinking among students. Collaborating with fellow academics, He actively refined teaching methods, emphasising innovative approaches to enhance the learning experience. Dr. Rabby is also an active member of several professional bodies, including the Australian Computer Society, IEEE and the Australian Marketing Association.

Nasim Ahmed received the PhD degree in Computer Science from Massey University, New Zealand, in 2022, and MSc in Computer Engineering from the University Malaysia Perlis, Malaysia, in 2009. His research interests span diverse areas including data science applications in healthcare and agriculture, big data clusters, artificial intelligence's role in education and management (specifically policymaking and industry), and data security.

Amandeep Sehmi, is an experienced academic leader and an industry professional with a deep commitment to advancing business education, integrating innovative technologies, and fostering continuous improvement in teaching practices. Holding a Ph.D. from the School of Business, Law, and Entrepreneurship at Swinburne University of Technology, Dr Sehmi has built a distinguished career in higher education, focusing on quality teaching, academic leadership, and the transformative potential of Artificial Intelligence (AI) in reshaping education. Currently serving as the Assistant Dean and Program Director (Management Undergraduate) at the Melbourne campus of Canterbury Institute of Management, Dr Sehmi works in close collaboration with the Dean to engage innovative academic approach for teaching, learning, assessments, and course evaluations. She has a particular research focus on Artificial Intelligence (AI) and its transformative potential in education. Before transitioning into academic leadership, Dr. Amandeep gained valuable experience in teaching at CIM, University of New South Wales, University of Adelaide College, Swinburne University of Technology and Kaplan Business School. Dr. Sehmi 's experience extends beyond academia. As a Market Research Consultant, she led research projects for major brands like Kmart, Target, and Australia Post. She Dr. Amandeep's professional background includes extensive experience in marketing and communications, having worked as a Senior Marketing & Communications Manager and Relationship Manager in New Zealand.

Rohit Bansal is working at Rockford College, Australia. He is a perseverant, passionate academician cum seasoned professional. He obtained Ph.D. in Management from Maharshi Dayanand University, Rohtak. With a rich experience of 17 years, he has achieved growth through robust and proactive academic initiatives. He is also working as online Instructor on Udemy.com (an online learning and teaching platform), launched six online courses with 15,000 students enrolled till date. He has authored & edited 51 books with renowned national & international publishers including Springer Nature, IGI Global USA, Scrivener-Wiley Publishing, De Gruyter, Germany, Central West Publishing, Australia etc. In addition to, Dr. Rohit has published 190 research papers and chapters in journals of repute including Scopus indexed as well as edited books. His research work is published in leading publishers like Springer, MDPI, HIndawi, IGI Global, Cell Press, De Gruyter, Elseveir, Inderscience, Wiley etc. He has also presented papers in 60 conferences and seminars including IIM Indore, IIM Ranchi, IIM Jammu and IIM Kozhikode. His area of interest includes marketing management, organizational behavior, customer engagement, digital marketing, influencer marketing, human resource management, mental health & well-being, emerging technologies, industry 5.0, sustainable development, e-learning and organizational development. He has proficiency in handling graduate and post graduate students by using interactive discussions and "hands-on" approaches to help them undergo experiential learning. He is on Editorial Advisory Board as a member in 120 national and international peer reviewed journals. He is Managing Editor of International Journal of 360° Management Review & International Journal of Techno – Management Research. He has served as member of technical committee in many international conferences. He has reviewed more than 600 articles for international peer-reviewed journals. Dr. Rohit has acted as Session Chair and speaker in many international conferences. He has been awarded many times for contribution to academics and research. Apart from this, he has been awarded one Australian Patent and have published thirteen Indian Patents. He is a member of many national and international organizations.

Nishita Pruthi is an Assistant Professor at the Asian School of Business, part of the Asian Education Group, Noida. Awarded with Ph.D. at the Institute of Management Studies and Research, Maharshi Dayanand University, Rohtak, Haryana, she exemplifies academic excellence and scholarly dedication. Dr. Pruthi has an impressive portfolio of nine research papers published in prestigious journals, with six indexed in Scopus, Web of Science, UGCCARE, and ABDC. Her contributions extend to authoring thirteen chapters in edited volumes published by globally renowned publishers such as Emerald, Bentham Science, and IGI Global. Additionally, she has co-edited three books published by IGI Global, USA, reflecting her versatility and scholarly rigor. Her academic journey is further distinguished by her active participation in twelve national and international conferences hosted by premier institutions, including IIM Indore, IIM Jammu, IIM Kozhikode, and Deakin University. Dedicated to fostering a transformative learning environment, Dr. Pruthi inspires critical thinking, ad creativity, ensuring her students are well-prepared to excel in dynamic academic and professional landscapes

***

Kirti K. is an Assistant Professor of Management at Chanakya National Law University, Patna, where she also serves as Chair of the Internship Cell. She holds a Ph.D. in Management from Patna University, with research focused on the impact of interpersonal conflict on human resource performance in public enterprises. Her academic interests span human resource management, organizational behavior, digital transformation, and strategic communication. Dr. Kirti's recent scholarly contributions include a co-authored Scopus-indexed article on blockchain implementation in supply chains (World Scientific) and several peer-reviewed book chapters with Springer on themes such as globalization, institutional change, and digital branding. She is currently leading a meta-analysis on poverty alleviation schemes across Indian states aligned with SDG 1 and has authored a chapter on Emotion AI and customer empathy for this volume. She is an active reviewer for Social Sciences & Humanities Open (Elsevier), The Singapore Economic Review (World Scientific Publishing) and Managing Editor of the UGC-CARE listed CNLU Law Journal. Her work reflects a commitment to academic excellence, ethical inquiry, and cross-disciplinary engagement.

Kirti Aggarwal Assistant Professor, MBA Department, Vaish College of Engineering, Rohtak Dr. Kirti Aggarwal is a dedicated academic professional with over a decade of teaching experience in higher education. Currently, she serves as an Assistant Professor in the MBA Department at Vaish College of Engineering, Rohtak, where she plays a crucial role in mentoring students and fostering academic excellence. Dr. Aggarwal earned her Ph.D. in Finance, focusing her research on "Human Resource Disclosure Practices in the Indian Corporate Sector". Her scholarly contributions have garnered recognition both nationally and internationally, with numerous publications in reputed journals. Her research interests lie in accounting and finance, and she has published various research papers, including 09 Scopus Indexed, 05 in UGC-CARE List, and 01 in UGC Approved journals. She has actively participated in several academic conferences, providing valuable insights into finance, business strategies, and management practices. Beyond her academic responsibilities, Dr. Aggarwal is deeply involved in editorial work. She serves as an Editorial Member of 21 national and international journals and is a Reviewer for 32 reputed journals (including some Scopus Indexed). In recognition of her contributions, she was felicitated with the 'Best Editor Award' at the 5th National Editors Summit, 2023, organized by Innovative Education and Scientific Research Foundation (IESRF) and sponsored by IP Innovative Publication Pvt. Ltd. (Journal of Management Research and Analysis), New Delhi. Dr. Aggarwal's teaching philosophy emphasizes creating an engaging and inclusive learning environment. She is particularly passionate about integrating real-world applications into her curriculum, thereby enhancing students' understanding and practical knowledge. Passionate about fostering a collaborative academic environment, Dr. Aggarwal is committed to ensuring the dissemination of high-quality research. Her extensive experience in research, teaching, and editorial work reflects her dedication to advancing knowledge and excellence in her field.

Chinmoy Bharadwaj holds a B. Tech in Computer Science and Engineering, M.Tech in Computer Science and Engineering, and a Ph.D. With over 10 years and 9 months of teaching experience at the Masters and Bachelors levels in Computer

Applications, he has developed expertise in several key areas, including Cloud Security, Edge Computing, Network Security, Wireless Sensor Networks, Internet of Things, Artificial Intelligence, and Machine Learning. His extensive teaching experience and research focus allow him to bring cutting-edge knowledge to his students and contribute to the advancement of these rapidly evolving fields.

Sippee Bharadwaj has a doctorate in speech signal processing with 7+ years of research experience in universities by conducting high-level scientific research for science-driven ventures. Proactive leader and team player integral to the success of numerous research initiatives. Dedicated research associate with good teaching experience and a strategic approach to problem solving. Diplomatic supervisor with sophisticated knowledge of multicultural and interdisciplinary teams. Resourceful, result-orientated, positive thinker, highly self-motivated, driven, and versatile with excellent communication and interpersonal skills, a highly developed work ethic, structured work, and a reputation for integrity.Her areas of interest include neuromarketing, AI and ML.

Bhupinder Chahal is a Professor at Yorkville University, Canada.

Saurabh Chandra is an accomplished legal scholar and currently serves as an Associate Professor & Assistant Dean at the School of Law, Bennett University, India. With a robust academic foundation, he earned his Bachelor of Arts and Bachelor of Laws (B.A. LL.B.) from the prestigious Aligarh Muslim University (A.M.U.). He further specialized in Business Laws, completing his Master of Laws (LL.M.) at the renowned National Law School of India University (NLSIU), Bangalore. Dr. Chandra's academic journey culminated in a Doctorate in Law (Ph.D.) from the esteemed National Law University (NLU) Delhi. In addition to his impressive legal education from some of India's most prestigious institutions, Dr. Saurabh Chandra also holds a master's degree in Management with specialization in Human Resources in Management and another in Journalism and Mass Communication. With 16 years of teaching experience, he has established himself as an expert in Business and Corporate Laws, with numerous publications to his credit. Dr. Saurabh Chandra is a distinguished expert in Business and Corporate Laws, with a profound understanding of the intricate legal frameworks that govern the business world. His expertise is complemented by a robust portfolio of scholarly publications, which underscores his commitment to advancing knowledge in his field. Throughout his career, Dr. Chandra has consistently produced high-impact research that addresses critical issues in Business and Corporate Law. His publications are widely recognized for their depth of analysis, innovative perspectives, and practical relevance. He has contributed extensively to leading journals, where his work has often set new standards for legal scholarship. His publications not only reflect his scholarly rigor but also his dedication to addressing contemporary challenges faced by businesses and policymakers. By bridging the gap between theoretical insights and real-world applications, Dr. Chandra's work serves as a valuable resource for academics, legal practitioners, and industry professionals alike. Dr. Chandra brings a wealth of experience in Academic Administration, having successfully held various key administrative positions. His diverse qualifications and extensive experience make him a distinguished figure in the academic and legal communities.

Nila Chotai is a dedicated academic leader and the Director of Executive Education and PGDM, with a strong background in the higher education industry. Holding a Doctor of Philosophy in Economics, she has 16 years of teaching experience and a proven track record in research, leadership, consulting, curriculum development, and mentoring. She has successfully managed management institutes, demonstrating her expertise in strategic planning and academic administration. Her contributions to executive education and postgraduate management programs have significantly enhanced learning outcomes and institutional excellence.

Admassu Huluka, Associate Professor of Development Studies, Department of Development Management, Ethiopian Civil Service University, Ethiopia Email: admassutesso@gmail.com

Zahid Hussain currently in his postgraduate degree in Marketing from Khadim Hussain Shah Institute of Technology, Karachi, Pakistan (KASBIT). He has done several research publications, conference papers and book chapters related Marketing, Food Marketing, Metaverse Marketing, Customer Satisfaction and Artificial Intelligence in marketing related fields in different impact factor journals at national and international level.

Ashima Jain, is Assistant Professor, School of Law, Manipal University Jaipur since 2022. She has been actively presenting various papers at ICSSR conferences and has published many paper. She has been affiliated to various prestigious organizations as lifetime member too like Life Member Certificate of the South Asian Society of Criminology & Victimology (SASCV), Sociological Society. She has participated in many FDP and given lectures. She has more than 2 years of experience in academics and has been actively taking participations in various conferences and publications.

Shashi Kant, Assistant Professor, College of Business and Economics, Department of Management, Bule Hora University, Ethiopia.

Manish Kothari is a visionary leader in business education with over 20 years of experience in academia and institution building. As the Founder and Managing Director of ISBR Business School, Founder Chairman of Oxford Group of Colleges, and Chairman of BET College, he has been instrumental in delivering quality education and career opportunities across disciplines. Holding a PhD in Business Administration and Management from the University of Mysore, he specializes in Human Resources, Organizational Development, and Leadership Development. His expertise spans strategic planning, curriculum design, faculty development, accreditation, and collaboration. Committed to education, innovation, and entrepreneurship, he strives to create a positive societal impact by fostering a culture of learning, growth, and empowerment with a strong emphasis on diversity, integrity, and excellence.

Pawan Kumar, Working as a Professor, Graphic Era deemed to be a University India

Larisa Mistrean has more than 28 years of scientific-didactic and research experience in the area of economics. She is an Associate Professor at the Academy of Economic Studies of Moldova, in the Investments and Banking Activity Department. Her research focuses on banking, consumer behavior, corporate governance, monetary policy, inter-, trans-, and multidisciplinary research, and neuromarketing. She was a Fulbright scholar for the period January-April 2024. She is the project director and principal investigator of the Moldovan state research project "Behavioral Evolutions of Consumers of Financial-Banking Services in the New Economic Configuration" (2021-2024). She is the national manager of CEEPUS "Supply Chains Management – From the Perspective of National Industries," PL-1910-01-2425 (2024-2026). She is an expert member in international Erasmus projects - JW CPW "Protecting Working Conditions in the Platform Economy: Moldovan-Lithuanian Social Dialogue," No. 101126470: SOCPL-2022-IND-REL-l (2023-2025), and a member of ERASMUS+ CBHE "UPGRADE - Enhancing Graduates' Employability Tracking in Moldova" (No. 101129166) (2024-2026). She is an MC member in COST ACTION CA21129, COST ACTION CA21102, COST ACTION 19103, and COST ACTION 17125. She is an editorial member and member of the committee of reviewers of the international and national journals of Finance and Banking Research; a congress president member, an executive board member, a member of the scientific committee, and a member of the organizing committee of the international scientific conferences organized in Turkey, Romania, and Moldova.

Vipin Mittal Associate Professor and Head, MBA Department, Vaish College of Engineering, Rohtak Dr. Vipin Mittal is a dedicated academic professional with over a decade of teaching experience in higher education. Currently, he serves as an Associate Professor and Head in the MBA Department at Vaish College of Engineering, Rohtak, where he plays a crucial role in mentoring students and fostering academic excellence. Dr. Mittal earned his Ph.D. in Marketing, focusing his research on "Customer Perception towards Plastic Money with Special Reference to Haryana". His scholarly contributions have garnered recognition both nationally and internationally, with numerous publications in reputed journals. He has published more than 20 research papers in national and international journals of repute. He has also presented papers at over 25 conferences and seminars. His research interests lie in accounting and finance. He has actively participated in several academic conferences, providing valuable insights into finance, business strategies, and management practices. Beyond his academic responsibilities, Dr. Mittal plays a significant role in several academic activities across various universities. He is actively involved in setting question papers for different courses, ensuring comprehensive coverage of the curriculum and maintaining academic standards. In addition, he contributes to the evaluation process by assessing answer sheets with accuracy and fairness. Dr. Mittal is also engaged in conducting practical examinations, providing students with opportunities to demonstrate their skills and practical knowledge. Dr. Mittal's teaching philosophy emphasizes creating an engaging and inclusive learning environment. He is particularly passionate about integrating real-world applications into his curriculum, thereby enhancing students' understanding and practical knowledge. Passionate about fostering a collaborative academic environment, Dr. Mittal is committed to ensuring the dissemination of high-quality research. His extensive experience in research and teaching work reflects his dedication to advancing knowledge and excellence in his field.

Mukul Mukul, working as Assistant Professor, Commerce Department, Graphic Era deemed to be a University India

Dereje Negeri, PhD Scholar Marketing Management Department, Bule Hora University, Ethiopia Email: derejedinsa12@gmail.com

Timilehin Olasoji OLUBIYI is the Dean of Management and Social Sciences at West Midlands Open University in Lagos State, Nigeria, and also a visiting lecturer at Babcock University in Ogun State, Nigeria. A seasoned professional, business management expert, senior government adviser, author, columnist and scholar with many years of hands-on experience and background in investment banking, public sector investment management, stockbroking, consulting, research, and entrepreneurial development. His main expertise and research interests include small business development and management, entrepreneurship, enterprise continuity, sustainability, green management, and innovation. His research has appeared in the Management and Marketing Journal, International Journal of Professional Business Review, Revista de Gestão Social e Ambiental, Emerald Publishing, Palgrave Macmillian, Springer Nature, Malaysian Management Journal, Routledge, and IGI Global. The significance of his research is emphasized by the publication of more than one hundred (100) academic articles in prestigious and reputable national and international journals, several of which are indexed in Scopus, UGC CARE, EBSCO, and the Directory of Open Access Journals (DOAJ).

Abhijit Pandit, holding M.Sc., M.B.A. (Marketing) and Ph.D. (Marketing) degrees along with professional affiliations such as MIMA and IIMCW, possesses over 20 years of teaching experience, including 13 years post-PhD. A dedicated academician and researcher, he has presented papers at prestigious national and international conferences, delivered keynote lectures at six such events, and received six best paper awards. His extensive scholarly contributions include 6 SCOPUS, 9 ABDC-B and 16 ABDC-C indexed publications, alongside 11 books, 12 book chapters and 7 patents, demonstrating his expertise in computational research techniques. Actively engaged in mentoring doctoral scholars, he has served at esteemed institutions such as ICFAI University Tripura, Amity University Kolkata, MAKAUT and West Bengal Health University, while also participating in numerous workshops, faculty development programs and philanthropic initiatives.

Bhupinder Singh working as Professor at Sharda University, India. Also, Honorary Professor in University of South Wales UK and Santo Tomas University Tunja, Colombia. His areas of publications as Smart Healthcare, Medicines, fuzzy logics, artificial intelligence, robotics, machine learning, deep learning, federated learning, IoT, PV Glasses, metaverse and many more. He has 3 books, 139 paper publications, 163 paper presentations in international/national conferences and seminars, participated in more than 40 workshops/FDP's/QIP's, 25 courses from international universities of repute, organized more than 59 events with international and national academicians and industry people's, editor-in-chief and co-editor in journals, developed new courses. He has given talks at international universities, resource person in international conferences such as in Nanyang Technological University Singapore, Tashkent State University of Law Uzbekistan; KIMEP University Kazakhstan, All'ah meh Tabatabi University Iran, the Iranian Association of International Criminal law, Iran and Hague Center for International Law and Investment, The Netherlands, Northumbria University Newcastle UK,

Neha Singla is currently working as a Research Analyst at Milestone Global Moneymart (P) Limited where she plays a crucial role of analyzing the market movements and different mutual funds. She had also served as an Assistant Professor at Delhi School of Professional Studies and Research. She has earned her Ph.D. in Finance, focusing her research on “Earnings Management Practices in Indian Banking Sector” from Guru Jambheshwar University of Science & Technology, Hisar, Haryana. Her scholarly contribution has garnered recognition both nationally and internationally, with numerous publications in reputed journals. Her research interest lies in accounting and finance, and she has published 04 research papers which are all Scopus Indexed. She is actively engaged in research work.

Sanjay Taneja is currently an Associate Professor in Research at Graphic Era University, Dehradun, India. His significant thrust areas are Banking Regulations, Banking and Finance (Fin Tech, Green Finance), Risks, Insurance Management, Green Economics and Management of Innovation in Insurance. He holds a double master’s degree (MBA &M.Com.) in management with a specialization in Finance and Marketing. He received his PG degrees in Management (Gold Medalist) from Chaudhary Devi University, Sirsa, India in 2012. He earned his Doctor of Philosophy (Sponsored By ICSSR) in Banking and Finance entitled “An Appraisal of financial performance of Indian Banking Sector: A Comparative study of Public, Private and Foreign banks in 2016 from Chaudhary Devi University, Sirsa, India. He received his Post Doctoral Degree from faculty of Social Sciences, Department of Banking and Insurance, Usak University, Turkey entitled on “Impact of the European Green Deal on Carbon (CO2) Emission in Turkey” in 2023. He has published research papers in reputed SCOPUS/Web of Science/SCI/ABDC/UGC Care Journals. Prof. Taneja has more than fifty publications in total (Scopus/ABDC/Web of Science- 27)

Index

A

B

C

D

E

F

K

L

V

www.ingramcontent.com/pod-product-compliance
Lightning Source LLC
LaVergne TN
LVHW080341011026
849161LV00009B/2153